Lecture Notes in Artificial Intelligence   8774

Subseries of Lecture Notes in Computer Science

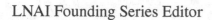

Lecture Notes in Artificial Intelligence 8794

Subseries of Lecture Notes in Computer Science

LNAI Series Editors

Randy Goebel
University of Alberta, Edmonton, Canada
Yuzuru Tanaka
Hokkaido University, Sapporo, Japan
Wolfgang Wahlster
DFKI and Saarland University, Saarbrücken, Germany

LNAI Founding Series Editor

Joerg Siekmann
DFKI and Saarland University, Saarbrücken, Germany

Neamat El Gayar   Friedhelm Schwenker
Ching Suen (Eds.)

Artificial Neural Networks in Pattern Recognition

6th IAPR TC 3 International Workshop, ANNPR 2014
Montreal, QC, Canada, October 6-8, 2014
Proceedings

 Springer

Volume Editors

Neamat El Gayar
Cairo University
Faculty of Computers and Information
Orman, Giza, Egypt
E-mail: elgayar.neamat@gmail.com

Friedhelm Schwenker
Ulm University
Institute for Neural Information Processing
Ulm, Germany
E-mail: friedhelm.schwenker@uni-ulm.de

Ching Suen
Concordia University
Department of Computer Science and Software Engineering
Montral, QC, Canada
E-mail: suen@cse.concordia.ca

ISSN 0302-9743   e-ISSN 1611-3349
ISBN 978-3-319-11655-6   e-ISBN 978-3-319-11656-3
DOI 10.1007/978-3-319-11656-3
Springer Cham Heidelberg New York Dordrecht London

Library of Congress Control Number: 2014948673

LNCS Sublibrary: SL 7 – Artificial Intelligence

Typesetting: Camera-ready by author, data conversion by Scientific Publishing Services, Chennai, India

Printed on acid-free paper

Springer is part of Springer Science+Business Media (www.springer.com)

Preface

This volume contains the papers presented at the 6th IAPR TC3 Workshop on Artificial Neural Networks for Pattern Recognition (ANNPR 2014), held at the Concordia University, Montreal, Canada during October 6–8, 2014. ANNPR 2014 follows the success of the ANNPR workshops of 2003 (Florence), 2006 (Ulm), 2008 (Paris), 2010 (Cairo), and 2012 (Trento). The series of ANNPR workshops has acted as a major forum for international researchers and practitioners from the communities of pattern recognition and machine learning based on neural networks.

The Program Committee of the ANNPR 2014 workshop selected 24 papers out of 37 for the scientific program, organized in regular oral presentations and a poster session. Three IAPR Invited Sessions given by Dr. Yoshua Bengio, University of Montreal, Canada, Dr. Michael J. Herrmann, University of Edinburgh, United Kingdom, and Dr. Zhi-Hua Zhou Nanjing University, China enriched the workshop.

This workshop would not have been possible without the help of many people and organizations. First of all, we are grateful to all the authors who submitted their contributions to the workshop. We thank the members of the Program Committee and the additional reviewers for performing the difficult task of selecting the best papers from a large number of high-quality submissions. We hope that readers of this volume may enjoy it and be inspired by its contributions. ANNPR 2014 was supported by the International Association for Pattern Recognition (IAPR), by the IAPR Technical Committee on Neural Networks and Computational Intelligence (TC3), by the University of Ulm, Germany, by Concordia University, Montreal, Canada, and by IMDS (International Medias Data Services), Montreal, Canada. Special thanks to the people of the local organization, in particular to Marleah Blom, Leila Kosseim, and Nicola Nobile.

Finally, we wish to express our gratitude to Springer for publishing our workshop proceedings within their LNCS/LNAI series.

July 2014

Neamat El Gayar
Friedhelm Schwenker
Ching Suen

Organization

Organization Committee

Neamat El Gayar   Montreal, Canada
Friedhelm Schwenker   Ulm University, Germany
Ching Suen   Montreal, Canada

Program Committee

Shigeo Abe (Japan)   Simone Marinai (Italy)
Amir Atiya (Egypt)   Heiko Neumann (Germany)
Andreas Fischer (Canada)   Günther Palm (Germany)
Markus Hagenbuchner (Australia)   Lionel Prevost (France)
Hans A. Kestler (Germany)   Edmondo Trentin (Italy)
Adam Krzyak (Canada)   Ah-Chung Tsoi (Macau)
Louisa Lam (Canada)

Local Arrangements

Marleah Blom
Leila Kosseim
Nicola Nobile

Sponsoring Institutions

International Association for Pattern Recognition (IAPR)
Technical Committee 3 (TC3) of the IAPR
Concordia University, Montreal, Canada
Ulm University, Ulm, Germany
IMDS (International Medias Data Services) of Montreal, Canada

Table of Contents

Invited Paper

Learning Algorithms and Architectures

Applications

Large Margin Distribution Learning

Zhi-Hua Zhou

National Key Laboratory for Novel Software Technology
Nanjing University, Nanjing 210023, China
zhouzh@lamda.nju.edu.cn

Abstract. Support vector machines (SVMs) and Boosting are possibly the two most popular learning approaches during the past two decades. It is well known that the *margin* is a fundamental issue of SVMs, whereas recently the margin theory for Boosting has been defended, establishing a connection between these two mainstream approaches. The recent theoretical results disclosed that the *margin distribution* rather than a single margin is really crucial for the generalization performance, and suggested to optimize the margin distribution by maximizing the margin mean and minimizing the margin variance simultaneously. Inspired by this recognition, we advocate the *large margin distribution learning*, a promising research direction that has exhibited superiority in algorithm designs to traditional large margin learning.

1 Introduction

Support vector machines (SVMs) and Boosting have both been very popular during the past two decades. SVMs belong to the family of *large margin methods* [18] whereas Boosting belongs to the family of *ensemble methods* [22]. The former roots in the statistical learning theory [19], exploiting the kernel trick explicitly to handle nonlinearity with linear classifiers; the latter comes from the proof construction [13] to the theoretical problem that whether weakly learnable equals strongly learnable [8]. It is clearly that these two approaches were born with apparent differences.

The *margin* [19] is a fundamental issue of SVMs as an intuitive understanding of the behavior of SVMs is to search for a large margin separator in a RKHS (reproducing kernel Hilbert space). It is worth noting that there is also a long history of research trying to explain Boosting with a margin theory. Though there were twists and turns in this line of studies, recently the margin theory for Boosting has finally been defended [5], establishing a connection between these two mainstream learning approaches. It is interesting that in contrast to large margin methods that focus on the maximization of a single margin, the recent theoretical results disclosed that the *margin distribution* rather than a single margin is really crucial for the generalization performance, and suggested to optimize the margin distribution by maximizing the margin mean and minimizing the margin variance simultaneously. Inspired by this recognition, we advocate *large margin distribution learning*, a promising research direction that has already exhibited superiority in algorithm designs [21].

N. El Gayar et al. (Eds.): ANNPR 2014, LNAI 8774, pp. 1–11, 2014.

In this article, we will first briefly introduce the efforts on establishing the margin theory of Boosting, and then explain the basic idea of large margin distribution learning. After that, we will show some simple implementation of large margin distribution learning, followed by concluding remarks.

2   The Long March of Margin Theory for Boosting

Overfitting is among the most serious obstacles for learning approaches to achieve strong generalization performances, and great efforts have been devoted to mechanisms that help reduce overfitting risk, such as decision tree pruning, neural networks early stopping, minimum description length constraint, structural risk minimization, etc. It is typically believed that when the training error reaches zero (even much before that), the training process should be terminated because the further training will unnecessarily increase the model complexity and therefore, leading to overfitting. Indeed, according to the Occam's razor, if we have multiple hypotheses consistent with observations, then the simpler, the better.

However, for `AdaBoost`, the most famous representative of Boosting, it has been observed that the generalization performance can be improved further if the training process continues even after the training error reaches zero, though the ensemble model becomes more complicated owing to the inclusion of more base learners. This seems contradictory to previous knowledge, and thus, to understand why `AdaBoost` seems resistant to overfitting is the most fascinating fundamental theoretical issue in Boosting studies.

To explain this phenomenon, Schapire et al. [14] presented the margin theory for Boosting. Let $\mathcal{X}$ and $\mathcal{Y}$ denote the input and output spaces, respectively. A training set of size m is an *i.i.d.* sample $S = \{(\boldsymbol{x}_1, y_1), \cdots, (\boldsymbol{x}_m, y_m)\}$ drawn according to D, an unknown underlying probability distribution over $\mathcal{X} \times \mathcal{Y}$. Denote $\Pr_D[\cdot]$ and $\Pr_S[\cdot]$ as the probability w.r.t. D and w.r.t. uniform distribution over S, respectively. Let $\mathcal{H}$ be a hypothesis space, and a base learner is a function $h\colon \mathcal{X} \to \mathcal{Y}$. Here, we focus on binary classification, i.e., $\mathcal{Y} = \{+1, -1\}$. Let $\mathcal{C}(\mathcal{H})$ denote the convex hull of $\mathcal{H}$, i.e., the ensemble model $f \in \mathcal{C}(\mathcal{H})$ is of the form

$$f = \sum_i \alpha_i h_i \text{ with } \sum_i \alpha_i = 1 \text{ and } \alpha_i \geq 0. \tag{1}$$

We call this ensemble model a voting classifier because the base learners are combined via voting (also called *additive model* in statistical literatures). Given an example $(\boldsymbol{x}, y)$, the *margin* w.r.t. the voting classifier $f = \sum \alpha_i h_i(\boldsymbol{x})$ is defined as $yf(\boldsymbol{x})$; in other words,

$$yf(\boldsymbol{x}) = \sum_{i\colon y=h_i(\boldsymbol{x})} \alpha_i - \sum_{i\colon y\neq h_i(\boldsymbol{x})} \alpha_i, \tag{2}$$

which shows the difference between the weights of base learners that classify $(\boldsymbol{x}, y)$ correctly and the weights of base learners that classify $(\boldsymbol{x}, y)$ incorrectly.

Based on the concept of margin, Schapire et al. [14] proved the first margin theorem for `AdaBoost` and upper bounded the generalization error as follows, where $\theta > 0$ is a threshold of margin over the training sample S.

Theorem 1. *(Schapire et al., 1998) For any $\delta > 0$ and $\theta > 0$, with probability at least $1 - \delta$ over the random choice of sample S with size m, every voting classifier $f \in C(\mathcal{H})$ satisfies the following bound:*

$$\Pr_D[yf(x) < 0] \leq \Pr_S[yf(x) \leq \theta] + O\left(\frac{1}{\sqrt{m}}\left(\frac{\ln m \ln |\mathcal{H}|}{\theta^2} + \ln\frac{1}{\delta}\right)^{1/2}\right). \quad (3)$$

This theorem implies that, when other variables are fixed, the larger the margin over the training sample, the better the generalization performance; this offers an explanation to why AdaBoost tends to be resistant to overfitting: It is able to increase the margin even after the training error reaches zero.

The margin theory looks intuitive and reasonable, and thus, it attracted a lot of attention. Notice that Schapire et al.'s bound (3) depends heavily on the smallest margin, because $\Pr_S[yf(x) \leq \theta]$ will be small if the smallest margin is large. Thus, Breiman [3] explicitly considered the *minimum margin*, $\hat{y}_1 f(\hat{x}_1) = \min_{i \in \{1..m\}}\{y_i f(x_i)\}$, and proved the following margin theorem:

Theorem 2. *(Breiman, 1999) For any $\delta > 0$, if $\theta = \hat{y}_1 f(\hat{x}_1) > 4\sqrt{\frac{2}{|\mathcal{H}|}}$ and $R \leq 2m$, with probability at least $1 - \delta$ over the random choice of sample S with size m, every voting classifier $f \in C(\mathcal{H})$ satisfies the following bound:*

$$\Pr_D[yf(x) < 0] \leq R\left(\ln(2m) + \ln\frac{1}{R} + 1\right) + \frac{1}{m}\ln\frac{|\mathcal{H}|}{\delta}, \quad (4)$$

where $R = \frac{32 \ln 2|\mathcal{H}|}{m\theta^2}$.

Breiman's minimum margin bound (4) is in $O(\ln m/m)$, sharper than Schapire et al.'s bound (3) that is in $O(\sqrt{\ln m/m})$. Thus, it was believed that the minimum margin is essential. Breiman [3] designed the `arc-gv` algorithm, a variant of `AdaBoost`, which directly maximizes the minimum margin. The margin theory would appear to predict that `arc-gv` should perform better than `AdaBoost`; however, empirical results show that though `arc-gv` does produce uniformly larger minimum margin than `AdaBoost`, its generalization error increases drastically in almost every case.[1] Thus, Breiman raised serious doubt about the margin theory, and almost sentenced the margin theory to death.

Seven years later, Reyzin and Schapire [12] found that, amazingly, Breiman had not controlled the model complexity well in experiments. To study the margin, one must fix the model complexity of base learners as it is meaningless to compare the margins of models with different complexities. In his experiments, Breiman [3] used `CART` decision trees, and considering that each decision tree leaf corresponds to an equivalent class in the instance space, Breiman tried to fix the model complexity by using trees with fixed number of leaves. Reyzin and Schapire found that the trees of `arc-gv` are generally deeper than that

---

[1] Similar empirical evidences have been reported by other researchers such as [7].

of AdaBoost, and they argued that trees with different heights may be with different model complexities. Then, they repeated Breiman's experiments using *decision stumps* with two leaves and observed that, comparing to AdaBoost, arc-gv is with larger minimum margin but smaller margin distribution. Thus, they claimed that the minimum margin is not essential, while the margin distribution characterized by the average or median margin is important.

Though Reyzin and Schapire showed that the empirical attack of Breiman is not deadly, it is far from validating the essentiality of margin distribution, because Breiman's generalization bound based on the minimum margin is quite tight. To enable the margin theory to gets renascence, it is crucial to have a sharper bound based on margin distribution.

For this purpose, Wang et al. [20] presented a sharper bound in term of the *Emargin*, i.e., $\arg\inf_{q\in\{q_0,q_0+\frac{1}{m},\cdots,1\}} KL^{-1}(q;u[\hat{\theta}(q)])$, as follows:

Theorem 3. *(Wang et al., 2008) For any $\delta > 0$, if $8 < |\mathcal{H}| < \infty$, with probability at least $1 - \delta$ over the random choice of sample S with size $m > 1$, every voting classifier $f \in C(\mathcal{H})$ satisfies the following bound:*

$$\Pr_D[yf(\boldsymbol{x}) < 0] \leq \frac{\ln|\mathcal{H}|}{m} + \inf_{q\in\{q_0,q_0+\frac{1}{m},\cdots,1\}} KL^{-1}(q;u[\hat{\theta}(q)]), \quad (5)$$

where $q_0 = \Pr_S\left[yf(\boldsymbol{x}) \leq \sqrt{8/|\mathcal{H}|}\right] < 1$, $u[\hat{\theta}(q)] = \frac{1}{m}\left(\frac{8\ln|\mathcal{H}|}{\hat{\theta}^2(q)}\ln\frac{2m^2}{\ln|\mathcal{H}|} + \ln|\mathcal{H}| + \ln\frac{m}{\delta}\right)$, $\hat{\theta}(q) = \sup\left\{\theta \in \left(\sqrt{8/|\mathcal{H}|},1\right] : \Pr_S[yf(\boldsymbol{x}) \leq \theta] \leq q\right\}$.

Here $KL^{-1}(q;u) = \inf_w\{w: w \geq q \text{ and } KL(q||w) \geq u\}$ is the inverse of the KL divergence $KL(q||\cdot)$ for a fixed q. Notice that the factors considered by (5) are different from that considered by (3) and (4). Though (5) was believed to be a generalization bound based on margin distribution, the Emargin is too unintuitive to inspire algorithm design.

Several years later, Gao and Zhou [5] revealed that both the minimum margin and Emargin are special cases of the *k-th margin*, which is still a single margin. Fortunately, they proved a sharper generalization bound based on margin distribution as follows by considering the same factors as in (3) and (4).

Theorem 4. *(Gao and Zhou, 2013) For any $\delta > 0$, with probability at least $1-\delta$ over the random choice of sample S with size $m \geq 5$, every voting classifier $f \in C(\mathcal{H})$ satisfies the following bound:*

$$\Pr_D[yf(\boldsymbol{x}) < 0] \leq \frac{2}{m} + \inf_{\theta\in(0,1]}\left[\Pr_S[yf(\boldsymbol{x}) < \theta] + \frac{7\mu + 3\sqrt{3\mu}}{3m} + \sqrt{\frac{3\mu}{m}\Pr_S[yf(\boldsymbol{x}) < \theta]}\right],$$
$$(6)$$

where $\mu = \frac{8}{\theta^2}\ln m\ln(2|\mathcal{H}|) + \ln\frac{2|\mathcal{H}|}{\delta}$.

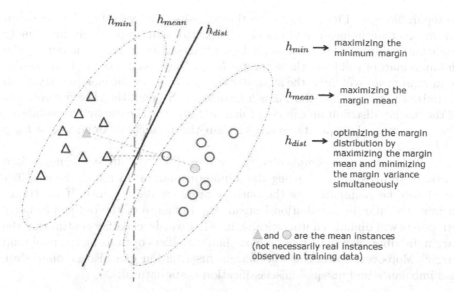

Fig. 1. A simple illustration of linear separators optimizing the minimum margin, margin mean and margin distribution, respectively

This result proves the essentiality of margin distribution to generalization performance. Thus, the margin theory for Boosting finally stands.[2]

Now, it is clear that the margin distribution can be improved further even after the training error reaches zero, and therefore, the generalization performance of AdaBoost can be improved further if the training process continues. This also implies that overfitting will finally occur, although very late, since the margin distribution cannot be improved endlessly. As for the contradictory to the Occam's razor, now our understanding is that the complexity of ensemble models is related to not only the number of learners but also the structural relation between the learners; thus, including more base learners in an ensemble does not necessarily lead to a higher model complexity. This is likely to be relevant to the diversity issue of ensemble methods [22], and theoretical exploration of this point may offer model complexity some new comprehension.

3   Optimizing Margin Distribution

Fig. 1 provides a simple illustration. Suppose we are trying to separate two categories of data points, i.e., red circles and blue triangles. For simplicity, consider

---

[2] Notice that instead of considering the whole function space, there are some studies about data-dependent margin-based generalization bounds, based on techniques such as the empirical cover number [15], empirical fat-shattering dimension [2] and Rademacher and Gaussian complexities [9, 10]. Some of these bounds are proven to be sharper than (3), but hard to show sharper than (4)-(6). Moreover, they fail to explain the resistance of AdaBoost to overfitting.

the separable case. First, we can see that classifiers maximizing the minimum margin, the margin mean[3] and the margin distribution, respectively, are usually significantly different. For example, in Fig. 1 the classifier trying to maximize the minimum margin will favor the separator h_{min}, the classifier trying to maximize the margin mean will favor the separator h_{mean}, whereas the classifier trying to maximize the margin distribution will favor h_{dist}. Second, the classifier optimizing the margin distribution can be intuitively better as the predictive confidence of h_{dist} on most data points are larger than the predictive confidence of h_{min} and h_{mean}.

Fig. 2 shows a more complicated case where there are outliers or noisy data points. If we insist on optimizing the minimum margin, in Fig. 2 the classifier will almost be dominated by the outliers or noisy data points. If we try to optimize the margin distribution instead, the influence of the outliers or noisy data points will diminish automatically. In other words, classifiers optimizing the margin distribution will be more robust than classifiers optimizing the minimum margin. Moreover, optimizing the margin distribution can also accommodate class imbalance and unequal misclassification costs naturally.

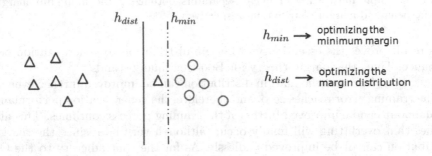

Fig. 2. Another illustration of linear separators with outliers or noisy data points

Notice that though the theoretical results proving the essentiality of margin distribution in Section 3 were derived for Boosting, the implications are far beyond Boosting. There are many learning approaches trying to optimize actually a single margin, particularly the minimum margin; the most famous representatives are SVMs.

For SVMs, $f(x) = w^\top \phi(x)$ where w is a linear predictor, $\phi(x)$ is a feature mapping of x induced by a kernel k, i.e., $k(x_i, x_j) = \phi(x_i)^\top \phi(x_j)$. Given an example (x, y), similar to that in Section 2, the margin γ w.r.t. f is defined as $yf(x)$ [4, 19]:

$$\gamma = yf(x) = yw^\top \phi(x). \tag{7}$$

---

[3] Notice that the mean instances are not necessarily observed in training data.

The SVMs formulation for separable case (hard-margin SVMs) is indeed a maximization of the minimum margin, i.e., $\min\{\gamma_i\}_{i=1}^m$:

$$\min_{\boldsymbol{w}} \frac{1}{2}\boldsymbol{w}^\top \boldsymbol{w} \tag{8}$$
$$\text{s.t. } y_i \boldsymbol{w}^\top \phi(\boldsymbol{x}_i) \geq 1$$
$$i = 1, \cdots, m.$$

The formulation for non-separable case (soft-margin SVMs) introduces the slack variables $\boldsymbol{\xi} = [\xi_1, \cdots, \xi_m]^\top$ to measure the losses of different instances, where C is a trading-off parameter:

$$\min_{\boldsymbol{w},\boldsymbol{\xi}} \frac{1}{2}\boldsymbol{w}^\top \boldsymbol{w} + C\sum_{i=1}^m \xi_i \tag{9}$$
$$\text{s.t. } y_i \boldsymbol{w}^\top \phi(\boldsymbol{x}_i) \geq 1 - \xi_i,$$
$$\xi_i \geq 0, \ i = 1, \cdots, m.$$

There exists a constant $\bar{C}$ such that (9) can be equivalently reformulated as follows, showing that the soft-margin SVMs are maximizing the k-th margin (i.e., the k-th smallest margin) [5]:

$$\max_{\boldsymbol{w}} \ \gamma_0 - \bar{C}\sum_{i=1}^m \xi_i \tag{10}$$
$$\text{s.t. } \gamma_i \geq \gamma_0 - \xi_i,$$
$$\xi_i \geq 0, \ i = 1, \cdots, m.$$

Hence, both hard-margin and soft-margin SVMs are indeed trying to optimize a single margin. It is very likely that they can be improved by replacing the optimization of a single margin by the optimization of margin distribution, while keeping the other parts of their solution strategies unchanged; this also applies to other large margin methods. Thus, the large margin distribution learning offers a promising way to derive more powerful learning approaches by simple adaptations.

To accomplish large margin distribution learning, we need to understand how to optimize the margin distribution. Reyzin and Schapire [12] suggested to maximize the average or median margin, and there are also efforts on maximizing the average margin or weighted combination margin [1,6,11]. These arguments, however, are all heuristics without theoretical justification.

In addition to (6), Gao and Zhou [5] proved anther form of their margin theorem, disclosing that the average or median mean is not enough, and to characterize the margin distribution, it is important to consider not only the *margin mean* but also the *margin variance*. This suggests a new direction for algorithm design, i.e., to optimize the margin distribution by maximizing the margin mean and minimizing the margin variance simultaneously. This argument has got supported empirically by some recent Boosting studies [16,17].

4 A Simple Implementation of Large Margin Distribution Learning

For a straightforward implementation of large margin distribution learning, as an example, we adapt the simple SVMs formulation (8) to the optimization of margin distribution [21].

Denote $X = [\phi(x_1), \cdots, \phi(x_m)]$ as the matrix whose i-th column is $\phi(x_i)$, $y = [y_1, \cdots, y_m]^\top$, and Y as a $m \times m$ diagonal matrix whose diagonal elements are $y_1, \cdots, y_m$. According to the definition in (7), the margin mean is

$$\bar{\gamma} = \frac{1}{m}\sum_{i=1}^{m} y_i w^\top \phi(x_i) = \frac{1}{m}(Xy)^\top w, \tag{11}$$

and the margin variance is

$$\hat{\gamma} = \frac{1}{m^2}\sum_{i=1}^{m}\sum_{j=1}^{m}(y_i w^\top \phi(x_i) - y_j w^\top \phi(x_j))^2$$
$$= \frac{2}{m^2}(m w^\top X X^\top w - w^\top X y y^\top X^\top w). \tag{12}$$

By incorporating into (8) the maximization of margin mean and the minimization of margin variance simultaneously, we get the hard-margin LDM (Large Margin distribution Machine) formulation [21]:

$$\min_{w} \frac{1}{2} w^\top w + \lambda_1 \hat{\gamma} - \lambda_2 \bar{\gamma} \tag{13}$$
$$\text{s.t.}\ \ y_i w^\top \phi(x_i) \geq 1$$
$$i = 1, \cdots, m,$$

where λ_1 and λ_2 are trading-off parameters. It is evident that (8) is a special case of (13) when λ_1 and λ_2 equal zero.

Similarly, we have the soft-margin LDM which degenerates to (10) when λ_1 and λ_2 equals zero:

$$\min_{w,\xi} \frac{1}{2} w^\top w + \lambda_1 \hat{\gamma} - \lambda_2 \bar{\gamma} + C\sum_{i=1}^{m} \xi_i \tag{14}$$
$$\text{s.t.}\ \ y_i w^\top \phi(x_i) \geq 1 - \xi_i,$$
$$\xi_i \geq 0,\ i = 1, \cdots, m.$$

Notice that in (14) the influence of the $C\sum_{i=1}^{m}\xi_i$ term can be subsumed by the λ_1 and λ_2 terms, whereas we keep it to let (14) and (10) look similar such that it is easy to perceive that adapting the soft-margin SVMs to the optimization of margin distribution is quite straightforward.

Solving (13) and (14) is not difficult. For example, by substituting (11)-(12), (14) leads to a quadratic programming problem:

$$\min_{\boldsymbol{w},\boldsymbol{\xi}} \frac{1}{2}\boldsymbol{w}^\top\boldsymbol{w} + \frac{2\lambda_1}{m^2}(m\boldsymbol{w}^\top\boldsymbol{X}\boldsymbol{X}^\top\boldsymbol{w} - \boldsymbol{w}^\top\boldsymbol{X}\boldsymbol{y}\boldsymbol{y}^\top\boldsymbol{X}^\top\boldsymbol{w})$$

$$- \lambda_2\frac{1}{m}(\boldsymbol{X}\boldsymbol{y})^\top\boldsymbol{w} + C\sum_{i=1}^{m}\xi_i \tag{15}$$

$$\text{s.t.} \ \ y_i\boldsymbol{w}^\top\phi(\boldsymbol{x}_i) \geq 1 - \xi_i, \tag{16}$$

$$\xi_i \geq 0, \ i = 1, \cdots, m.$$

A dual coordinate descent method for kernel LDM and an average stochastic gradient descent method for large-scale linear kernel LDM have been developed, with details in [21]. Table 1 shows some experimental results of comparing LDM to SVM, where it can be seen that LDM is significantly better on more than half of the experimental datasets and never worse than SVM. Such a simple implementation of large margin distribution learning also exhibits superior performance to many other related methods [1, 6, 11] in experiments [21].

Table 1. Comparing predictive accuracy (mean±std.) of SVM and LDM. •/∘ indicates the performance of LDM is significantly better/worse than SVM (paired t-tests at 95% significance level). The win/tie/loss counts are summarized in the last row.

Data sets	Linear kernel		RBF kernel	
	SVM	LDM	SVM	LDM
promoters	.723±.071	.721±.069	.684±.100	.715±.074•
planning-relax	.683±.031	.706±.034•	.708±.035	.707±.034
colic	.814±.035	.832±.026•	.822±.033	.841±.018•
parkinsons	.846±.038	.865±.030•	.929±.029	.927±.029
colic.ORIG	.618±.027	.619±.042	.638±.043	.641±.044
sonar	.725±.039	.736±.036	.842±.034	.846±.032
vote	.934±.022	.970±.014•	.946±.016	.968±.013•
house	.942±.015	.968±.011•	.953±.020	.964±.013•
heart	.799±.029	.791±.030	.808±.025	.822±.029•
breast-cancer	.717±.033	.725±.027•	.729±.030	.753±.027•
haberman	.734±.030	.738±.020	.727±.024	.731±.027
vehicle	.959±.012	.959±.013	.992±.007	.993±.006
clean1	.803±.035	.814±.019•	.890±.020	.891±.024
wdbc	.963±.012	.968±.011•	.951±.011	.961±.010•
isolet	.995±.003	.997±.002•	.998±.002	.998±.002
credit-a	.861±.014	.864±.013•	.858±.014	.861±.013
austra	.857±.013	.859±.015	.853±.013	.857±.014•
australian	.844±.019	.866±.014•	.815±.014	.854±.016•
fourclass	.724±.014	.723±.014	.998±.003	.998±.003
german	.711±.030	.738±.016•	.731±.019	.743±.016•
w/t/l (SVM vs. LDM)	0/8/12		0/10/10	

5   Conclusion

Recently the margin theory for Boosting has been defended [5], showing that the *margin* is not only a fundamental issue of SVMs but also an essential factor of Boosting. In contrast to previous belief on single margins such as the minimum margin optimized by SVMs, the recent theoretical results disclosed that the *margin distribution* rather than a single margin is crucial for the generalization performance. Inspired by this recognition, in this article we advocate *large margin distribution learning*. We also briefly introduce how the SVMs can be easily adapted to large margin distribution learning by maximizing the margin mean and minimizing the margin variance simultaneously, while such a simple implementation leads to the LDMs that exhibit superior performance to SVMs [21]. Overall, large margin distribution learning exhibits a promising direction to derive powerful learning approaches.

Acknowledgments. This article summarizes the author's keynote talk at the ANNPR'2014, Montreal, Canada. The author was supported by the National Science Foundation of China (61333014, 61321491).

References

1. Aiolli, F., San Martino, G., Sperduti, A.: A kernel method for the optimization of the margin distribution. In: Proceedings of the 18th International Conference on Artificial Neural Networks, Prague, Czech, pp. 305–314 (2008)
2. Antos, A., Kégl, B., Linder, T., Lugosi, G.: Data-dependent margin-based generalization bounds for classification. Journal of Machine Learning Research 3, 73–98 (2002)
3. Breiman, L.: Prediction games and arcing classifiers. Neural Computation 11(7), 1493–1517 (1999)
4. Cortes, C., Vapnik, V.: Support-vector networks. Machine Learning 20(3), 273–297 (1995)
5. Gao, W., Zhou, Z.-H.: On the doubt about margin explanation of boosting. Artificial Intelligence 199-200, 22–44 (2013) (arXiv:1009.3613, September 2010)
6. Garg, A., Roth, D.: Margin distribution and learning algorithms. In: Proceedings of the 20th International Conference on Machine Learning, Washington, DC, pp. 210–217 (2003)
7. Grove, A.J., Schuurmans, D.: Boosting in the limit: Maximizing the margin of learned ensembles. In: Proceedings of the 15th National Conference on Artificial Intelligence, Menlo Park, CA, pp. 692–699 (1998)
8. Kearns, M., Valiant, L.G.: Cryptographic limitations on learning boolean formulae and finite automata. In: Proceedings of the 21st Annual ACM Symposium on Theory of Computing, Seattle, WA, pp. 433–444 (1989)
9. Koltchinskii, L., Panchanko, D.: Empirical margin distributions and bounding the generalization error of combined classifiers. Annuals of Statistics 30(1), 1–50 (2002)
10. Koltchinskii, L., Panchanko, D.: Complexities of convex combinations and bounding the generalization error in classification. Annuals of Statistics 33(4), 1455–1496 (2005)

11. Pelckmans, K., Suykens, J., Moor, B.D.: A risk minimization principle for a class of parzen estimators. In: Platt, J.C., Koller, D., Singer, Y., Roweis, S. (eds.) Advances in Neural Information Processing Systems 20, pp. 1137–1144. MIT Press, Cambridge (2008)
12. Reyzin, L., Schapire, R.E.: How boosting the margin can also boost classifier complexity. In: Proceeding of 23rd International Conference on Machine Learning, Pittsburgh, PA, pp. 753–760 (2006)
13. Schapire, R.E.: The strength of weak learnability. Machine Learning 5(2), 197–227 (1990)
14. Schapire, R.E., Freund, Y., Bartlett, P.L., Lee, W.S.: Boosting the margin: A new explanation for the effectives of voting methods. Annuals of Statistics 26(5), 1651–1686 (1998)
15. Shawe-Taylor, J., Williamson, R.C.: Generalization performance of classifiers in terms of observed covering numbers. In: Proceedings of the 4th European Conference on Computational Learning Theory, Nordkirchen, Germany, pp. 153–167 (1999)
16. Shen, C., Li, H.: Boosting through optimization of margin distributions. IEEE Transactions on Neural Networks 21(4), 659–666 (2010)
17. Shivaswamy, P.K., Jebara, T.: Variance penalizing AdaBoost. In: Shawe-Taylor, J., Zemel, R.S., Bartlett, P.L., Pereira, F.C.N., Weinberger, K.Q. (eds.) Advances in Neural Information Processing Systems 24, pp. 1908–1916. MIT Press, Cambridge (2011)
18. Smola, A.J., Bartlett, P.L., Schölkopf, B., Schuurmans, D. (eds.): Advances in Large Margin Classifiers. MIT Press, Cambridge (2000)
19. Vapnik, V.: The Nature of Statistical Learning Theory. Springer, New York (1995)
20. Wang, L., Sugiyama, M., Yang, C., Zhou, Z.-H., Feng, J.: On the margin explanation of boosting algorithm. In: Proceedings of the 21st Annual Conference on Learning Theory, Helsinki, Finland, pp. 479–490 (2008)
21. Zhang, T., Zhou, Z.-H.: Large margin distribution machine. In: Proceedings of the 20th ACM SIGKDD Conference on Knowledge Discovery and Data Mining, New York, NY (2014)
22. Zhou, Z.-H.: Ensemble Methods: Foundations and Algorithms. CRC Press, Boca Raton (2012)

A Decorrelation Approach for Pruning of Multilayer Perceptron Networks

Hazem M. Abbas

The German University in Cairo
Faculty of Media Engineering and Technology
New Cairo, Cairo, Egypt
hazem.abbas@guc.edu.eg,
http://met.guc.edu.eg

Abstract. In this paper, the architecture selection of a three–layer non-linear feedforward network with linear output neurons and sigmoidal hidden neurons is carried out. In the proposed method, the conventional back propagation (BP) learning algorithm is used to train the network by minimizing the representation error. A new pruning algorithm employing statistical analysis can quantify the importance of each hidden unit. This is accomplished by providing lateral connections among the neurons of the hidden layer and minimizing the variance of the hidden neurons. Variance minimization has resulted in decorrelated neurons and thus the learning rule for the lateral connections in the hidden layer becomes a variation of the anti-Hebbian learning. The decorrelation process minimizes any redundant information transferred among the hidden neurons and therefore enables the network to capture the statistical properties of the required input-output mapping using the minimum number of hidden nodes. Hidden nodes with least contribution to the error minimization at the output layer will be pruned. Experimental results show that the proposed pruning algorithm correctly prunes irrelevant hidden units.

Keywords: Neural Networks, Backpropagation Learning, Optimal Network Architecture, Pruning Algorithms, Statistical Learning.

1 Introduction

Finding an optimal architecture of feedforward neural networks is a very important issue for both classification and approximation problems. A small architecture will be unable to capture the internal representation required to perform the required input-output mapping. On the other hand, a large architecture will tend to over-fit the training data which leads to poor generalization capabilities of the designed network. It is therefore necessary to design the network with the smallest architecture and still can perform satisfactorily with unseen data. The generalization of a neural network architecture can be assessed by changing the the network size, i.e., the number of nodes and/or weights. There are many approaches to tackle the network design problem. Network construction algorithms starts with a small number of hidden nodes and then grows additional hidden

N. El Gayar et al. (Eds.): ANNPR 2014, LNAI 8774, pp. 12–22, 2014.

nodes and/or weights until a satisfactory design is found [1, 2, 3, 4]. Pruning algorithms start with a seemingly large network that is trained until an acceptable performance is achieved. Based on certain criteria, some hidden units or weights can be removed if they are considered useless [5, 6, 7, 8, 9, 10]. A recent survey on pruning methods can be found in [11] and older ones in [12, 13]. The third approach employs regularization techniques, which involves the addition of a penalty term to the objective function to be minimized [14, 15, 16, 17]. Algorithms that combine both constructive and pruning methods have been proposed in [18, 19].

The algorithm proposed in this work belongs to the pruning algorithms family. Normally, individual weights, hidden units and/or input units are the parameters that can be considered for removal. Two methods are normally employed to remove any of these candidate parameters: sensitivity analysis and role interpretation of the node. Sensitivity analysis techniques quantify the relevance of a network parameter as how important a slight deviation in a network parameter on the network performance [20, 21, 22]. The sensitivity measures could fail to identify possible correlations among nodes and possess high computational complexity. Node pruning techniques are post-training algorithms where the correlations among nodes in the hidden and output layers are exploited to decide which node to remove [23, 24, 9].

In this paper, a pruning method for a three-layer feedforward network is proposed. The method relies on reducing the variance of hidden layer nodes. Lateral connections among the hidden nodes are provided to accommodate such variance minimization. By doing so, the resulting updating rule will provide a kind of anti-Hebbian learning mechanism that will eventually lead to removal of nodes with least variance or contribution to the network mapping performance.

The paper is organized as follows. Section 2 introduces the proposed network architecture. The proposed decorrelational cost function and the learning rules required to train the network are presented in Section 3. Some necessary conditions on the nature of the node activation functions are discussed in Section 4. The training and pruning algorithm is detailed in Section 5. In Section 6, some simulation results of applying the proposed network and the conventional BP net to some benchmark problems are analyzed.

2   Network Structure

The neural network structure we are concerned with here is the three-layer network: an input layer, I, a hidden layer, H, and an output layer, O (Fig. 1). The neurons in the hidden layer have the sigmoid function while the output layer can have either sigmoidal or linear neurons. Neurons of the input layer can feed into neurons in the following layer through linking weights. Lateral connections are introduced in the hidden layer to decorrelate the output of its neurons. The net input to each node is the sum of the weighted outputs of the nodes feeding into this node.

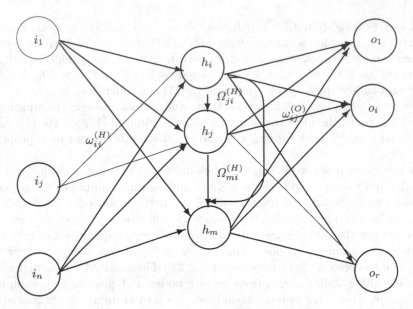

Fig. 1. The proposed network structure

The net input to the ith node in the output layer is calculated as

$$net_i^{(O)} = \sum_j^H \omega_{ij}^{(O)} a_i^{(H)} + \omega_{i0}^{(O)} \tag{1}$$

where $\omega_{ij}^{(O)}$ is the weight connecting the jth hidden node and the ith output node, $\omega_{i0}^{(O)}$ is the threshold (bias) of the node and $a_j^{(H)}$ is the output of the jth node in the hidden layer. Similarly, the net input to the jth node in the hidden layer is

$$net_j^{(H)} = \sum_i^I \omega_{ji}^{(H)} a_i^{(I)} + \omega_{j0}^{(H)} + \sum_{k<j}^H \Omega_{jk}^{(H)} a_k^{(H)} \tag{2}$$

where $\Omega_{jk}^{(H)}$ is a lateral connection from the kth hidden node to the jth hidden node. It should be noted that lateral connections in the hidden layer are only allowed from lower order to higher order nodes as indicated in the second summation. This is to ensure that the network is strictly feedforward. The output of sigmoidal node, j is

$$a_j^{(H)} = f(net_j) = \frac{1}{2}\left(\frac{1-e^{-net_j}}{1+e^{-net_j}}\right) \tag{3}$$

whereas the activation of the linear output nodes is simply

$$a_j^{(O)} = net_j.$$

3 Cost Functions and Training Method

Assume that there are P distinct training patterns, each consisting of a pair of input and target patterns $\{\mathbf{a}^{(I,p)}, \mathbf{t}^{(O,p)}\}$. The network connections, $\{\omega^{(H)}, \omega^{(O)}, \Omega^{(H)}\}$, are to be adjusted in order to satisfy two objectives:

1. all the input/output pairs are mapped within a certain acceptable representation error
2. the variance of hidden nodes are minimized so that neurons with smallest influence on the representation error can be removed

Hence, the overall cost function, $\mathcal{J}$, that needs to be minimized can be expressed as follows:

$$\mathcal{J} = \mathcal{J}_o + \mathcal{J}_h \tag{4}$$

The mapping cost function, $\mathcal{J}_o$ is defined as

$$\mathcal{J}_o(\omega^{(H)}, \omega^{(O)}, \Omega^{(H)}) = \frac{1}{P} \sum_p^P \sum_k^O \left(t_k^{O,p} - a_k^{O,p} \right)^2 \tag{5}$$

where $t_k^{O,p}$ is the kth target component and $a_k^{O,p}$ is the output of the kth node of the output layer when the p pattern pair is presented. The hidden layer variance cost function, $\mathcal{J}_h$ is defined as

$$\mathcal{J}_h(\omega^{(H)}, \Omega^{(H)}) = \frac{1}{P} \sum_p^P \sum_i^H \left(a_i^{H,p} - \overline{a}_i^{H,p} \right)^2 \tag{6}$$

where $\overline{a}_i^{H,p}$ is the average value of hidden node i. The gradient descent learning rule is used to update the all weights of this network, i.e., the weight γ_{kj} is updated along the gradient direction using the equation

$$\Delta \gamma_{ij} = -\eta_\gamma \frac{\partial \mathcal{J}}{\partial \gamma_{ij}}$$

where $\gamma_{ij} \in \{\omega^{(H)}, \omega^{(O)}, \Omega^{(H)}\}$ and η_γ is a suitable learning rate that might differ with the type of γ_{kj}. It is evident that the backpropagation (BP) learning rules [25] can be applied to the three sets of weights when the $\mathcal{J}_o$ is minimized. Therefore, and dropping the variable p, the BP algorithm is applied as follows:

$$\Delta \gamma_{ij} = -\eta_\gamma \frac{\partial \mathcal{J}_o}{\partial \gamma_{ij}} = -\eta_\gamma \frac{\partial \mathcal{J}_o}{\partial net_i} \cdot \frac{\partial net_i}{\partial \gamma_{ij}} = \eta_\gamma \, \delta_i \, a_j \tag{7}$$

where η_γ is the learning rate and $\delta_i = -\frac{\partial \mathcal{J}_o}{\partial net_i}$ which is defined as

$$\delta_i = \begin{cases} \left(t_i^{(O)} - a_i^{(O)} \right) f'\left(net_i^{(O)} \right) & \text{if } i \text{ is an output node} \\ f'\left(net_i^{(H)} \right) \sum_l \delta_l^{(O)} \, \omega_{li}^{(O)} & \text{if } i \text{ is a hidden node.} \end{cases} \tag{8}$$

In addition, the two sets of weights, $\{\omega^{(H)}, \Omega^{(H)}\}$, will be also updated to minimize the hidden layer cost function, $\mathcal{J}_h$. Assuming that $\phi \in \{\omega^{(H)}, \Omega^{(H)}\}$, the update rule for ϕ is defined as

$$\Delta\phi_{ij} = -\eta_\phi \frac{\partial \mathcal{J}_h}{\partial \phi_{ij}} = -\eta_\phi \frac{\partial \mathcal{J}_h}{\partial a_i^{(H)}} \cdot \frac{\partial a_i^{(H)}}{\partial net_i^{(H)}} \cdot \frac{\partial net_i^{(H)}}{\partial \phi_{ij}} = -\eta_\phi \, a_i^{(H)} \, f' \left(net_i^{(H)} \right) a_j^{(H/I)} \quad (9)$$

Here, $a_j^{(H/I)}$ is either the output of an input node, $j \in I$ or a hidden node with $j \in H$, and $j < i$.

Although the BP rules (7,8) will reduce the system error to one of its local minima, it does not guarantee that the smallest number of hidden nodes will be employed in the representation task. Actually the error reduction task at the hidden layer is distributed over the all nodes and different nodes can contribute to the same task. Nothing in the BP rules imply that the correlation between the nodes is minimized. By inspecting the update rule, (9), one can observe that the weight update is proportional to the negative product of pre- and post-synaptic activations of the two nodes. This is a form of the anti-Hebbian learning rule [26] that is normally employed to decorrelate the output of two neurons. This rule has been employed in networks that extract the principal components of the presented data [27]. An anti-Hebbian learning rule can accomplish this decorrelation process. Hebb [28] had suggested that an excitatory connection between two neurons be strengthened if their activities are correlated and weakened otherwise. Similarly, an inhibitory connection will be strengthened if the activities were uncorrelated and weakened otherwise. This leads to the anti-Hebbian learning rule

$$\Delta\Omega_{ij}^{(H)} \propto -\beta \, a_i^{(H)} \, a_j^{(H)}. \quad (10)$$

Ideally, at convergence, when the average change in $\Omega_{ij}^{(H)}$ is zero, i.e., $E\{\Delta\Omega_{ij}^{(H)}\}$ $\rightarrow 0$, the correlation, $E\{a_i^{(H)} a_j^{(H)}\}$, must vanish. This ensures that the hidden layer activations are orthogonal. The rule (10) differs from the conventional Hebb rule only in the sign before its learning rate, β.

4   Conditions on the Neuron Activations and Network Stability

The application of this decorrelation rule (9) dictates that the neurons are of a certain sigmoidal shape. Also, since the network employs lateral connections in the hidden layer, the stability of the performance should be investigated.

If the hidden layer were composed of sigmoidal neurons whose activation values take any real positive value in the range $[0,1]$, then the condition (9) will be satisfied only when all hidden neurons activations are at zero level, i.e., the trivial solution of this function minimization. Hence, there will be no internal representation at the hidden layer and thus the system error, $\mathcal{J}$, will be at a very poor local minimum. This makes it necessary that the hidden nodes should produce both positive and negative values. Therefore, a bioploar sigmoid should be utilized.

The major problem with Hebbian learning is that the weights can grow indefinitely. This has led some researchers to modify the Hebb rule either by doing weight normalization [28] or by adding a forgetting factor [29]. Fortunately, the anti-Hebbian rule, (10), is stable without providing any modification and the weights, $\{\Omega\}$, will be bounded. This can be shown as follows. Assume that the weight Ω_{ij} increases, then using (2) and (3), the activation of node i, will increase. By virtue of the updating rule, (10), Ω_{ij} will decrease.

It should be also noted that decorrelation process performed at the hidden layer is similar to the Gram-Schmidt (GS) orthogonalization [30] applied to network pruning [31, 32]. However, there are many differences both in architecture and learning procedure. The GS nets generate a decorrelated version of the hidden layer output by introducing another layer after the hidden layer. The proposed architecture directly generates a decorrelated hidden layer output by simply adding and training lateral connections. Also, the GS orthogonalization is carried out using linear mapping. In this work, the anti-Hebbian decorrelation is embedded into the nonlinear operation of the sigmoid. Moreover, the plain BP rules cannot be applied directly with the GS nets due to the existence of the decorrelation layers.

5   The Training and Pruning Algorithm

The Algorithm (1) describes the procedure of training the network using the BP algorithm and how to select hidden neurons to remove from the hidden layer. First, the weights and biases of an $(n - m - r)$ three–layer network are initialized to small random values. Here, n, m and r stand for the number of neurons in the three layers, respectively. Then, the BP learning algorithm is applied to perform the training task. After every sweep of the whole set of data patterns, the error, J_o (7), is calculated and compared to the error threshold, ϵ. If, $J_o > \epsilon$, the training phase should be resumed till the error becomes less than ϵ.

At this point, the network is able to find the set of weights that can represent the required input-output mapping within a certain threshold. There are two possibilities regarding the hidden layer. The first is that all hidden neurons are fully utilized in mapping process. The other possibility is that the number of active hidden neurons could be reduced without (probably) violating the specified error threshold. A selection criterion needs to be applied to choose which neuron to remove or to become *inactive*. A reasonable criterion is *to excise the hidden node which has the least contribution in the error reduction process*. Alternatively, it is the node which, if removed, will result in the smallest error increase. A suitable way to perform this is to calculate the *average correlation coefficient*, $\overline{\rho}_i$, between the ith hidden neuron and the error at the output layer provided that the ith hidden neuron is not contributing to the output. The correlation coefficient, $\rho(i, j)$, between the output of ith hidden node, h_i, and the error element at jth output neuron, $e_i = t_i - o_i$, assuming that neuron i is inactive, is defined as:

$$\rho(i, j) = \frac{\mathrm{cov}\,(i, j)}{\sqrt{[\mathrm{var}(i)\,\mathrm{var}(j)]}} \tag{11}$$

Algorithm 1. Backpropagation Training and Pruning Algorithm

Initialize the network weights with small random value
while Minimum hidden nodes is not reached **do**
   repeat
      for Pattern $= 1 : P$ **do**
         1. Present input pattern at the input layer
         2. Starting from the input layer, use Eqns. (1,2,3) to compute
the activities of the neurons at each layer.
         3.Calculate the error, $\mathcal{J}_o$, at the output layer
         4. Compute the variable, δ_i (8), for all nodes in
the ouput and hidden layer, respectively.
         5. Compute the change of weights of the three set of weights using

$$\Delta\omega_{ij}^{(O)} = \eta_\gamma \delta_i^{(O)} a_j^{(H)}$$

$$\Delta\omega_{ij}^{(H)} = \eta_\gamma \delta_i^{(H)} a_j^{(I)} - \eta_\phi a_i^{(H)} a_i^{\prime(H)} a_j^{(I)}$$

$$\Delta\Omega_{ij}^{(H)} = \eta_\gamma \delta_i^{(H)} a_j^{(H)} - \eta_\phi a_i^{(H)} a_i^{\prime(H)} a_j^{(I)}$$

         6. Updates the weights
      end for
      Calculate Representation error, $\mathcal{J}_o$, after the update
    until Representation error is within threshold, $\mathcal{J}_o < \epsilon$
    Calculate the contribution of each hidden node in reducing $\mathcal{J}_o$
    Remove the hidden node with the least contribution, $\overline{\rho_i}$ (12)
end while

---

where $\text{cov}(i,j) = E\left\{([t_j - o_j] + \omega_{ji}h_i)(h_i)\right\}$, with $\text{var}(i) = E\left\{h_i^2\right\}$ and var
$(j) = E\left\{([t_j - o_j] + \omega_{ji}h_i)^2\right\}$. The average correlation coefficient, $\overline{\rho_i}$, then becomes

$$\overline{\rho_i} = \frac{1}{n}\sum_{j=1}^{r} \rho(i,j). \tag{12}$$

The criterion guarantees that the neurons with maximal error reduction will be left intact, while improving the convergence by removing the less contributing neurons. It should be noted that the removal of hidden nodes requires a readjustment of the bias vector of the remaining nodes. Thus, an increase in $\mathcal{J}_o$ will be noticed for the first few sweeps after the excision. However, further training will fix this problem.

6 Simulation Results

In this section, we present some simulations of the conventional BP algorithm and the proposed model when they were applied to two benchmark problems: the XOR and the three-bit parity problems. For both models, a momentum term was added to the weight change formula, (7), i.e.,

$$\Delta\omega_{kj}(l) = \eta\,\delta_k(n)\,a_j(l) + \alpha\,\Delta\omega_{kj}(l-1)$$

where α is the momentum factor and l is the current iteration number. To compare the performance of the two networks, they had to be initialized to the same values. The parameters of each network were updated after every training epoch, a sweep of the presentation of the entire training set. The learning is considered complete when the performance index, $\mathcal{J}_o$, went below 0.0001.

6.1   The XOR Problem

Using a conventional BP algorithm, it was found that only two hidden nodes were needed to solve the XOR classification. Hence, the weights of a larger network with a 2–4–1 structure was initialized with small random values. The network is trained using the proposed BP Pruning algorithm. Learning rates were set to: $\eta_\gamma = 0.7$, $\alpha = 0.2$, and $\eta_\phi = 0.3$. The initial values of $\Omega_{ij}^{(H)}$ were set to zero in order to ensure the proposed network will have the same initial start as the BP network. The network was allowed to train till the error threshold was reached after 354 epochs. The calculated values of $\overline{\rho_i}, i = 1, \cdots, 4$ for the 4 hidden nodes were reported to be $+0.2260, +\mathbf{0.0032}, +0.5714$, and $+0.7889$, respectively. The correlation matrix, $C_H = E\{(h - \bar{h})(h - \bar{h})\}$, with h being the activation values of hidden layer nodes, is found to be

$$
\begin{pmatrix}
0.0139 & 0 & 0 & 0 \\
+0.0001 & \mathbf{0.0000} & 0 & 0 \\
-0.0046 & -0.0004 & 0.0068 & 0 \\
-0.0015 & +0.0005 & -0.0036 & 0.0054
\end{pmatrix}
$$

When the last step in Algorithm (1) is reached and decision about which node to remove, it is quite obvious that node 2 with variance equal to zero and minimum $\overline{\rho_2}$ is to be removed. The matrix clearly shows that the hidden nodes were highly decorrelated. When the algorithm is resumed after removing the second hidden node, the target error was reached after further 26 epochs of training. The $\overline{\rho_i}$ values are $(\mathbf{0.2307}, 0, +0.6184, +0.7513)$ and

$$
C_H =
\begin{pmatrix}
0.0139 & 0 & 0 & 0 \\
0 & 0 & 0 & 0 \\
-0.0038 & 0 & 0.0069 & 0 \\
-0.0010 & 0 & -0.0043 & 0.0043
\end{pmatrix}
$$

Here, node 1 will have the least effect on reducing the representation error and is chosen to be excised. After training for extra for extra 400 epochs, the reported values were $\overline{\rho_i} = (0, 0, +0.4317, +0.9020)$ and

$$
C_H =
\begin{pmatrix}
0 & 0 & 0 & 0 \\
0 & 0 & 0 & 0 \\
0 & 0 & 0.0131 & 0 \\
0 & 0 & +0.0035 & 0.0289
\end{pmatrix}
$$

Evidently, the variance of the two remaining nodes have increased to account for the lost representation by the removed node while the anti-Hebbian sort of learning kept the remaining two nodes largely uncorrelated. A very important observation has been reported when a conventional BP network is trained. While the BP network tends to almost equally distribute the internal representation task between the hidden nodes (nearly equal autocorrelation values), the proposed network, on the other hand, forced the hidden nodes to discover only the necessary features. This is represented by the unequal autocorrelation values in $\mathbf{C}_H$. It is worth mentioning that in implementing the algorithm, no matrix calculation is required. The computation of the correlation coefficients, $\rho(i,j)$, is carried out using local information and can be easily performed in a recursive fashion. The matrices, $\mathbf{C}_H$, shown above are for sole purpose of demonstrating the decorrelation capabilities of the proposed method.

6.2   The 3-Bit Parity Problem:

A 3-bit parity problem needs only a two hidden nodes to accomplish a 100% correct mapping. A 3-7-1 network was initialized and trained using the proposed BP pruning algorithm with the same learning parameters used with the XOR problem. After 1600 epochs, the correlation variable $\overline{\rho_i}$ had the following values for the 7 nodes: $+0.1428, +0.9469, +0.2721, +\mathbf{0.0019}, +0.0866, +0.0381, +\mathbf{0.0046}$. Clearly, nodes 7 and 10 have the least contribution and both have zero variance. Node 10 was chosen to be removed. At this point, both nodes could have been removed simultaneously, but selecting one node at a time will help to better understand the learning behaviour. As expected, further training for 100 epochs resulted in removing node 7. Correlation between the removed node and its predecessors are all null which suggests total independence between the nodes in solving the problem. Continuing with the training and pruning process, nodes 9, 8, and 5 were excised at epoch 1786, 2030, and 2155, respectively. The remaining two nodes had a very small correlation value of 0.0004.

The algorithm still needs to be tested with a larger classification data set such as the UCI Machine Learning database and be compared with other pruning algorithms mentioned in Section 1. Also, the generalization capabilities of the resulting network has to be assessed and compared with the a network of the same size that is trained with the conventional BP algorithm. Finally, the proposed method needs to be compared to a simple cross-validation approach in terms of complexity and performance, i.e., how the resulting optimal NN structure, trained on a subset of the data, will perform over the majority of the subsets.

7   Conclusions

In this work, a pruning algorithm for the hidden neurons of a three–layer network was investigated. The network has linear output neurons and bipolar sigmoidal hidden neurons. The algorithm works by providing lateral connection among the hidden nodes in such a way that the nodes are only connected to earlier ones in the same hidden layer. Minimizing the variance of hidden nodes resulted in a learning rule that is of anti-Hebbian nature. This anti-Hebbian learning rule has been used to train these lateral connections to orthogonalize the outputs of all hidden nodes. The BP rules have been employed to train all forward connections. Test results indicate that the proposed method managed to find the optimal number of hidden nodes for both the XOR and 3-parity problems which fully decorrelates the hidden layer outputs.

References

[1] Ash, T.: Dynamic node creation in backpropagation networks. Connection Sciences 1, 365–375 (1989)

[2] Christian, S.F., Lebiere, C.: The cascade-correlation learning architecture. In: Advances in Neural Information Processing Systems 2, pp. 524–532. Morgan Kaufmann (1990)

[3] Yau Kwok, T., Yeung, D.Y.: Constructive algorithms for structure learning in feedforward neural networks for regression problems. IEEE Transactions on Neural Networks 8, 630–645 (1997)

[4] Platt, J.: A resource-allocating network for function interpolation. Neural Comput. 3(2), 213–225 (1991)

[5] Han, H.G., Qiao, J.F.: A structure optimisation algorithm for feedforward neural network construction. Neurocomput. 99, 347–357 (2013)

[6] Yau Kwok, T., Yeung, D.Y.: Constructive algorithms for structure learning in feedforward neural networks for regression problems. IEEE Transactions on Neural Networks 8, 630–645 (1997)

[7] Xu, J., Ho, D.W.: A new training and pruning algorithm based on node dependence and jacobian rank deficiency. Neurocomputing 70, 544–558 (2006)

[8] Engelbrecht, A.P.: A new pruning heuristic based on variance analysis of sensitivity information. Trans. Neur. Netw. 12(6), 1386–1399 (2001)

[9] Sietsma, J., Dow, R.J.F.: Creating artificial neural networks that generalize. Neural Network 4(1), 67–79 (1991)

[10] Hassibi, B., Stork, D.G.: Second order derivatives for network pruning: Optimal brain surgeon. In: Advances in Neural Information Processing Systems 5, [NIPS Conference], pp. 164–171. Morgan Kaufmann Publishers Inc., San Francisco (1993)

[11] Augasta, M.G., Kathirvalavakumar, T.: Pruning algorithms of neural networks - a comparative study. Central Europ. J. Computer Science 3(3), 105–115 (2013)

[12] Reed, R.: Pruning algorithms-a survey. Trans. Neur. Netw. 4(5), 740–747 (1993)

[13] Castellano, G., Fanelli, A.M., Pelillo, M.: An iterative pruning algorithm for feedforward neural networks. Trans. Neur. Netw. 8(3), 519–531 (1997)

[14] Girosi, F., Jones, M., Poggio, T.: Regularization theory and neural networks architectures. Neural Comput. 7(2), 219–269 (1995)

[15] Schittenkopf, C., Deco, G., Brauer, W.: Two strategies to avoid overfitting in feedforward networks. Neural Networks 10(3), 505–516 (1997)

[16] Miller, D.A., Zurada, J.M.: A dynamical system perspective of structural learning with forgetting. Trans. Neur. Netw. 9(3), 508–515 (1998)

[17] Ishikawa, M.: Structural learning with forgetting. Neural Netw. 9(3), 509–521 (1996)

[18] Islam, M.M., Murase, K.: A new algorithm to design compact two-hidden-layer artificial neural networks. Neural Netw. 14(9), 1265–1278 (2001)

[19] Ma, L., Khorasani, K.: New training strategies for constructive neural networks with application to regression problems. Neural Netw. 17(4), 589–609 (2004)

[20] Cun, Y.L., Denker, J.S., Solla, S.A.: Advances in neural information processing systems 2, pp. 598–605. Morgan Kaufmann Publishers Inc., San Francisco (1990)

[21] Setiono, R.: A penalty-function approach for pruning feedforward neural networks. Neural Comput. 9(1), 185–204 (1997)

[22] Suzuki, K., Horiba, I., Sugie, N.: A simple neural network pruning algorithm with application to filter synthesis. Neural Process. Lett. 13(1), 43–53 (2001)

[23] Kanjilal, P., Dey, P., Banerjee, D.: Reduced-size neural networks through singular value decomposition and subset selection. Electronic Letters 17, 1515–1518 (1993)

[24] Fletcher, L., Katkovnik, V., Steffens, F., Engelbrecht, A.: Optimizing the number of hidden nodes of a feedforward artificial neural network. In: IEEE World Congress on Computational Intelligenc, pp. 1608–1612 (1998)

[25] Rumelhart, D., Hintont, G., Williams, R.: Learning representations by back-propagating errors. Nature 323(6088), 533–536 (1986)

[26] Carlson, A.: Anti-Hebbian learning in a non-linear neural network. Biol. Cybern. 64(2), 171–176 (1990)

[27] Chen, Z., Haykin, S., Eggermont, J.J., Becker, S.: Correlative Learning: A Basis for Brain and Adaptive Systems (Adaptive and Learning Systems for Signal Processing, Communications and Control Series). Wiley-Interscience (2007)

[28] Rubner, J., Tavan, P.: A self-organizing network for principal-component analysis. EPL (Europhysics Letters) 10(7), 693 (1989)

[29] Oja, E.: A simplified neuron model as a principal component analyzer. Journal of Mathematical Biology 15, 267–273 (1982)

[30] Cheney, W., Kincaid, D.R.: Linear Algebra: Theory and Applications, 1st edn. Jones and Bartlett Publishers, Inc., USA (2008)

[31] Maldonado, F., Manry, M.: Optimal pruning of feedforward neural networks based upon the schmidt procedure. In: The Thirty-Sixth Asilomar Conference on Signals, Systems and Computers, vol. 2 (1998)

[32] Orfanidis, S.: Gram-schmidt neural nets. Neural Computation 2, 116–126 (1990)

Unsupervised Active Learning of CRF Model for Cross-Lingual Named Entity Recognition

Mohamed Farouk Abdel Hady, Abubakrelsedik Karali, Eslam Kamal, and Rania Ibrahim

Microsoft Research
Cairo, Egypt
mohabdel@microsoft.com

Abstract. Manual annotation of the training data of information extraction models is a time consuming and expensive process but necessary for the building of information extraction systems. Active learning has been proven to be effective in reducing manual annotation efforts for supervised learning tasks where a human judge is asked to annotate the most informative examples with respect to a given model. However, in most cases reliable human judges are not available for all languages. In this paper, we propose a cross-lingual unsupervised active learning paradigm (XLADA) that generates high-quality automatically annotated training data from a word-aligned parallel corpus. To evaluate our paradigm, we applied XLADA on English-French and English-Chinese bilingual corpora then we trained French and Chinese information extraction models. The experimental results show that XLADA can produce effective models without manually-annotated training data.

Keywords: Information extraction, named entity recognition, cross-lingual domain adaptation, unsupervised active learning.

1 Introduction

Named Entity Recognition (NER) is an information extraction task that identifies the names of locations, persons, organizations and other named entities in text, which plays an important role in many Natural Language Processing (NLP) applications such as information retrieval and machine translation. Numerous supervised machine learning algorithms such as Maximum Entropy, Hidden Markov Model and Conditional Random Field (CRF) [1] have been adopted for NER and achieved high accuracy. They usually require large amount of manually annotated training examples. However, it is time-consuming and expensive to obtain labeled data to train supervised models. Moreover, in sequence modeling like NER task, it is more difficult to obtain labeled training data since hand-labeling individual words and word boundaries is really complex and need professional annotators. Hence, the shortage of annotated corpora is the obstacle of supervised learning and limits the further development, especially for languages for which such resources are scarce. Active learning is the method which,

N. El Gayar et al. (Eds.): ANNPR 2014, LNAI 8774, pp. 23–34, 2014.
© Springer International Publishing Switzerland 2014

instead of relying on random sampling from the large amount of unlabeled data, it reduces the cost of labeling by actively participates in the selection of the most informative training examples then an oracle is asked for labeling the selected sample. There are two settings depending on the oracle type: supervised setting [2] where requires human annotators as oracle for manual annotation and the unsupervised setting where the oracle is an automation process. Using different settings, active learning may determine much smaller and most informative subset from the unlabeled data pool. The difference between unsupervised active learning and semi-supervised learning [3] is that the former depends on an oracle to automatically annotate the most informative examples with respect to the underlying model. The later depends on the underlying model to automatically annotate some unlabeled data, to alleviate mislabeling noise the model selects the most confident examples.

For language-dependent tasks such as information extraction, to avoid the expensive re-labeling process for each individual language, cross-lingual adaptation, is a special case of domain adaptation, refers to the transfer of classification knowledge from one source language to another target language.

In this paper, we present a framework for incorporating unsupervised active learning in the cross-lingual domain adaptation paradigm ($XLADA$) that learns from labeled data in a source language and unlabeled data in the target language. The motivation of $XLADA$ is to collect large-scale training data and to train an information extraction model in a target language without manual annotation but with the help of an effective information extraction system in a source language, bilingual corpus and word-level alignment model.

2   Related Work

Yarowsky et al. [4] used word alignment on parallel corpora to induce several text analysis tools from English to other languages for which such resources are scarce. An NE tagger was transferred from English to French and achieved good classification accuracy. However, Chinese NER is more difficult than French and word alignment between Chinese and English is also more complex because of the difference between the two languages.

Some approaches have exploited Wikipedia as external resource to generate NE tagged corpus. Kim et al. [5] build on prior work utilizing Wikipedia metadata and show how to effectively combine the weak annotations stemming from Wikipedia metadata with information obtained through English-foreign language parallel Wikipedia sentences. The combination is achieved using a novel semi-CRF model for foreign sentence tagging. The model outperforms both standard annotation projection methods and methods based solely on Wikipedia metadata. $XLADA$does not leverage Wikipedia because its content is poor in some languages like Chinese.

Fu et al. [6] presents an approach to generate large-scale Chinese NER training data from an English-Chinese discourse level aligned parallel corpus. It first employs a high performance NER system on one side of a bilingual corpus.

And then, it projects the NE labels to the other side according to the word level alignment. At last, it selects labeled sentences using different strategies and generate an NER training corpus. This approach can be considered as passive domain adaptation while *XLADA* is active learning framework that filters out the auto-labeled data and selects the most informative training sentences.

Muslea et al. [7] introduced *Co-Testing*, a multi-view active learning framework, where two models are trained on two independent and sufficient sets of features. The most informative sentences are the points of disagreement between the two models that could improve their performance and a human judge is asked for labeling them. On the other hand, *XLADA* looks for the most informative sentences for the target model and we don't have judges.

Jones et al. [3] adapted semi-supervised learning *Co-EM* to information extraction tasks to learn from both labeled and unlabeled data that makes use of two distinct feature sets (training document's noun phrases and context). It is interleaved in the supervised active learning framework *Co-Testing*. *XLADA* differs in that cross-lingual label propagation on a parallel corpus is interleaved for automatic annotation instead of using *Co-EM* approach and that it adopts an unsupervised active learning strategy.

XLADA is more practical than the framework proposed by Li et al. [8] that depends on cross-lingual features extracted from the word-aligned sentence pair in training the target language CRF model. Hence, it isn't possible to extract named entities from a sentence in the target language unless it is aligned with a sentence in the source language.

3   Algorithmic Overview

The architecture of the proposed combination of cross-lingual domain adaptation and active learning paradigm *XLADA* is shown in Figure 1.

3.1   Initial Labeling

Source Language NER. An effective source language NER is applied on the source-side of the bilingual corpus U_S to identify named entities such as person, location, organization names, denote the output L_S. In our experiments, the source language is English and English Stanford NER[1] is used. The system is based on linear chain CRF [1] sequence models that can recognize three types of named entities (Location, Person and Organization).

Word Alignment of Parallel Corpus. Sentence alignment and word alignment is performed on the given unlabeled bilingual corpus U_S and U_T. First, sentence level alignment is performed then we applied word dependent transition model based HMM (WDHMM) for word alignment [9].

---

[1] http://nlp.stanford.edu/software/CRF-NER.shtml

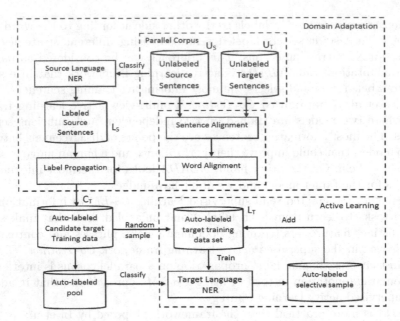

Fig. 1. Architecture of cross-lingual active domain adaptation (*XLADA*)

Label Propagation. We project the NE labels to the target side of the parallel corpus to automatically annotate target language sentences, according to the result of word alignment, as shown in Figure 2. The output is a set of candidate training sentences. A target sentence is filtered out from the set of candidate training sentences if the number of named entities after label propagation is less than the number of named entities in the source sentence.

3.2   Unsupervised Active Learning

The amount of auto-labeled sentences in the target language training is too huge to be used for training the information extraction model. Also they are noisy because of the errors in source language NER or word-level alignment. Unsupervised active learning is adopted for selecting high quality training sentences used to train CRF model. The manual annotation of the selected sentences by human judges is replaced with the alignment-based automatic annotation.

We randomly select a set of auto-labeled training sentences L_T. An initial CRF model is trained with L_T. Since a random set of auto-labeled sentences is not sufficient to train a good prediction model in the target language, additional labeled data is required to reach a reasonable prediction model. Afterward, *XLADA* will proceed in an iterative manner.

A pool CP_T of the large amount of auto-labeled sentences is selected. There are two ways to select the sentences in the pool, either a random sample or by assigning a score for each target sentence and finally choose sentences with the highest score (most confident sentences).

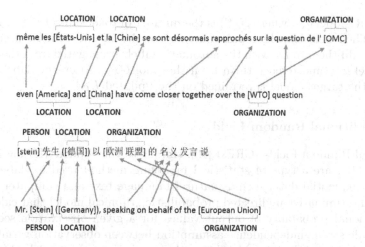

Fig. 2. Projection of named-entity tags from English to Chinese and French sentences

The score of each target sentence depends on the score given to its corresponding source sentence in the parallel corpus, as follows:

$$score(S) = \min_{w_i \in S} \max_{c_j \in classes} P(c_j|w_i, \theta_{src}) \tag{1}$$

The source NER model θ_{src} assigns probability for each token of how likely it belongs to each entity type: person, location, organization or otherwise. Then, the entity type for each token is the class with maximum probability $P(c_j|w_i, \theta_{src})$. We apply the forward-backward algorithm to compute them.

In each round of active learning, the current target NER model θ_{tgt} tags each target sentence in the auto-labeled pool CP_T. The critical challenge lies in how to select the most informative sentences for labeling. Based on different measurements of target sentence informativeness, we propose the following metric to measure how informative is a given sentence S.

$$inform(S) = \frac{1}{N(S)} \sum_{w_i \in S} \hat{y}(w_i) P(\hat{y}_{tgt}(w_i)|w_i, \theta_{tgt}) \tag{2}$$

where $\hat{y}(w_i) = I(\hat{y}_{src}(w_i) \neq \hat{y}_{tgt}(w_i))$ the indicator boolean function between

$$\hat{y}_{src}(w_i) = \arg \max_{c_j \in classes} P(c_j|w_i, \theta_{src}) \tag{3}$$

the NE label propagated from the source NER model θ_{src} through alignment information and

$$\hat{y}_{tgt}(w_i) = \arg \max_{c_j \in classes} P(c_j|w_i, \theta_{tgt}) \tag{4}$$

the NE tag assigned by the current target NER model θ_{tgt} to the ith word in S.

The most informative sentences are the ones that the target NER model θ_{tgt} didn't learn yet (least confident on its NE prediction) and mismatch with the source NER model θ_{src} where $N(S)$ is the number of tokens in S the two models disagree. Then we select the top N sentences or the ones less than a predefined threshold, add them to L_T with the automatic labels propagated from the source NER model θ_{src} and remove them from the pool CP_T. After new labels being acquired, the target model is retrained on the updated L_T.

3.3   Conditional Random Field

Conditional Random Fields (CRFs) [10], similar to the Hidden Markov Models (HMMs) [11] , are a type of statistical modeling method used for labeling or parsing of sequential data, such as natural language text and computer vision. CRF is a discriminative undirected probabilistic graphical model that calculates the conditional probability of output values for a given observation sequence. HMMs made strong independence assumption between observation variables, in order to reduce complexity, which hurts the accuracy of the model while CRF does not make assumptions on the dependencies among observation variables.

Figure 3 shows the graphical representation of liner chain CRFs. Because of its linear structure, linear chain CRF is frequently used in natural language processing to predict sequence of labels Y for a given observation sequence X. The inference of a linear-chain CRF model is that given an observation sequence X, we want to find the most likely sequence of labels Y. The probability of Y given X is calculated as follows:

$$P(Y|X) = \frac{1}{Z(X)} \exp(\sum_{t=1}^{T} \sum_{i=1}^{n} w_i f_i(y_{t-1}, y_t, X, t)) \tag{5}$$

where

$$Z(X) = \sum_{Y'} \exp(\sum_{t=1}^{T} \sum_{i=1}^{n} w_i f_i(y'_{t-1}, y'_t, X, t)) \tag{6}$$

In the equation, the observation sequence $X = (x_1, \ldots, x_T)$, the label sequence $Y = (y_1, \ldots, y_T)$ where y_t is the label for position t, state feature function is concerned with the entire observation sequence, the transition feature between labels of position $t-1$ and t on the observation sequence is also considered. Each feature f_i can either be state feature function or transition feature function. The coefficients w_is are the weights of features and can be estimated from training data. $Z(X)$ is a normalization factor.

4   Experiments

4.1   Datasets

The performance of *XLADA* is evaluated on the unsupervised learning of Chinese and French NER for named entity recognition of three entity types, person (PER), location (LOC) and organization (ORG). To achieve this goal, unlabeled

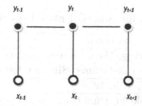

Fig. 3. Graphical representation of linear-chain CRF

training data set and labeled test data set is required for each target language. As unlabeled training data, two bilingual parallel corpora is used. The English-Chinese corpus is 20 million parallel sentences and the English-French corpus contains 40 million parallel sentences. The corpora involve a variety of publicly available data sets including United Nations proceedings[2], proceedings of the European Parliament[3], Canadian Hansards[4] and web crawled data. Both sides of each corpus were segmented (in Chinese) and tokenized (in English and French).

Table 1. Corpora used for performance evaluation

test set	Chinese	French
#sentences	5,633	9,988
#Person	2,807	3,065
#Location	7,079	3,153
#Organization	3,827	1,935

Table 1 shows a description of the corpora used as labeled test data for *XLADA*. One is the Chinese OntoNotes Release 2.0 corpus [5] and the second is a French corpus manually labeled using crowd sourcing. A group of five human annotators was asked to label each sentence then the majority NE tag is assigned to each token.

4.2 Setup

A widely used open-source NER system, Stanford Named Entity Recognizer is employed to detect named entities in the English side of the English-Chinese and English-French parallel corpora. The number of sentences that has at least one named entity detected by the Stanford NER is around 4 million sentences for Chinese and 10 million sentences for French. The features used to train the CRF model are shown in Figure 2. It's worth mentioning that the trainer used here is a local implementation of CRF (not Stanford's implementation) since Stanford's implementation is very slow and memory consuming.

---

[2] http://catalog.ldc.upenn.edu/LDC94T4A
[3] http://www.statmt.org/europarl/
[4] http://www.isi.edu/natural-language/download/hansard/
[5] http://catalog.ldc.upenn.edu/LDC2008T04

Table 2. Features used for Named Entity Recognition CRF model

type	extracted features
Shape Features	WordString, WordShape, StartsWithCapital, AllCapital Character N-Grams, Shape Character N-Grams
Brown Clusters [12]	Levels 0, 4, 8, 12
Bags of Words	Date Tokens, Punctuations, Personal Titles, Stop Words
Contextual	Previous 3 words and next 2 word features

As baselines for comparison, we have studied the following data selection techniques:

- *random sample*: The first NER model was trained on randomly sample of 340,000 and 400,000 sentences from the four million auto-labeled sentences and ten million sentences for Chinese and French language, respectively. (upper horizontal dashed lines in Figure 4)
- *most confident sample*: the second NER model was trained on the set of the top 340,000 and the top 400,000 most confident sentences (based on the min-max score function defined on section 3) for Chinese and French, respectively. (lower horizontal dashed lines in Figure 4)

For active learning, we have randomly chosen 100,000 auto-labeled sentences to train the initial NER for Chinese and French, respectively. And then, we have created a pool(set) of two million sentences where we have two experiments:

- *random pool*: one with a pool of randomly chosen sentences regardless of tagging confidence.
- *most confident pool*: another experiment with a pool of target sentences corresponding to the most confident source sentences selected by min-max score function.

The initial NER is applied on the pool and the informativeness of each sentence is measured using the function defined in section 3.

- *informative addition*: The most informative sentences are the sentences with score less than 0.9. At the end of the first iteration, the labeled training set is augmented with the newly-selected most informative sentences and the target NER is re-trained, this process is repeated for 20 iterations where the final NER for Chinese and French has been trained on 340,000 sentences and 400,000 sentences, respectively.
- *random addition*: another baseline for comparison where in each iteration, a number of auto-labeled sentences in the target language, Chinese or French, is randomly selected, equals to the number of most informative sentences selected in the same iteration at the *informative addition* experiment.

4.3   Results

The performance of unsupervised Chinese and French NER systems is reported in Table 4 and Table 3, respectively where the best performing data selection

Table 3. The performance of unsupervised French NER models trained using *XLADA* compared to baselines

Entity type	Pool type	XLADA		Baseline			
		informative addition		*random addition*		most confident sample	random sample
		most confident	*random*	*most confident*	random		
PER	Precision	78.1	**79.7**	78.6	79.0	77.9	77.0
	Recall	**82.0**	79.6	78.8	75.4	71.9	76.2
	F1	**80.0**	79.7	78.7	77.1	74.7	76.6
LOC	Precision	82.9	82.9	83.2	83.8	73.0	**84.6**
	Recall	65.9	**66.6**	64.0	64.9	59.2	63.8
	F1	73.4	**73.9**	72.3	73.2	65.4	72.7
ORG	Precision	54.1	50.1	52.0	51.7	**59.1**	49.5
	Recall	50.0	**52.2**	48.7	49.6	25.0	48.1
	F1	**51.9**	51.1	50.3	50.6	35.2	48.8

technique is bold faced. Figure 4 shows the learning curve of target NER models using the different training data selection techniques for Chinese and French, respectively. The F1 measure of both *random sample* NER and *most confident sample* NER is drawn as a horizontal dashed line.The results show that *XLADA* outperforms the random sample baseline.

For Chinese NER. For person NE, *XLADA* with *informative addition* using *most confident pool* achieves the highest F1-score 80.4% compared to 59.5% for *most confident sample* and 75.1% for random sample. This is attributed to the increase in person recall from 43.6% and 63.2% to 69.7% and 68.0% respectively. For location NE, *XLADA* with *informative addition* using *most confident pool* achieves the highest F1-score 83.1% compared to 73.3% for *most confident sample* and 81.7% for random sample. This is attributed to the increase in location recall from 64.6% and 74.0% to 76.4% and 75.0% respectively. For organization NE, *XLADA* with *informative addition* using *random pool* achieves the highest F1-score 65.9% compared to 44.5% for *most confident sample* and 62.6% for random sample. This is attributed to the increase in organization recall from 29.4% and 50.3% to 55.2%, respectively.

For French NER. For person NE, *XLADA* with *informative addition* using *most confident pool* achieves the highest F1-score 80.0% compared to *most confident sample* with 74.7% and *random sample* with 76.6% . This is attributed to the increase in person recall from 71.9% and 76.2% to 82.0%. For location NE, *XLADA* with *informative addition* using *random pool* achieves the highest F1-score 73.9% compared to domain adaptation without active learning: *most confident sample* of 65.4% and *random sample* of 72.7%. This is attributed to the increase in location recall from 59.2% and 63.8% to 66.6%. For organization NE, *XLADA* with *informative addition* using *most confident pool* achieves the highest

Table 4. The performance of unsupervised Chinese NER models trained using *XLADA* compared to baselines

	selection	XLADA		Baseline			
	method	*informative addition*		*random addition*		most	random
Entity type	Pool type	*most confident*	*random*	*most confident*	random	confident sample	sample
	Precision	**94.9**	93.5	92.9	91.8	93.4	92.4
PER	Recall	**69.7**	68.0	62.7	65.0	43.6	63.2
	F1	**80.4**	78.8	74.9	76.1	59.5	75.1
	Precision	91.1	**92.2**	90.9	91.1	84.9	91.1
LOC	Recall	**76.4**	75.0	73.6	73.9	64.6	74.0
	F1	**83.1**	82.7	81.3	81.6	73.3	81.7
	Precision	80.8	81.7	87.2	78.6	**91.4**	83.0
ORG	Recall	51.9	**55.2**	47.7	51.6	29.4	50.3
	F1	63.2	**65.9**	61.6	62.3	44.5	62.6

F1-score 51.9% compared to *most confident sample* of 35.2% and *random sample* of 48.8%. This is attributed to the significant improvement of organization recall from 25.0% and 48.1% to 50.0%.

4.4   Discussion

The improvement in recall means increase in the coverage of the trained NER model. This is attributed to the high quality of the training sentences selected by the proposed selective sampling criterion compared to random sampling. In addition, it is better than selecting target sentences where the English NER model is most confident about their corresponding English ones. The reason is that although the English NER model is most confident, this does not alleviate the passive nature of the target NER model as it has no control on the selection of its training data based on its performance. That is, it implies that the selected sentences do not carry new discriminating information with respect to the target NER model. In all cases, the *random sample* outperforms the *most confident sample*. The reason that selecting only the most confident sentences tends to narrow the coverage of the constructed NER. Figure 4 shows that *XLADA* achieves the most significant performance improvement in the early iterations, then the learning curve starts to saturate.

In general, the results of *organization* NE type are lower than the results of *Person* and *Location*. The reason is that ORG names are more complex than *Person* and *Location* names. They usually consist of more words, which may result in more word alignment errors and then lead to more training sentences being filtered out. Another reason behind this is that ORG names mostly consist of a combination of common words. gNot only for French and Chinese but also English ORG entity recognition is more difficult, which also results in more noise among the ORG training sentences.

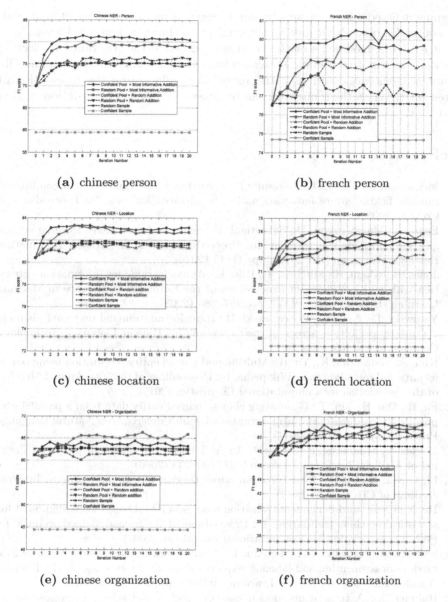

Fig. 4. Performance of unsupervised Chinese and French NER

5 Conclusions

The manual annotation of training sentences to build an information extraction system for each language is expensive, error-prone and time consuming. We introduced an unsupervised variant of active learning in the cross-lingual automatic annotation framework that replaces the manual annotation with the alignment-based automatic annotation. It depends on the existence of high quality source language NER model, bilingual parallel corpus and word-level alignment model. A modified score function is proposed as the criterion for selecting the most informative training sentences from the huge amount of automatically annotated sentences. Although the reported results are on the recognition of three entity types, the framework can be generalized to any information extraction task.

References

1. McCallum, A., Li, W.: Early results for named entity recognition with conditional random fields, feature induction and web-enhanced lexicons. In: Proceedings of CoNLL (2003)
2. Esuli, A., Marcheggiani, D., Sebastiani, F.: Sentence-based active learning strategies for information extraction. In: Proceedings of the 2nd Italian Information Retrieval Workshop (IIR 2010), pp. 41–45 (2010)
3. Jones, R., Ghani, R., Mitchell, T., Rilo, E.: Active learning for information extraction with multiple view. In: Proceedings of the European Conference in Machine Learning (ECML 2003), vol. 77, pp. 257–286 (2003)
4. Yarowsky, D., Ngai, G., Wicentowski, R.: Inducing multilingual text analysis tools via robust projection across aligned corpora. In: Human Language Technology Conference, pp. 109–116 (2001)
5. Kim, S., Toutanova, K., Yu, H.: Multilingual named entity recognition using parallel data and metadata from Wikipedia. In: Proceedings of the 50th Annual Meeting of the Association for Computational Linguistics (2012)
6. Fu, R., Qin, B., Liu, T.: Generating chinese named entity data from a parallel corpus. In: Proceedings of the 5th International Joint Conference on Natural Language Processing, pp. 264–272 (2011)
7. Muslea, I., Minton, S., Knoblock, C.A.: Active learning with multiple views. Journal of Artificial Intelligence Research 27, 203–233 (2006)
8. Li, Q., Li, H., Ji, H.: Joint bilingual name tagging for parallel corpora. In: Proceedings of CIKM 2012 (2012)
9. He, X.: Using word-dependent transition models in HMM based word alignment for statistical machine translation. In: Proceedings of the Second Workshop on SMT (WMT). Association for Computational Linguistics (2007)
10. Lafferty, J., McCallum, A., Pereira, F.: Conditional random fields: Probabilistic models for segmenting and labeling sequence data. In: Proceedings of the International Conference on Machine Learning (ICML), pp. 282–289 (2001)
11. Rabiner, L.: A tutorial on hidden markov models and selected applications in speech recognition. Proceedings of the IEEE 77, 257–286 (1989)
12. Brown, P.F., de Souza, P.V., Mercer, R.L., Pietra, V.J.D., Lai, J.C.: Class-based n-gram models of natural language. Computational Linguistics 18(4) (1992)

Incremental Feature Selection
by Block Addition and Block Deletion
Using Least Squares SVRs

Shigeo Abe

Kobe University
Rokkodai, Nada, Kobe, Japan
abe@kobe-u.ac.jp
http://www2.kobe-u.ac.jp/~abe

Abstract. For a small sample problem with a large number of features, feature selection by cross-validation frequently goes into random tie breaking because of the discrete recognition rate. This leads to inferior feature selection results. To solve this problem, we propose using a least squares support vector regressor (LS SVR), instead of an LS support vector machine (LS SVM). We consider the labels (1/-1) as the targets of the LS SVR and the mean absolute error by cross-validation as the selection criterion. By the use of the LS SVR, the selection and ranking criteria become continuous and thus tie breaking becomes rare. For evaluation, we use incremental block addition and block deletion of features that is developed for function approximation. By computer experiments, we show that performance of the proposed method is comparable with that with the criterion based on the weighted sum of the recognition error rate and the average margin error.

Keywords: Backward feature selection, feature ranking, forward feature selection, incremental feature selection, pattern classification, support vector machines, support vector regressors.

1 Introduction

To realize a classifier with high generalization ability, feature selection, which eliminates redundant and irrelevant features, is especially important for a small sample problem with a large number of features (SSPLF). In such a problem, to avoid deleting important features for classification, wrapper methods [1–3], which use recognition rate-based criteria, are preferable to filter methods, which use more simpler criteria [4–6].

For kernel-based classifiers, imbedded methods, in which feature selection and training are done simultaneously are also used [7, 8].

For wrapper methods, forward selection and backward selection are often used. In forward selection, a feature is sequentially added to an initially empty set, and in backward selection, a feature is sequentially deleted from the set initialized with all the features. Because forward selection is faster than backward selection

N. El Gayar et al. (Eds.): ANNPR 2014, LNAI 8774, pp. 35–46, 2014.

if the number of selected features is small, but less stable, the combination of forward selection and backward selection is also used [3, 9–11].

There are several approaches to speed up wrapper methods: some feature selection methods combine filter methods and wrapper methods and use filter methods as a preselector [12–14]. In [3], instead of sequential forward selection and backward selection, block addition (BA) of features followed by block deletion (BD) of features is proposed.

Incremental selection has also been proposed [15–19] to speed up feature selection. In [19], BABD for input variable selection is extended to incremental selection and speedup was shown for the small sample problems with a large number of input variables (SSPLV).

In applying a wrapper method to an SSPLF, frequently we need to break ties in feature selection and feature ranking, because the feature selection/ranking criterion is discrete. In addition, the number of selected features is very small because the 100% recognition rate is easily obtained for the validation data set. This worsens the generalization ability. To avoid this, we used the weighted sum of the recognition error rate and the average margin error [3]. This led to more stable feature selection for microarray data sets.

In this paper, instead of the weighted sum of error rate and the average margin error used in [3], we propose using the mean absolute error by the least squares support vector regressor (LS SVR), assuming the labels $(1/ - 1)$ as the targets of regression. Because, unlike the regular SVM, for the LS SVM, classifiers and regressors have the same form, training for the LS SVM and that for the LS SVR are the same. The only difference is whether the recognition error is calculated or the mean absolute error is calculated. Thus, a classification problem is easily converted into the associated regression problem, whose absolute error is continuous. Therefore, unlike the LS SVM, tie breaking rarely happens for the LS SVR.

The procedure for feature selection is based on incremental block addition and block deletion [3, 19]. Starting from the empty set, we repeat adding multiple features at a time to the set. We stop addition when the generalization ability of the set is no longer improved. Then from the set of selected features, we delete multiple features at a time until the generalization ability is not improved.

In Section 2, we discuss the idea of feature selection and selection criteria. Then in Section 3 we discuss the proposed methods based on incremental block addition and block deletion, and in Section 4, we show the results of computer experiments using two-class benchmark data sets including microarray data sets.

2   Idea of Feature Selection and Selection Criteria

For an SSPLF such as microarray data sets, the optimal set of features that realizes the generalization ability comparable to that of the original set of features is usually not so large. In such a situation, forward selection is faster than backward selection. Therefore, by forward selection we select a set of features whose generalization ability is comparable to that of the original set of features.

But because an added feature may become redundant after another feature is added, we perform backward selection for the set of features selected by forward selection.

To speedup feature selection, we use multiple feature addition (block addition) and multiple feature deletion (block deletion) and combine BABD with incremental feature selection.

To avoid frequent tie breaking in feature selection and feature ranking, we use a continuous selection criterion.

Let the decision function for a two class problem be

$$z = f(\mathbf{x}) = \mathbf{w}^\top \phi(\mathbf{x}_i) + b, \tag{1}$$

where $\mathbf{x}$ and z are the feature vector and the decision output, respectively, $\mathbf{w}$ is the coefficient vector of the separating hyperplane in the feature space, $\phi(\mathbf{x})$ is the mapping function that maps $\mathbf{x}$ into the feature space, and b is the bias term.

For M training input-output pairs $\{\mathbf{x}_i, y_i\}$ $(i = 1, \ldots, M)$, the LS SVM is given by

$$\text{minimize} \quad \frac{1}{2} \mathbf{w}^\top \mathbf{w} + \frac{C}{2} \sum_{i=1}^{M} \xi_i^2 \tag{2}$$

$$\text{subject to} \quad y_i f(\mathbf{x}_i) = 1 - \xi_i \quad \text{for} \quad i = 1, \ldots, M, \tag{3}$$

where C is the margin parameter, $y_i = 1$ for Class 1 and -1 for Class 2, and ξ_i is the slack variable associated with $\mathbf{x}_i$.

Multiplying y_i to both sides of (3) and replacing $y_i \xi_i$ with ξ_i, we obtain

$$\text{minimize} \quad \frac{1}{2} \mathbf{w}^\top \mathbf{w} + \frac{C}{2} \sum_{i=1}^{M} \xi_i^2 \tag{4}$$

$$\text{subject to} \quad f(\mathbf{x}_i) = y_i - \xi_i \text{ for } i = 1, \ldots, M. \tag{5}$$

The above LS SVM is the same as the LS SVR.

In a wrapper method, we use the recognition error rate E_C. For the training data set it is given by

$$E_\mathrm{C} = \frac{1}{M} \sum_{i=1}^{M} e_i \quad \text{for} \quad e_i = \begin{cases} 0 & \text{for} \quad y_i f(\mathbf{x}_i) \geq 0, \\ 1 & \text{for} \quad y_i f(\mathbf{x}_i) < 0. \end{cases} \tag{6}$$

Because the recognition error rate is discrete, for an SSPLF, frequent tie breaking occurs for feature selection and feature ranking. Therefore, in [3] we proposed the following MM criterion:

$$E_{\mathrm{M}_\mathrm{C}} = E_\mathrm{C} + r\, E_\mathrm{M}, \tag{7}$$

where r is a positive parameter and $r = 1/M$, and E_M is the mean margin error given by

$$E_\mathrm{M} = \frac{1}{M} \sum_{i=1}^{M} \xi_i \quad \text{where} \quad \xi_i = \begin{cases} 0 & \text{for} \quad y_i f(\mathbf{x}_i) \geq 1, \\ 1 - y_i f(\mathbf{x}_i) & \text{for} \quad y_i f(\mathbf{x}_i) < 1. \end{cases} \tag{8}$$

Because the LS SVM can also be used as a regressor, we consider the classification problem as a function approximation problem: we assume that the class labels $(1/-1)$ are target values of a function approximation problem. Then, training the LS SVM is equivalent to training the associated LS SVR.

Thus, instead of (7), we consider using the mean absolute error:

$$E_{\mathrm{MAE}} = \frac{1}{M} \sum_{i=1}^{M} |y_i - f(\mathbf{x}_i)|. \tag{9}$$

Because $y_i = 1$ or -1, minimization of (9) leads to minimization of the recognition error. But model selection by cross-validation using (9) does not necessarily lead to the same model obtained by cross-validation using the recognition error or the MM criterion given by (7).

3   Feature Selection by Incremental Block Addition and Block Deletion

We use incremental BABD for function approximation discussed in [19]. The algorithm for pattern classification is essentially the same. In the following we explain incremental BABD.

In incremental BABD, initially we select a subset from the set of initial features and select features from the subset by BABD. Then we add features that are not yet processed to the set of selected features and repeat BABD until all the features are processed. This procedure is called one-pass incremental BABD.

By this method, important features may be discarded before the new features are added. To prevent this, we repeat one-pass BABD until no further improvement in the selection criterion is obtained. This procedure is called multi-pass incremental BABD.

Now we explain incremental BABD more in detail referencing the corresponding steps in Algorithm 1, which is an extension of iterative BABD discussed in [20].

Let $I^m = \{1, \ldots, m\}$ be the set of the original m features. Initially, we select the set of m' features, $I^{m'}$, from I^m as the initial set of features (Step 1), and calculate the MAE for $I^{m'}$, $E^{m'}$, by cross-validation. This is used as the threshold of feature selection for $I^{m'}$, $T^{m'}$ (Step 2):

$$T^{m'} = E^{m'}. \tag{10}$$

By BA, we iterate feature ranking and feature addition until

$$E^j \leq T^{m'} \leq E^j + \varepsilon_{\mathrm{M}} \tag{11}$$

is satisfied, where ε_{M} is a positive value, I^j is the set of selected j features, $j \leq m'$, and $I^j \subseteq I^{m'}$. The right-hand side inequality is to control the number of selected features, and as the value of ε_{M} is decreased, the number of selected features is increased. Then if $E^j < T^{m'}$, we update the threshold by

$$T^{m'} = E^j. \tag{12}$$

In the feature ranking we rank features in $I^{m'}$ in the ascending order of MAEs, which are evaluated by temporarily adding a feature to the set of selected features. Then we add, to the set of selected features, from the top ranked to the 2^kth ranked features, where $k = 1, \ldots, 2^A$ and A is a user defined parameter, and evaluate the MAE by cross-validation (Step 3). If the minimum MAE for $k \in \{1, \ldots, 2^A\}$ is smaller than or equal to $T^{m'}$, we permanently add the associated features, and update the threshold. If the right-hand side inequality in (11) is satisfied, finish BA. If not, repeat BA. Otherwise, if the minimum MAE is smaller than that at the previous BA step, we permanently add the associated features, update the threshold, and repeat BA. Otherwise, we add the top ranked feature and repeat BA (Step 4).

Because redundant features may be added by BA, we delete these features by BD repeating feature ranking and deletion of features.

For each feature in I^j we evaluate the MAE by cross-validation temporarily deleting the feature (Step 5).

We generate set S^j that includes features whose MAE is not larger than $T^{m'}$. If S^j is empty we terminate BD. If only one element is in S^j, delete this feature and iterate BD (Step 6). Otherwise, we temporarily delete all the features in S^j and evaluate the MAE by cross-validation. If it is not larger than $T^{m'}$, we permanently delete these features and update j, and repeat BD (Step 7). If not, we rank features in S^j and temporarily delete the top half and evaluate the MAE by cross-validation. We repeat this until feature deletion is succeeded (Step 8).

After BD is succeeded, E^j for the resulting set of features I^j satisfies

$$E^j \leq T^{m'}. \tag{13}$$

Then we update the threshold by $T^{m'} = E^j$ and repeat BD.

The above procedure guarantees that the MAE for the selected features is not larger than that for $I^{m'}$, i.e., $E^j \leq E^{m'}$.

Let i_{Inc} be the number of features that are added at the incremental step. We add i_{Inc} features from $I^m - I^{m'}$ to I^j,

Let the resulting set of features be $I^{j+i_{\mathrm{Inc}}}$. Then the MAE for $I^{j+i_{\mathrm{Inc}}}$ is $E^{j+i_{\mathrm{Inc}}}$. We set the threshold $T^{m'+i_{\mathrm{Inc}}}$ by $T^{m'+i_{\mathrm{Inc}}} = E^{j+i_{\mathrm{Inc}}}$. Here, we must notice that

$$T^{m'+i_{\mathrm{Inc}}} \leq T^{m'}. \tag{14}$$

is not always satisfied.

We iterate the above BABD for $I^{j+i_{\mathrm{Inc}}}$. Let the resulting set of features be I^o, where $o \leq j + i_{\mathrm{Inc}}$. Then

$$E^o \leq T^{m'+i_{\mathrm{Inc}}} \tag{15}$$

is satisfied. If (14) is satisfied,

$$E^o \leq T^{m'} \tag{16}$$

is also satisfied. But otherwise, there is no guarantee that the above inequality is satisfied.

If (16) is satisfied, we repeat BABD adding the variables not processed. Otherwise, we consider that the BABD for this step failed and undo the feature selection at this step; namely, we restart BABD with threshold $T^{m'}$ and I^j, and add remaining features to I^j.

In one-pass incremental BABD, we repeat the BABD until all the variables are processed. In multi-pass incremental BABD, to reduce the absolute error further, we repeat the above procedure until the selection criterion does not change (Step 9).

Algorithm 1 (Incremental BABD).

Initialization

Step 1 Set $I^{m'}(\subseteq I^m)$, $j = 0$, and $E^j = \infty$.

Block Addition

Step 2 Calculate $E^{m'}$ for $I^{m'}$. Set $T^{m'} = E^{m'}$.

Step 3 Add feature i in $I^{m'} - I^j$ temporarily to I^j, calculate $E^j_{i_{\text{add}}}$, where i_{add} denotes that feature i is temporarily added, and generate feature ranking list V^j. Set $k = 1$.

Step 4 Calculate E^{j+k} $(k = 1, 2^1, \ldots, 2^A)$. If $E^{j+k} < T^{m'}$, set $j \leftarrow j+k, T^{m'} \leftarrow E^j$. And if $T^{m'} \leq E^j + \varepsilon_{\text{M}}$, go to Step 5; if not, go to Step 3. Otherwise, if $E^{j+k} < E^j$ is satisfied, set $j \leftarrow j + k$ and go to Step 3. Otherwise, if $E^j \leq T^{m'}$, go to Step 5; otherwise, set $j \leftarrow j + 1, T^{m'} \leftarrow E^j$ and go to Step 3.

Block Deletion

Step 5 Delete temporarily feature i in I^j and calculate $E^j_{i_{\text{del}}}$, where i_{del} denotes that feature i is temporarily deleted.

Step 6 Calculate S^j. If S^j is empty, $I^o = I^j$ and go to Step 9. If only one feature is included in S^j, set $I^{j-1} = I^j - S^j$, set $j \leftarrow j - 1$ and go to Step 5. If S^j has more than two features, generate V^j and go to Step 7.

Step 7 Delete all the features in V^j from I^j: $I^{j'} = I^j - V^j$, where $j' = j - |V^j|$ and $|V^j|$ denotes the number of elements in V^j. Then, calculate $E^{j'}$ and if $E^{j'} > T^{m'}$, go to Step 8. Otherwise, update j with j', $T^{m'} \leftarrow E^{j'}$, and go to Step 5.

Step 8 Let V'^j include the upper half elements of V^j. Set $I^{j'} = I^j - \{V'^j\}$, where $\{V'^j\}$ is the set that includes all the features in V'^j and $j' = j - |\{V'^j\}|$. Then, if $E^{j'} \leq T^{m'}$, delete features in V'^j and go to Step 5 updating j with j' and $T^{m'}$ with $E^{j'}$. Otherwise, update V^j with V'^j and iterate Step 8 until $E^{j'} < T^{m'}$ is satisfied.

Step 9 If E^o is larger than $T^{m'}$ in the previous step, undo current BABD. If some features in I^m are not added, $I^{m'} = I^o \cup I^{i_{\text{Inc}}}$, $m' \leftarrow o + i_{\text{Inc}}$, $j = 0$, $E^j = \infty$, and go to Step 2. Otherwise, if one-pass, terminate feature selection; otherwise if $T^{m'}$ decreases from previous $T^{m'}$, go to Step 1. If not, stop feature selection.

4   Performance Evaluation

Because feature selection based on the E_C criterion performed poorly for a large number of features [3], in this section, we compare the MAE criterion with the

MM criterion and incremental BABD with batch BABD using two kinds of data sets: data sets with small numbers of features and microarray data sets with large numbers of features. We set $A = 5$ and $\varepsilon_M = 10^{-5}$ as in [3]. In incremental feature selection, we set $m' = i_{Inc}$ and add features from the first to the last.

4.1 Data Sets with Small Numbers of Features

We used the ionosphere and WDBC data sets [21]. We divided each data set randomly into training and test data sets and generated 20 pairs.

For these data sets, in [3] we showed that the recognition rates of the test data sets and the numbers of selected features by batch BABD were comparable to those shown in [2, 8, 13]. Therefore, here, we only compare the proposed method with batch BABD.

We used the RBF kernels: $K(\mathbf{x}, \mathbf{x}') = \boldsymbol{\phi}^\top(\mathbf{x}) \boldsymbol{\phi}(\mathbf{x}') = \exp(-\gamma \|\mathbf{x} - \mathbf{x}'\|^2/m)$, where γ is a positive parameter. Using all the features we determined the γ and C values by fivefold cross-validation changing $\gamma = \{0.001, 0.01, 0.5, 1.0, 5.0, 10, 15, 20, 50, 100\}$ and $C = \{1, 10, 50, 100, 500, 1000, 2000\}$. During and after feature selection we fixed the γ and C values to the determined values.

We measured the average feature selection time per data set using a personal computer with 3GHz CPU and 2GB memory.

Table 1 shows the results for the ionosphere and WDBC data sets. The upper part for each data set shows the result for the MM criterion and the lower part, the MAE criterion. In the "Data (Tr/Te/In)" column, the first row of each data set shows the name of the data set followed by the numbers of training data, test data, and features. The first column also includes performance with the standard deviation using all the features: the recognition rates for the test data sets and those for the validation data sets in the parentheses. For the MAE criterion, MAEs are shown in the parentheses.

In the second column, MM denotes batch BABD with the MM criterion and MAE, that with the MAE criterion. And for instance "20" denotes the one-pass incremental BABD with 20 features added, and "m" in 20m denotes the multi-pass incremental BABD. The third column shows the recognition (approximation) performance after feature selection. And the fourth and the fifth columns show the number of selected features and the feature selection time, respectively.

For each performance measure, the best performance is shown in bold face.

From the table, except for two cases by one-pass incremental BABD, the recognition rates (MAEs) by cross-validation were improved by feature selection, but for the test data sets, the recognition rates were decreased. This was caused by overfitting.

Now compare the MM and MAE criteria. Using all the features, the recognition rates of the test data sets by the MAE criterion were better for both data sets. This means that different γ and C values were selected by cross-validation. But the differences including those after feature selection were small.

As for the effect of incremental BABD, although multi-pass incremental BABD improved the recognition rates (MAEs) by cross-validation, in some cases

Table 1. Comparison of selection methods

Data (Tr/Te/In)	Method	Test Rate (CV Rate/MAE)	Selected	Time [s]
Ionosphere (281/70/34)	MM	**93.93**±2.59(97.10±0.80)	15.20±5.0	**14.70**±2.12
94.21±1.89(95.57±0.67)	20	92.64±3.07(96.57±0.84)	13.9±3.3	15.55±2.82
	20m	92.79±2.73(97.12±0.55)	13.1±3.7	37.80±13.06
	10	91.86±3.50(96.51±0.86)	11.3±3.1	17.70±1.31
	10m	92.29±3.04(**97.17**±0.71)	11.4±3.3	45.15±17.36
	1	91.29±2.78(95.14±1.51)	**5.7**±1.7	35.90±5.84
	1m	91.71±3.55(96.05±1.63)	7.2±2.3	142.9±66.60
95.29±2.31(0.2640±1.16)	MAE	**94.21**±2.57(0.2278±0.0120)	13.5±2.4	**13.75**±1.41
	20	93.14±3.25(0.2315±0.0134)	10.7±2.7	14.15±1.42
	20m	93.43±3.33(**0.2267**±0.0127)	10.9±2.6	33.95±9.86
	10	92.21±2.77(0.2321±0.0145)	7.8±2.9	15.80±1.29
	10m	92.14±3.11(0.2274±0.0145)	8.6±3.2	34.45±9.46
	1	91.50±3.85(0.2406±0.0140)	**4.8**±0.7	32.90±3.99
	1m	91.36±3.51(0.2344±0.0112)	5.4±1.2	91.90±41.29
WDBC(455/114/30)	MM	**97.11**±1.15(98.41±0.33)	16.6±4.4	**40.45**±8.99
97.41±0.98(98.09±0.34)	20	97.02±1.09(98.32±0.38)	14.4±2.7	41.50±6.34
	20m	96.93±1.13(**98.57**±0.24)	14.7±3.6	100.5±27.00
	10	97.06±1.01(98.26±0.35)	13.2±3.4	42.90±4.38
	10m	96.71±1.24(98.56±0.34)	12.7±3.4	126.5±25.51
	1	96.14±1.02(98.01±0.38)	**6.6**±1.4	114.3±9.81
	1m	95.96±1.16(98.33±0.34)	7.5±2.0	381.7±134.0
97.72±1.22(0.2335±0.0067)	MAE	96.14±1.40(0.1622±0.0058)	5.3±1.0	**30.45**±2.42
	20	96.10±1.43(0.1622±0.0058)	5.2±1.0	34.20±2.27
	20m	95.92±1.62(0.1619±0.0058)	5.0±1.2	62.10±9.72
	10	96.10±1.43(0.1622±0.0058)	5.2±1.0	35.55±2.31
	10m	95.92±1.62(0.1619±0.0058)	5.0±1.2	71.25±10.50
	1	**96.19**±1.60(0.1617±0,0051)	**4.2**±1.0	86.05±8.23
	1m	96.05±1.58(**0.1616**±0.0052)	4.4±1.2	366.7±810.1

one-pass incremental BABD showed better recognition rates for the test data
sets. Except for the WDBC data set with the MAE criterion, the recognition
rates for the test data sets decreased as i_{Inc} was decreased.

The numbers of selected features decreased as i_{Inc} was decreased and they
were minimum when $i_{Inc} = 1$ both for one- and multi-pass feature selection.

Feature selection time by batch BABD was shortest for all four cases. This
means that because the numbers of features were not so large, incremental feature selection did not contribute in speeding up feature selection.

4.2 Microarray Data Sets

We compared BABD with the MM criterion and BABD with the MAE criterion for microarray data sets (see [22] for details of data sets), each of which
consisted of 100 pairs of training and test data sets. Because microarray data
sets have a small number of samples and a large number of features, they are
linearly separable and overfitting occurs easily. Therefore, we used linear kernels:
$K(\mathbf{x}, \mathbf{x}') = \mathbf{x}^\top \mathbf{x}'$ and fixed $C = 1$.

To measure feature selection time, we used a personal computer with 3.4GHz CPU and 16GB memory.

To determine the number of added features (i_{Inc}), we carried out incremental BABD with the MAE criterion for the breast cancer data set (1) changing i_{Inc}. Figure 1 shows the result for one- and multi-pass BABD. As shown in Fig. (a), the MAE for the training data by multi-pass BABD was better than that by one-pass BABD. But there was not much difference in the recognition rates of the test data by one- and multi-pass BABD (Fig. (b)), although by one-pass BABD the feature selection time was shorter and the number of selected features was smaller. From Figs. (b) and (c), we set $i_{Inc} = 500$ in the following experiments.

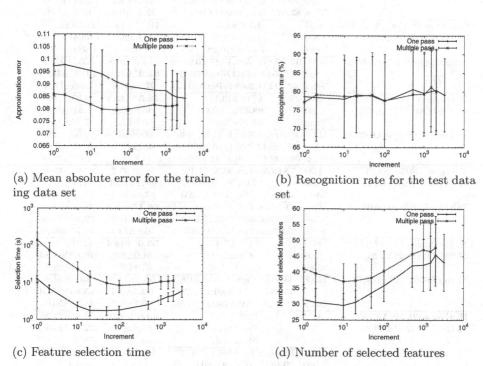

(a) Mean absolute error for the training data set

(b) Recognition rate for the test data set

(c) Feature selection time

(d) Number of selected features

Fig. 1. Feature selection for the breast cancer (1) data set

Table 2 shows the results. In the table if 100% recognition rates were obtained, they are not shown. The triplet in the "Summary" row shows from the left the numbers that the best/second best/third performance were obtained. In the "Selected" and "Time" columns, the average value with the asterisk shows that it is statistically significant between the values for the MM criterion and one-pass incremental method with $i_{inc} = 500$ by the Welch t-test with the confidence interval of 95%.

Comparing the results for the MM and MAE criteria, there is not much difference of the recognition rates of the test data sets (statistically comparable). In some cases (e.g., the breast cancer (3) and hepatocellular carcinoma data sets),

Table 2. Performance comparison of incremental BABD and batch BABD

Data (Tr/Te/In)	Method	Test Rate (CV Rate/MAE)	Selected	Time [s]
B. cancer (1) (14/8/3226)	MM	**80.50**±11.36	**40.5***±11.9	4.04±2.08
73.87±11.47 (76.50±7.09)	500	78.25±10.55	44.7±8.3	**2.20***±0.57
	500m	79.25±10.93	47.7±7.3	6.71±2.58
73.87±11.47(0.6215±0.0709)	MAE	79.12±9.85(0.0843±0.0104j	43.1±9.0	5.85±1.88
	500	**80.63**±11.37(0.0875±0.0101)	**42.1**±6.7	**2.51***±0.59
	500m	79.25±12.15(**0.0816**±0.0102)	45.8±7.7	8.98±3.52
B. cancer (2) (14/8/3226)	MM	**83.38**±13.12	**43.9***±12.4	4.34±2.08
91.88±10.21(83.50±7.93)	500	82.63±13.45	50.1±7.9	**2.39***±0.63
	500m	82.13±13.02	55.0±9.7	7.45±2.60
91.88±10.21(0.6356±0.0729)	MAE	82.00±11.64(0.0950±0.0138)	49.9±12.8	7.16±2.35
	500	83.50±12.10(0.0982±0.0117)	**47.9**±9.3	**3.21***±0.65
	500m	**83.87**±12.03(**0.0904**±0.0117)	52.9±10.8	10.21±3.69
B. Cancer (3) (78/19/24188)	MM	63.37±9.93	**70.7***±15.5	847.0±358.0
67.32±9.42(66.96±4.58)	500	62.95±9.29	82.6±8.5	**555.4***±74.74
	500m	**64.58**±10.28	84.6±8.1	1999±673.8
67.32±9.42(0.8167±0.0474)	MAE	**63.47**±10.39(**0.1547**±0.0122)	115.8±15.8	3557±915.2
	500	62.05±8.82(0.1701±0.0095)	**94.3***±12.0	**1463***±107.4
	500m	62.79±11.25(0.1576±0.0117)	97.7±11.8	5595±1724
B. cancer (s) (14/8/3226)	MM	67.00±13.17	**39.5***±12.4	3.85±2.09
69.12±10.82(72.79±9.30)	500	**68.87**±11.92	46.5±7.9	**2.31***±0.69
	500m	68.75±12.69	50.9±7.5	7.47±2.61
69.13±10.82(0.7248±0.0816)	MAE	67.37±13.33(0.1051±0.0149)	46.0±10.5	6.56±2.39
	500	67.50±12.75(0.1110±0.0139)	**43.9**±7.8	**3.03***±0.67
	500m	**69.13**±13.16(**0.1012**±0.0121)	49.3±8.8	10.25±4.30
C. cancer (40/20/2000)	MM	**81.05**±6.68(99.53±1.10)	91.8±35.7	47.22±43.42
79.64±6.54(79.67±6.21)	500	80.82±7.06(99.70±0.89)	**84.1**±23.4	**30.17***±11.83
	500m	80.86±7.05(**99.95**±0.35)	87.0±16.2	73.55±45.44
79.64±6.54(0.6819±0.0880)	MAE	**81.82**±6.49(0.2423±0.0319)	**66.1***±19.9	28.17±10.93
	500	81.50±6.40(0.2357±0.0268)	71.1±15.6	**22.68***±4.68
	500m	81.23±6.88(**0.2223**±0.0275)	76.9±13.7	72.78±26.70
H. Carcinoma (33/27/7129)	MM	64.63±7.45	**53.0***±14.4	42.21±20.45
67.96±7.00(66.21±7.34)	500	64.70±7.81	61.0±8.8	**26.52***±3.84
	500m	**64.74**±7.80	66.4±8.2	84.05±28.07
67.96±7.00(0.8263±0.0708)	MAE	63.56±8.14(0.1538±0.0196)	65.3±13.9	101.5±34.46
	500	**65.04**±8.24(0.1601±0.0192)	**63.5**±9.6	**45.64***±5.34
	500m	64.78±7.99(**0.1480**±0.0176)	66.4±9.0	153.3±56.05
H. glioma (21/29/12625)	MM	70.07±8.46	**49.6***±13.6	78.66±36.30
75.59±7.58(72.71±10.23)	500	70.38±8.39	61.6±9.7	**22.66***±3.03
	500m	**70.52**±8.58	66.3±9.4	79.46±27.45
75.59±7.58(0.7718±0.0124)	MAE	**71.17**±8.63(0.1364±0.0232)	**52.5**±12.9	131.1±38.38
	500	70.10±8.60(0.1409±0.0217)	54.6±10.2	**30.74***±4.26
	500m	70.41±9.13(**0.1286**±0.0192)	58.9±9.2	110.6±40.56
Leukemia (38/34/7129)	MM	94.38±3.88	**47.9***±12.2	43.76±20.11
94.44±4.70(92.45±3.32)	500	**94.41**±3.87	56.6±8.6	**25.93***±4.94
	500m	94.29±3.90	62.3±7.3	74.31±24.85
94.44±4.70(0.4866±0.0392)	MAE	94.06±3.58(0.0883±0.0129)	66.3±14.5	126.3±42.17
	500	94.32±3.80(0.0896±0.0110)	**62.1***±10.0	**66.78***±7.83
	500m	**94.59**±3.83(**0.0829**±0.0095)	64.9±10.5	196.6±59.68
P. cancer (102/34/12600)	MM	**84.65**±6.08(99.18±1.64)	350.5±243.8	33970±34855
87.03±4.56(88.52±2.27)	500	83.74±6.75(99.77±0.44)	**251.3***±134.1	**9625***±5279
	500m	84.29±6.54(99.88±0.32)	288.2±115.6	39593±36943
87.03±4.56(0.8757±0.0429)	MAE	80.68±6.30(0.4039±0.0385)	**105.4***±26.4	2490±1037
	500	82.62±6.18(0.3988±0.0235)	135.3±21.8	**1974***±176.6
	500m	**83.38**±6.45(**0.3662**±0.0213)	153.7±20.3	9157±3141
	MM	7/2/9	10/4/4	0/17/1
Summary	500	4/8/6	8/10/0	18/0/0
	500m	7/8/3	0/4/14	0/1/17

the MAE criterion selected more features and thus feature selection time was longer. But for the colon cancer data sets, the opposite was true. The above results confirm that the MAE criterion is comparable to the MM criterion.

From the "Summary" rows, we found that multi-pass incremental BABD showed the best recognition rates for the test data sets, but the numbers of selected features were the largest and also feature selection was slowest. The recognition rates by one-pass incremental BABD were comparable with those by batch BABD and feature selection was the fastest, but the numbers of selected features were the second to batch BABD. Therefore, one-pass BABD can be an alternative to the batch BABD.

The reason why one-pass BABD performed well for the microarray data sets although it was not for the ionosphere and WDBC data sets is as follows: because the numbers of features are very large and the number of training samples are very small, there exist many alternative subsets of features that realize best generalization performance. In addition, because the number of added features was usually much larger than the number of selected features, during incremental BABD, optimal features were not deleted, or even if deleted, alternative features remained.

5   Conclusions

In this paper, we proposed using the MAE (mean absolute error) criterion in selecting features of small sample problems with a large number of features. Setting class labels $(1/-1)$ as the targets of regression, we train the least squares SVM and calculate the MAE. Because the MAE is continuous, tie breaking, which is a problem for a discrete criterion, does not occur frequently. Therefore, feature selection is stabilized.

We evaluate the MAE criterion by incremental block addition and block deletion (BABD) using the microarray data sets. The results show that the MAE criterion is comparable with the MM criterion, which is the weighted sum of the recognition error rate and the average margin error, and that the one-pass incremental BABD is comparable in generalization abilities to batch BABD with faster feature selection.

Acknowledgment. This work was supported by JSPS KAKENHI Grant Number 25420438.

References

1. Abe, S.: Modified backward feature selection by cross validation. In: Proc. ESANN 2005, pp. 163–168 (2005)
2. Maldonado, S., Weber, R.: A wrapper method for feature selection using support vector machines. Information Sciences 179(13), 2208–2217 (2009)
3. Nagatani, T., Abe, S.: Feature selection by block addition and block deletion. In: Mana, N., Schwenker, F., Trentin, E. (eds.) ANNPR 2012. LNCS, vol. 7477, pp. 48–59. Springer, Heidelberg (2012)

4. Guyon, I., Weston, J., Barnhill, S., Vapnik, V.: Gene selection for cancer classification using support vector machines. Machine Learning 46(1-3), 389–422 (2002)
5. Peng, H., Long, F., Dingam, C.: Feature selection based on mutual information: Criteria of max-dependency, max-relevance, and min-redundancy. IEEE Trans. Pattern Analysis and Machine Intelligence 27(8), 1226–1238 (2005)
6. Herrera, L.J., Pomares, H., Rojas, I., Verleysen, M., Guilén, A.: Effective input variable selection for function approximation. In: Kollias, S.D., Stafylopatis, A., Duch, W., Oja, E. (eds.) ICANN 2006, Part I. LNCS, vol. 4131, pp. 41–50. Springer, Heidelberg (2006)
7. Bradley, P.S., Mangasarian, O.L.: Feature selection via concave minimization and support vector machines. In: Proc. ICML 1998, pp. 82–90 (1998)
8. Neumann, J., Schnörr, C., Steidl, G.: Combined SVM-based feature selection and classification. Machine Learning 61(1-3), 129–150 (2005)
9. Stearns, S.D.: On selecting features for pattern classifiers. In: Proc. ICPR, pp. 71–75 (1976)
10. Pudil, P., Novovičová, J., Kittler, J.: Floating search methods in feature selection. Pattern Recognition Letters 15(11), 1119–1125 (1994)
11. Zhang, T.: Adaptive forward-backward greedy algorithm for sparse learning with linear models. In: NIPS 21, pp. 1921–1928 (2009)
12. Bi, J., Bennett, K.P., Embrechts, M., Breneman, C.M., Song, M.: Dimensionality reduction via sparse support vector machines. J. Machine Learning Research 3, 1229–1243 (2003)
13. Liu, Y., Zheng, Y.F.: FS_SFS: A novel feature selection method for support vector machines. Pattern Recognition 39(7), 1333–1345 (2006)
14. Nagatani, T., Ozawa, S., Abe, S.: Fast variable selection by block addition and block deletion. J. Intelligent Learning Systems & Applications 2(4), 200–211 (2010)
15. Liu, H., Setiono, R.: Incremental feature selection. Applied Intelligence 9(3), 217–230 (1998)
16. Perkins, S., Lacker, K., Theiler, J.: Grafting: Fast, incremental feature selection by gradient descent in function space. J. Machine Learning Research 3, 1333–1356 (2003)
17. Ruiz, R., Riquelme, J.C., Aguilar-Ruiz, J.S.: Incremental wrapper-based gene selection from microarray data for cancer classification. Pattern Recognition 39(12), 2383–2392 (2006)
18. Bermejo, P., Gamez, J.A., Puerta, J.M.: Speeding up incremental wrapper feature subset selection with naive Bayes classifier. Knowledge-Based Systems 55, 140–147 (2014)
19. Abe, S.: Incremental input variable selection by block addition and block deletion. In: ICANN 2014 (accepted, 2014)
20. Abe, S.: Feature selection by iterative block addition and block deletion. In: Proc. SMC 2013, pp. 2677–2682 (2013)
21. Asuncion, A., Newman, D.J.: UCI machine learning repository (2007), http://www.ics.uci.edu/~mlearn/MLRepository.html
22. Abe, S.: Support Vector Machines for Pattern Classification, 2nd edn. (2010)

Low-Dimensional Data Representation in Data Analysis

Alexander Bernstein[1,2] and Alexander Kuleshov[1,2]

[1] Kharkevich Institute for Information Transmission Problems RAS, Moscow, Russia
kuleshov@iitp.ru
[2] National Research University Higher School of Economics (HSE), Moscow, Russia
abernstein@hse.ru

Abstract. Many Data Analysis tasks deal with data which are presented in high-dimensional spaces, and the 'curse of dimensionality' phenomena is often an obstacle to the use of many methods, including Neural Network methods, for solving these tasks. To avoid these phenomena, various Representation learning algorithms are used, as a first key step in solutions of these tasks, to transform the original high-dimensional data into their lower-dimensional representations so that as much information as possible is preserved about the original data required for the considered task. The above Representation learning problems are formulated as various Dimensionality Reduction problems (Sample Embedding, Data Manifold embedding, Data Manifold reconstruction and newly proposed Tangent Bundle Manifold Learning) motivated by various Data Analysis tasks. A new geometrically motivated algorithm that solves all the considered Dimensionality Reduction problems is presented.

Keywords: Machine Learning, Representation Learning, Dimensionality Reduction, Manifold Learning, Tangent Learning, Tangent Bundle Manifold Learning, Kernel methods.

1 Introduction

The goal of Data Analysis, which is a part of Machine Learning, is to extract previously unknown information from a dataset. Thus, it is supposed that information is reflected in the structure of a dataset which must be discovered from the data. Many Data Analysis tasks, such as Pattern Recognition, Classification, Clustering, Prognosis, Function reconstruction, and others, which are challenging for machine learning algorithms, deal with real-world data that are presented in high-dimensional spaces, and the 'curse of dimensionality' phenomena is often an obstacle to the use of many methods for solving these tasks.

To avoid these phenomena, various Representation learning algorithms are used as a first key step in solutions of these tasks. Representation learning (Feature extraction) algorithms transform the original high-dimensional data into their lower-dimensional representations (or features) so that as much information as possible is preserved about the original data required for the considered Data Analysis task.

After that, the initial Data Analysis task may be reduced to the corresponding task for the constructed lower-dimensional representation of the original dataset.

N. El Gayar et al. (Eds.): ANNPR 2014, LNAI 8774, pp. 47–58, 2014.

Of course, construction of the low-dimensional data representation for subsequent using in specific Data Analysis task must depend on the considered task, and success of machine learning algorithms generally depends on the data representation [1].

Representation (Feature) learning problems that consist in extracting a low-dimensional structure from high-dimensional data can be formulated as various Dimensionality Reduction (DR) problems, whose different formalizations depend on Data Analysis tasks considered further.

This paper is about DR problems in Data Analysis tasks. We describe a few key Data Analysis tasks that lead to different formulations of the DR: Sample Embedding for Clustering, Data Space (Manifold) embedding for Classification, Manifold Learning for Forecasting, etc. We also present a new geometrically motivated algorithm that solves all the considered DR problems.

The rest of the paper is organized as follows. Sections 2-5 contain definitions of various DR problems motivated by their subsequent using in specific Data Analysis tasks. The proposed DR solution is described in Section 6.

2   Sample Embedding Problem

One of the key Data Analysis tasks related to unsupervised learning is Clustering, which consists in discovering groups and structures in data that contain 'similar' (in one sense or another) sample points. Constructing a low-dimensional representation of original high-dimensional data for subsequent solution of the Clustering problem may be formulated as a specific DR problem, which will be referred to as the **Sample Embedding** problem and is as follows: Given an input dataset

$$\mathbf{X}_n = \{X_1, X_2, \dots, X_n\} \subset \mathbf{X}$$

randomly sampled from an unknown Data Space (DS) $\mathbf{X}$ embedded in a p-dimensional Euclidean space R^p, find an 'n-point' Embedding mapping

$$h_{(n)}: \mathbf{X}_n \subset R^p \to \mathbf{Y}_n = h_{(n)}(\mathbf{X}_n) = \{y_1, y_2, \dots, y_n\} \subset R^q \qquad (1)$$

of the sample $\mathbf{X}_n$ to a q-dimensional dataset $\mathbf{Y}_n$ (feature sample), q < p, which 'faithfully represents' the sample $\mathbf{X}_n$ while inheriting certain subject-driven data properties like preserving the local data geometry, proximity relations, geodesic distances, angles, etc.

If the term 'faithfully represents' in the Sample Embedding problem corresponds to the 'similar' notion in the initial Clustering problem, we can solve the reduced Clustering problem for the constructed low-dimensional feature dataset $\mathbf{Y}_n$. After that, we can obtain some solution of the initial Clustering problem: clusters in the initial problem are images of clusters discovered in the reduced problem by using a natural inverse mapping from $\mathbf{Y}_n$ to the original dataset $\mathbf{X}_n$.

The term 'faithfully represents' is not formalized in general, and in various Sample Embedding methods it is different due to choosing some optimized cost function $L_{(n)}(\mathbf{Y}_n|\mathbf{X}_n)$ which defines an 'evaluation measure' for the DR and reflects desired properties of the n-point Embedding mapping $h_{(n)}$ (1). As is pointed out in some papers, a general view on the DR can be based on the 'concept of cost functions.'

There exist a number of methods (techniques) for the Sample Embedding. Linear methods are well known and use such techniques as the PCA [2]. Various nonlinear techniques are based on Auto-Encoder Neural Networks [3, 4, 5], Kernel PCA [6], and others.

A newly emerging direction in the field of the Sample Embedding, which has been a subject of intensive research over the last decades, consists in constructing a family of algorithms based on studying the local structure of a given sampled dataset that retains local properties of the data with the use of various cost functions. Examples of such 'local' algorithms are: Locally Linear Embedding (LLE), Laplacian Eigenmaps (LE), Hessian Eigenmaps, ISOMAP, Local Tangent Space Alignment (LTSA), etc., described in [7, 8, 9] and other works. Some of these algorithms (LLE, LE, ISOMAP) can be considered in the same framework based on the Kernel PCA applied to various data-based kernels.

Note that Sample Embedding algorithms are based on the sample only, and no assumptions about the DS X are required for their descriptions. However, the study of properties of the algorithms is based on assumptions about both the DS and a way for extracting the sample from the DS.

3   Data Space (Manifold) Embedding problem

Another key Data Analysis task related to supervised learning concerns the Classification problem in which the original dataset consists of labeled examples: outputs (labels) $\Lambda_n = \{\lambda_1, \lambda_2, \ldots, \lambda_n\}$ are known for the corresponding inputs $\{X_1, X_2, \ldots, X_n\}$ sampled from the DS X; each label λ belongs to a finite set $\{1, 2, \ldots, m\}$ with $m \geq 2$. The problem is to generalize a function or mapping from inputs to outputs which can then be used to generate an output for a previously unseen input $X \in X$.

In the case of high-dimensional original inputs X_n, it is possible to construct low-dimensional features $\{y_1, y_2, \ldots, y_n\}$ (1) by using the Sample Embedding algorithm. After that, we can consider the reduced sample $[Y_n, \Lambda_n]$ instead of the sample $[X_n, \Lambda_n]$. For the possibility of using the solution of the reduced classification problem built for the reduced dataset, it is necessary to construct a lower-dimensional representation for a new unseen (usually called Out-of-Sample, OoS) input $X \in X / X_n$. Thus, it is necessary to consider another specific DR problem which is an extension of the Sample Embedding and can be referred to as the Data Space Embedding (Parameterization) problem: Given an input dataset (sample) X_n from the DS $X \subset R^p$, construct a low-dimensional parameterization of the DS which produces an Embedding mapping

$$h: X \subset R^p \to Y = h(X) \subset R^q \qquad (2)$$

from the DS X, including the OoS points, to the Feature Space (FS) $Y \subset R^q$, $q < p$, which preserves specific properties of the DS X. The term 'preserves specific properties' is not formalized in general and can be different due to choosing various cost functions reflecting specific preserved data properties.

The definition of the Data Space Embedding problem uses values of the Embedding mapping h (2) at the OoS points too. Thus, to justify the problem solution and study properties of the solution, we must define a Data Model describing the DS and a Sampling Model offering a way for extracting both the sample X_n and the OoS points from the DS. The most popular models in the DR are Manifold Data Models, see [7, 8, 9] and others works, in which the DS X is a q-dimensional manifold embedded in an ambient p-dimensional Euclidean space R^p, q < p, and referred to as the Data Manifold (DM). In most studies, DM is modeled using a single coordinate chart.

The Sampling Model is typically defined as a probability measure μ on the DM X whose support Supp(μ) coincides with the DM X. In accordance with this model, the dataset X_n and OoS points $X \in X / X_n$ are selected from the DM X independently of each other according to the probability measure μ.

A motivation for using the Manifold Data model consists in the following empirical fact: as a rule, high-dimensional real-world data lie on or near some unknown low-dimensional Data Manifold embedded in an ambient high-dimensional 'observation' space. This assumption is usually referred to as the Manifold assumption.

Various non-linear DR problems applied to the data which are described by the Manifold Data Model are usually referred to as the Manifold Learning (ML) problem [7, 8, 9]; the above-defined Data Space Embedding problem under the Manifold Data Model will be referred to as the Manifold Embedding problem. In the introduced terms, the Manifold Embedding problem is to construct a parameterization of the DM (global low-dimensional coordinates on the DM) from a finite dataset sampled from the DM. Note that there is no generally accepted definition for the ML.

Manifold assumption allowed constructing a family of algorithms based on studying the local structure of a given sampled dataset that retains local properties of the data with the use of various cost functions. Examples of such 'local' algorithms are described in [7, 8, 9] and other works; an 'OoS extension' for some local algorithms has been found in [10].

4   Manifold Learning Problem as Data Manifold Reconstruction

Manifold Embedding is usually a first step in various Data Analysis tasks in which reduced q-dimensional features y = h(X) are used in the reduced learning procedures instead of initial p-dimensional vectors X. If the Embedding mapping h in the Manifold Embedding preserves only specific properties of high-dimensional data, then substantial data losses are possible when using a reduced vector y = h(X) instead of the initial vector X. To prevent these losses, the mapping h must preserve as much as possible available information contained in the high-dimensional data [11]. Thus, it is necessary to consider the Manifold Embedding problem, in which the term *'faithfully represents'* has a specified meaning reflecting the possibility for reconstructing the initial vector $X \in X$ from the feature y = h(X) with small reconstruction error. Note that this error can be considered as a valid evaluation measure ('universal quality criterion') for Manifold Embedding procedures describing a measure of preserving information contained in the high-dimensional data [11].

There is a natural reconstruction of the vector $X \in \mathbf{X}$ from its lower-dimensional feature $y = h(X)$ for feature sample points $y \in \mathbf{Y}_n$. But in some tasks there may arise the problem of accurately reconstructing the points $X \in \mathbf{X}$ from their low-dimensional features $y = h(X)$ for Feature-Out-of-Sample, FOoS, points $y = h(X) \in h(\mathbf{X}) / \mathbf{Y}_n$. This possibility is directly required in various Data Analysis tasks such as multidimensional time series prognosis [12], data-based approximation of function with high-dimensional inputs [13], etc.

As an example, consider the problem of Electricity price curve forecasting [12] which is as follows. Electricity 'daily-prices' are described by a multidimensional time series (electricity price curve) $X_t = (X_{t1}, X_{t2}, \ldots , X_{t,24})^T \in R^{24}$ consisting of 'hour-prices' in the course of day t. Based on given vectors $\mathbf{X}_{1:T} = \{X_1, X_2, \ldots , X_T\} \subset R^{24}$, it is required to construct a forecast $\widehat{X}_{T+1}$ for X_{T+1}. The forecasting algorithm [12] uses replacement of the vectors X_t by their low-dimensional features $Y_t = h(X_t) \in R^q$ (the LLE method is used; the value q = 4 is selected as an appropriate dimension of the features). Then the forecast $\widehat{Y}_{T+1}$ for $Y_{T+1} = h(X_{T+1})$ based on the feature sample $\mathbf{Y}_{1:T} = h(\mathbf{X}_{1:T})$ in the reduced low-dimensional problem is constructed by using standard forecasting techniques. But then it is necessary to reconstruct the daily-price forecast $\widehat{X}_{T+1}$ from the feature forecast $\widehat{Y}_{T+1}$ which is the FOoS point in the general case.

A newly direction in the field of Machine Learning is meta-modeling in which data-based models (called meta-models [14] or surrogate models [15]) are constructed by learning on a set of input and output data prototypes obtained as a result of full-scale and/or computational experiments with some original complicated time-consuming models. As a rule, surrogate models have higher computational efficiency and can be used to replace original complicated models for further study (forecasting, optimization, etc.) [14, 15].

Input data which are original descriptions of objects under modeling can have high dimensionality, and the DR technique in meta-modeling is used for constructing reduced 'low-dimensional' surrogate models [16]. Thereafter, optimization or forecasting problems for the 'full-dimensional' model amounts to the corresponding reduced problems in the low-dimensional Feature space.

For example, meta-modeling is used in the wing shape optimization problem in aircraft designing [17], in which the DR is used to construct a low-dimensional wing airfoil parameterization [13]. In this problem the FOoS points appear as a result of solving optimization problems in the Feature space; thus, the reconstruction possibility is required in the DR.

However, the most of popular Manifold Embedding methods have a common drawback: they do not allow reconstructing high-dimensional points X from low-dimensional features h(X). Thus, it is necessary to formulate the ML problem in such a way that its solution does not have the above drawbacks. In other words, a corresponding ML procedure must reconstruct the unknown DM together with its low-dimensional parameterization from the sample.

We consider the ML problem called the Data Manifold Reconstruction problem, in which a low-dimensional representation of the DM allows accurate reconstruction of the DM [18, 19].

A strict definition is as follows: Given an input dataset X_n sampled from a q-dimensional DM X embedded in an ambient p-dimensional space R^p, $q < p$, and covered by a single chart, construct an **ML-solution** $\theta = (h, g)$ consisting of two inter-related mappings: an Embedding mapping h (2) and a Reconstruction mapping

$$g: Y \subset R^q \to R^p,$$

which determine a reconstructed value $r_\theta(X) = g(h(X))$ as a result of successively applying the embedding and reconstruction mappings to a vector $X \in X$. The solution must ensure the approximate equality

$$g(h(X)) \approx X \quad \text{for all} \quad X \in X, \tag{3}$$

and the Reconstruction error $\delta_\theta(X) = |X - r_\theta(X)|$ is a measure of quality of the solution θ at a point $X \in X$.

The Reconstruction mapping g must be defined not only on the feature sample Y_n (with an obvious reconstruction), but also on the FOoS features $y = h(X) \in Y / Y_n$ obtained by embedding the OoS points X.

The solution θ determines also a q-dimensional Reconstructed Manifold (RM)

$$X_\theta = \{X = g(y) \in R^p: y \in Y_\theta \subset R^q\} \tag{4}$$

embedded in R^p and parameterized by the chart g defined on the FS $Y_\theta = h(X)$. The approximate equalities (3) can be considered as the Manifold proximity property

$$X_\theta \approx X, \tag{5}$$

meaning that the RM $X_\theta = r_\theta(X)$ accurately reconstructs the DM X from the sample.

Note that the Data manifold reconstruction solution $\theta = (h, g)$ allows reconstructing the unknown DM X by the parameterized RM X_θ, whereas the Embedding Manifold solution h reconstructs a parameterization of the DM only.

From the statistical point of view, the defined Data manifold reconstruction problem may be considered as a Statistical Estimation Problem: Given a finite dataset randomly sampled from a smooth q-dimensional Data Manifold X covered by a single chart, estimate X by data-based q-dimensional manifold also covered (parameterized) by a single chart.

It is natural to evaluate the quality of the estimator X_θ (4) (sample-based q-dimensional manifold in R^p also covered by a single chart) by the Hausdorff distance $H(X_\theta, X)$ between the DM and RM [20]; the following relation between the qualities of the Data Manifold Reconstruction and Estimation problems takes a place:

$$H(X_\theta, X) \leq \sup_{X \in X} \delta_\theta(X).$$

Note that the defined Data Manifold Reconstruction problem differs from the Manifold approximation problem, in which an unknown manifold embedded in a high-dimensional ambient space must be approximated by some geometrical structure with close geometry, without any 'global parameterization' of the structure. For the latter problem, some solutions are known such as approximations by a simplicial complex

[21], by finitely many affine subspaces called 'flats' [22], tangential Delaunay complex [23], k-means and k-flats [24], and others. However, the Manifold approximation methods have a common drawback: they do not find a low-dimensional representation (parameterization) of the DM approximation; such parameterization is usually required in Machine Learning tasks with high-dimensional data.

There are some (though limited number of) methods for reconstruction of the DM **X** from the FS h(**X**). For a specific linear DM, the reconstruction can easily be made with the PCA. For a nonlinear DM, Auto-Encoder Neural Networks [3, 4, 5] determine both the embedding and reconstruction mappings. The LLE and LTSA methods also allow some reconstruction of the original vectors from their features.

5   Tangent Bundle Manifold Learning

The Reconstruction error $\delta_\theta(X)$ can be directly computed at sample points $X \in \mathbf{X}_n$; for an OoS point X it describes the generalization ability of the considered Data Manifold Reconstruction solution θ at a specific point X. Local lower and upper bounds are obtained for the maximum reconstruction error in a small neighborhood of an arbitrary point $X \in \mathbf{X}$ [19]; these bounds are defined in terms of the distance between the tangent spaces $L(X)$ and $L_\theta(r_\theta(X))$ to the DM **X** and the RM $\mathbf{X}_\theta$ at the points X and $r_\theta(X)$, respectively. It follows from the bounds that the greater the distances between these tangent spaces, the lower the local generalization ability of the solution θ. Thus, it is natural to require that the MR-solution ensures not only Manifold proximity (5) but also Tangent proximity

$$L(X) \approx L_\theta(r_\theta(X)) \qquad \text{for all } X \in \mathbf{X} \qquad (6)$$

between these tangent spaces in some selected metric on the Grassmann manifold Grass(p, q) consisting of all q-dimensional linear subspaces in R^p (the tangent spaces are treated as elements of the Grass(p, q)).

The requirement of the Tangent proximity for the Data Manifold Reconstruction solution arises also in various applications in which the MR is an intermediate step for Intelligent Data Analysis problem solution. For example, to ensure closeness between specific iterative optimization processes in the original and reduced design spaces, which are induced by the same optimization gradient-based method, it is necessary to guarantee accurate reconstruction of not only the DM (design space) **X** but also its tangent spaces. In Image Analysis, Data (Image) manifold may be very curved in an ambient space [25], and accurate reconstruction of the differential structure of the Image manifold (first of all, reconstruction of the tangent spaces) is required [26].

A statement of the extended Data Manifold Reconstruction problem, which includes an additional requirement of the tangent spaces proximity, has been proposed in [18, 19] and is described below.

The set TB(**X**) = {(X, L(X)): $X \in \mathbf{X}$} composed of points X of the manifold **X** equipped by tangent spaces L(X) at these points, is known in the Manifold theory as the Tangent Bundle of the manifold **X**. Thus, accurate reconstruction of the DM **X** from the sample, which ensures accurate reconstruction of its tangent spaces too, can be considered as reconstruction of the Tangent Bundle TB(X). Therefore, the amplification of the ML consisting in accurate reconstruction of the tangent bundle TB(X) from the sample X_n may be referred to as the Tangent Bundle Manifold Learning.

A strict definition of the TBML is as follows: Given dataset X_n sampled from a q-dimensional DM X embedded in an ambient p-dimensional Euclidean space R^p, q < p, construct TBML-solution θ = (h, g) which provides Tangent Bundle proximity consisting in the Manifold proximity (5) and the Tangent proximity (6), where $L_\theta(r_\theta(X))$ = Span($J_g(h(X))$) is the tangent space to the RM X_θ at a point $r_\theta(X)$ spanned by columns of the Jacobian $J_g(y)$ of the mapping g(y) at a point y = h(X) ∈ Y_θ.

The TBML-solution θ determines the Reconstructed tangent bundle

$$RTB_\theta(X_\theta) = \{(g(y), Span(J_g(y))): y \in Y_\theta\} \tag{7}$$

of the RM X_θ, which is close to the TB(X), and the q-dimensional submanifold

$$L_\theta = \{Span(J_g(y)): y \in Y_\theta\} \subset Grass(p, q)$$

of the Grassmann manifold which reconstructs the Tangent Manifold

$$L = \{L(X): X \in X\} \subset Grass(p, q).$$

The next section briefly describes the TBML-solution called the Grassmann & Stiefel Eigenmaps (GSE) algorithm [18, 19], which also gives new solutions for all the DR problems specified in Sections 2-4 above.

6   Tangent Bundle Manifold Learning Solution

The GSE algorithm consists of three successively performed steps: Tangent Manifold Learning, Manifold Embedding, and Tangent Bundle reconstruction.

In the Tangent Manifold Learning Step, a sample-based family H = {H(X), X ∈ X} consisting of p×q matrices H(X) smoothly depending on X ∈ X is constructed to meet the relations

$$L_H(X) \approx L(X) \quad \text{for all} \quad X \in X;$$

here $L_H(X)$ are q-dimensional linear spaces in R^p spanned by columns $H^{(1)}(X)$, $H^{(2)}(X), \dots , H^{(q)}(X)$ of the matrices H(X).

The family H is constructed in such a way as to provide the additional property: vector fields $H^{(1)}(X), H^{(2)}(X), \dots , H^{(q)}(X) \in L_\theta(r_\theta(X))$ must be potential and, therefore, meet the following relations

$$\nabla_{H^{(i)}} H^{(j)}(X) = \nabla_{H^{(j)}} H^{(i)}(X) \tag{8}$$

for i, j = 1, 2, ... , q; here ∇_H denotes covariant differentiation with respect to the vector field H(X) ∈ $L_\theta(r_\theta(X))$.

Let us briefly describe the Tangent Manifold Learning Step of the GSE. At first, the tangent space L(X) for the points X ∈ X are estimated by the q-dimensional linear space $L_{PCA}(X)$ which is a result of the PCA applied to sample points from an ε_n-ball in R^p centered at X; here $\varepsilon_n = O(n^{-1/(q+2)})$ is a small parameter.

The data-based kernel $K(X, X')$, $X', X \in \mathbf{X}$, is constructed as a product

$$K_E(X, X') \times K_G(X, X'),$$

where K_E is the Euclidean 'heat' kernel introduced in the LE algorithm [27] and

$$K_G(X, X') = K_{BC}(L_{PCA}(X), L_{PCA}(X'))$$

is the Binet–Cauchy kernel [28] on the Grass(p, q); this aggregate kernel reflects not only geometrical nearness between the points X and X' but also nearness between the linear spaces $L_{PCA}(X)$ and $L_{PCA}(X')$, whence comes nearness between the tangent spaces L(X) and L(X').

The set $\mathbf{H}_n$ consisting of explicitly written p×q matrices H_i that approximate the matrices $H(X_i)$, meet the constraints $Span(H_i) = L_{PCA}(X_i)$ and satisfy the conditions (8) written in a form of finite differences, is constructed to minimize the quadratic form

$$\Delta_{H,n}(\mathbf{H}_n) = \tfrac{1}{2}\textstyle\sum_{i,j=1}^n K(X_i, X_j) \times \|H_i - H_j\|_F^2, \qquad (9)$$

under the normalizing condition $\sum_{i=1}^n K(X_i) \times (H_i^T \times H_i) = I_q$ required to avoid a degenerate solution; here $K(X) = \sum_{j=1}^n K(X, X_j)$ and $K = \sum_{i=1}^n K(X_i)$.

Given $\mathbf{H}_n$, the p×q matrix H(X) for an arbitrary point $X \in \mathbf{X}$ is chosen to minimize the form $\Delta_H(H, X) = \sum_{j=1}^n K(X, X_j) \times \|H(X) - H_j\|_F^2$ under the specified linear conditions.

The exact solution of the minimizing problem (9) under the conditions (8) is obtained as a solution of specified generalized eigenvector problems. The matrix H(X) which minimizes the quadratic form $\Delta_H(H, X)$ is written in an explicit form.

This Step gives a new solution for the Tangent Manifold Learning problem of estimating the tangent spaces L(X) in the form of a smooth function of the point $X \in \mathbf{X}$, which was considered in some previous works. The matrices whose columns approximately span the tangent spaces were constructed using Artificial Neural Networks with one hidden layer [29] or Radial Basis Functions [30]. The constructed linear spaces $\{L_H(X_i)\}$ are the result of an alignment of the PCA-based linear spaces $\{L_{PCA}(X_i)\}$; a similar alignment problem was studied in the LTSA [31] with using a cost function which differs from our cost function (9).

The mappings h and g will be constructed in the next parts to provide the proximities

$$g(h(X)) \approx X \quad and \quad J_g(h(X)) \approx H(X), \qquad (10)$$

whence comes the Tangent Bundle proximity (5), (6).

In the Manifold Embedding Step, given the family $\mathbf{H}$ already constructed, the Embedding mapping h(X) is constructed for $X \in \mathbf{X}$.

Taylor series expansions $g(y') - g(y) \approx J_g(y) \times (y' - y)$ for near points y, y', under the desired equalities (10) for mappings h and g specified further, imply the equalities:

$$X' - X \approx H(X) \times (h(X') - h(X)) \tag{11}$$

for near points X, X' $\in$ **X**.

Under the family **H** already constructed, these approximate equalities can be considered as regression equations for the features h(X). First, consider equations (11) written for near sample points, and compute a preliminary vector set $Y_n = \{y_1, y_2, \ldots, y_n\}$ as a standard least squares solution, which minimizes the weighted residual

$$\sum_{i,j=1}^{n} K(X_i, X_j) \times \left| X_j - X_i - H(x_i) \times (y_j - y_i) \right|^2$$

under the normalizing condition $y_1 + y_2 + \ldots + y_n = 0$.

Then, based on Y_n, choose a value y = h(X) for an arbitrary point X $\in$ **X** by minimizing over y the weighted residual

$$\sum_{j=1}^{n} K(X, X_j) \times \left| X_j - X - H(X) \times (y_j - y) \right|^2.$$

Thus, under Y_n, the value h(X) for an arbitrary point X $\in$ **X** (including sample points) is written as

$$h(X) = h_{KNR}(X) + v^{-1}(X) \times Q_{PCA}^T(X) \times \tau(X),$$

here $v(X) = Q_{PCA}^T(X) \times H(X)$, $\tau(X) = \frac{1}{K(X)} \sum_{j=1}^{n} K(X, X_j) \times (X - X_j)$ and

$$h_{KNR}(X) = \frac{1}{K(X)} \sum_{j=1}^{n} K(X, X_j) \times y_j$$

is standard Kernel Non-parametric Regression estimator for h(X) based on the preliminary values $y_j \in Y_n$ of the vector h(X) at the sample points.

The constrained mapping h determines the Feature space $Y_\theta = h(X)$. This Step gives a new solution for the Manifold Embedding problem.

In the Tangent Bundle reconstruction step, given the family **H** and the mapping h already constructed, the mapping g is constructed to meet the proximities (3) and (6). This step gives a new solution for the Data Manifold Reconstruction.

The data-based kernel k(y, y') on Y_θ and the linear spaces $L^*(y) \in$ Grass(p, q) depending on y $\in Y_\theta$ are constructed to provide the equalities

$$k(h(X), h(X')) \approx K(X, X')$$

and $L^*(h(X)) \approx L_{PCA}(X)$ for near points X $\in$ **X** and X' $\in X_n$.

The reconstruction function **g**(y) is constructed with using kernel nonparametric regression technique based on known values $X_i = g(y_i)$ at the points $y_i = h(X_i)$ with taking into account the known values $J_g(y_i) = H(X_i)$, i = 1, 2, ..., n.

In the as asymptotic $n \to \infty$ and under an appropriate choice of the algorithm parameters, the rate in proximities (3) and (6) is $O(n^{-2/(q+2)})$ and $O(n^{-1/(q+2)})$, respectively [32]; the first rate coincides with the asymptotically minimax lower bound [20] for the Hausdorff distance between the DM $\mathbf{X}$ and RM $\mathbf{X}_\theta$. Thus, the RM $\mathbf{X}_\theta$ estimates the DM $\mathbf{X}$ with the optimal rate of convergence.

Acknowledgments. This work is partially supported by the Russian Foundation for Basic Research, research projects 13-01-12447 and 13-07-12111.

References

1. Bengio, Y., Courville, A., Vincent, P.: Representation Learning: A Review and New Perspectives. arXiv:1206.5538v2, 1–64 (2012)
2. Jollie, T.: Principal Component Analysis. Springer, New-York (2002)
3. Hecht-Nielsen, R.: Replicator neural networks for universal optimal source coding. Science 269, 1860–1863 (1995)
4. Kramer, M.: Nonlinear Principal Component Analysis using autoassociative neural networks. AIChE Journal 37(2), 233–243 (1991)
5. Hinton, G.E., Salakhutdinov, R.R.: Reducing the dimensionality of data with neural networks. Science 313(5786), 504–507 (2006)
6. Schölkopf, B., Smola, A., Müller, K.: Nonlinear component analysis as a kernel eigenvalue problem. Neural Computation 10(5), 1299–1319 (1998)
7. Cayton, L.: Algorithms for manifold learning. Univ. of California at San Diego (UCSD), Technical Report CS2008-0923, pp. 541–555. Citeseer (2005)
8. Huo, X., Ni, X., Smith, A.K.: Survey of Manifold-based Learning Methods. In: Liao, T.W., Triantaphyllou, E. (eds.) Recent Advances in Data Mining of Enterprise Data, pp. 691–745. World Scientific, Singapore (2007)
9. Ma, Y., Fu, Y. (eds.): Manifold Learning Theory and Applications. CRC Press, London (2011)
10. Bengio, Y., Delalleau, O., Le Roux, N., Paiement, J.-F., Vincent, P., Ouimet, M.: Learning Eigenfunctions Link Spectral Embedding and Kernel PCA. Neural Computation 16(10), 2197–2219 (2004)
11. Lee, J.A., Verleysen, M.: Quality assessment of dimensionality reduction: Rank-based criteria. Neurocomputing 72(7-9), 1431–1443 (2009)
12. Chen, J., Deng, S.-J., Huo, X.: Electricity price curve modeling and forecasting by manifold learning. IEEE Transaction on Power Systems 23(3), 877–888 (2008)
13. Bernstein, A.V., Burnaev, E.V., Chernova, S.S., Zhu, F., Qin, N.: Comparison of Three Geometric Parameterization methods and Their Effect on Aerodynamic Optimization. In: Proceedings of International Conference on Evolutionary and Deterministic Methods for Design, Optimization and Control with Applications to Industrial and Societal Problems, Eurogen 2011, Capua, Italy, September 14-16 (2011)
14. Gary Wang, G., Shan, S.: Review of Metamodeling Techniques in Support of Engineering Design Optimization. J. Mech. Des. 129(3), 370–381 (2007)
15. Forrester, A.I.J., Sobester, A., Keane, A.J.: Engineering Design via Surrogate Modelling. A Practical Guide. Wiley, New-York (2008)
16. Kuleshov, A.P., Bernstein, A.V.: Cognitive Technologies in Adaptive Models of Complex Plants. Information Control Problems in Manufacturing 13(1), 1441–1452 (2009)

17. Bernstein, A., Kuleshov, A., Sviridenko, Y., Vyshinsky, V.: Fast Aerodynamic Model for Design Technology. In: Proceedings of West-East High Speed Flow Field Conference, WEHSFF-2007. IMM RAS, Moscow (2007), http://wehsff.imamod.ru/pages/s7.html

18. Bernstein, A.V., Kuleshov, A.P.: Tangent Bundle Manifold Learning via Grassmann & Stiefel Eigenmaps. In arXiv preprint: arXiv:1212.6031v1 [cs.LG], pp. 1–25 (December 2012)

19. Bernstein, A.V., Kuleshov, A.P.: Manifold Learning: generalizing ability and tangent proximity. International Journal of Software and Informatics 7(3), 359–390 (2013)

20. Genovese, C.R., Perone-Pacifico, M., Verdinelli, I., Wasserman, L.: Minimax Manifold Estimation. Journal Machine Learning Research 13, 1263–1291 (2012)

21. Freedman, D.: Efficient simplicial reconstructions of manifold from their samples. IEEE Transaction on Pattern Analysis and Machine Intelligence 24(10), 1349–1357 (2002)

22. Karygianni, Sofia, Frossard, Pascal. Tangent-based manifold approximation with locally linear models. In: arXiv:1211.1893v1 [cs.LG] (November 6, 2012)

23. Boissonnat, J.-D., Ghosh, A.: Manifold reconstruction using tangential Delaunay complexes. Discrete & Computational Geometry 51(1), 221–267 (2014)

24. Canas, G.D., Poggio, T., Rosasco, L.A.: Learning Manifolds with K-Means and K-Flats. In: Advances in Neural Information Processing Systems, NIPS (2012)

25. Pless, R., Souvenir, R.: A Survey of Manifold Learning for Images. IPSJ Transactions on Computer Vision and Applications 1, 83–94 (2009)

26. Kuleshov, A.P., Bernstein, A.V.: Tangent Bundle Manifold Learning for Image Analysis. In: Vuksanovic, B., Verikas, A., Zhou, J. (eds.) Proceedings of SPIE, Sixth International Conference on Machine Vision, ICMV 2013, London, The United Kingdom, November 16-17, vol. 9067, pp. 201–205 (2013)

27. Belkin, M., Niyogi, P.: Laplacian eigenmaps for dimensionality reduction and data representation. Neural Computation 15, 1373–1396 (2003)

28. Wolf, L., Shashua, A.: Learning over sets using kernel principal angles. J. Mach. Learn. Res. 4, 913–931 (2003)

29. Bengio, Y., Monperrus, M.: Non-local manifold tangent learning. In: Saul, L., Weiss, Y., Bottou, L. (eds.) Advances in Neural Information Processing Systems, vol. 17, pp. 129–136. MIT Press, Cambridge (2005)

30. Dollár, P., Rabaud, V., Belongie, S.: Learning to Traverse Image Manifolds. In: Schölkopf, B., Platt, J.C., Hoffman, T. (eds.) Advances in Neural Information Processing Systems, 19, pp. 361–368. MIT Press, Cambridge (2006)

31. Zhang, Z., Zha, H.: Principal Manifolds and Nonlinear Dimension Reduction via Local Tangent Space Alignment. SIAM Journal on Scientific Computing 26(1), 313–338 (2005)

32. Kuleshov, A., Bernstein, A., Yanovich, Y.: Asymptotically optimal method in Manifold estimation. In: Márkus, L., Prokaj, V. (eds.) Abstracts of the XXIX-th European Meeting of Statisticians, Budapest, July 20-25, p. 325 (2013)

Analyzing Dynamic Ensemble Selection Techniques Using Dissimilarity Analysis

Rafael M. O. Cruz, Robert Sabourin, and George D. C. Cavalcanti

École de Technologie Supérieure, Université du Québec,
1100 Notre-Dame Ouest, Montréal, Canada
Centro de Informática, Universidade Federal de Pernambuco
Recife, Brazil
`{cruz,Robert.Sabourin}@livia.etsmtl.ca`
`gdcc@cin.ufpe.br`
`http://www.livia.etsmtl.ca`

Abstract. In Dynamic Ensemble Selection (DES), only the most competent classifiers are selected to classify a given query sample. A crucial issue faced in DES is the definition of a criterion for measuring the level of competence of each base classifier. To that end, a criterion commonly used is the estimation of the competence of a base classifier using its local accuracy in small regions of the feature space surrounding the query instance. However, such a criterion cannot achieve results close to the performance of the Oracle, which is the upper limit performance of any DES technique. In this paper, we conduct a dissimilarity analysis between various DES techniques in order to better understand the relationship between them and as well as the behavior of the Oracle. In our experimental study, we evaluate seven DES techniques and the Oracle using eleven public datasets. One of the seven DES techniques was proposed by the authors and uses meta-learning to define the competence of base classifiers based on different criteria. In the dissimilarity analysis, this proposed technique appears closer to the Oracle when compared to others, which would seem to indicate that using different bits of information on the behavior of base classifiers is important for improving the precision of DES techniques. Furthermore, DES techniques, such as LCA, OLA, and MLA, which use similar criteria to define the level of competence of base classifiers, are more likely to produce similar results.

Keywords: Ensemble of classifiers, dynamic ensemble selection, dissimilarity analysis, meta-learning.

1 Introduction

In recent years, ensembles of Classifiers (EoC) have been widely studied as an alternative for increasing efficiency and accuracy in pattern recognition [1,2]. Classifier ensembles involve two basic approaches, namely, classifier fusion and dynamic ensemble selection. With classifier fusion approaches, each classifier in the ensemble is used, and their outputs are aggregated to give the final prediction. However, such techniques [1,3] present two main problems: they are based on the assumption that the base classifiers commit independent errors, which rarely occurs to find in real pattern recognition applications.

N. El Gayar et al. (Eds.): ANNPR 2014, LNAI 8774, pp. 59–70, 2014.

On the other hand, Dynamic Ensemble Selection (DES) techniques [4] rely on the assumption that each base classifier[1] is an expert in a different local region of the feature space. DES techniques work by measuring the level of competence of each base classifier, considering each new test sample. Only the most competent(s) classifier(s) is(are) selected to predict the class of a new test sample. Hence, the key issue in DES is defining a criterion for measuring the level of competence of a base classifier. Most DES techniques [5,6,7,8] use estimates of the classifier's local accuracy in small regions of the feature space surrounding the query instance as search criteria to carry out the ensemble selection. However, in our previous work [7], we demonstrated that this criterion is limited, and cannot achieve results close to the performance of the Oracle, which represents the best possible result of any combination of classifiers [2]. In addition, as reported by Ko et al. [5], addressing the behavior of the Oracle is much more complex than applying a simple neighborhood approach, and the task of figuring out its behavior based merely on the pattern feature space is not an easy one.

To tackle this issue, in [9] we proposed a novel DES framework in which multiple criteria regarding the behavior of a base classifier are used to compute its level of competence. In this paper, we conduct a dissimilarity analysis between different DES techniques in order to better understand their relationship. The analysis is performed based on the difference between the levels of competence of a base classifier estimated by the criterion embedded in each DES technique. All in all, we compare the DES criteria of seven state-of-the-art DES techniques, including our proposed meta-learning framework. In addition, we also formalize the Oracle as an ideal DES technique (i.e., a DES scheme which selects only the classifiers of the pool that predict the correct class for the query instance) to be used in the analysis.

The dissimilarities between different DES criteria are computed in order to generate a dissimilarity matrix, which is then, used to project each DES technique onto a two-dimensional space, called the Classifier Projection Space (CPS) [10]). In the CPS, each DES technique is represented by a point, and the distance between two points corresponds to their degree of dissimilarity. Techniques that appear close together present similar behavior (i.e., they are more likely to produce the same results), while those appearing far apart in the two-dimensional CPS can be considered different. Thus, a spatial relationship is achieved between different techniques. The purpose of the dissimilarity analysis is twofold: to understand the relationship between different DES techniques (i.e., whether or not the criteria used by DES techniques present a similar behavior), and in order to determine which DES technique presents a behavior that is closer to the behavior of the ideal DES scheme (Oracle).

This paper is organized as follows: Section 2 introduces the DES techniques from the literature that are used in the dissimilarity analysis. The proposed meta-learning framework is described in Section 3. Experiments are conducted in Section 4, and finally, our conclusion is presented in the last section.

---

[1] The term base classifier refers to a single classifier belonging to an ensemble or a pool of classifiers.

2   Dynamic Ensemble Selection Techniques

The goal of dynamic selection is to find an ensemble of classifiers, $C' \subset C$ containing the best classifiers to classify a given test sample x_j. This is different from static selection, where the ensemble of classifiers C' is selected during the training phase, and considering the global performance of the base classifiers over a validation dataset. In dynamic selection, the classifier competence is measured on-the-fly for each query instance x_j.

The following DES techniques are described in this section: Overall Local Accuracy (OLA) [6], Local Classifier Accuracy (LCA) [6], Modified Local Accuracy (MLA) [8], KNORA-Eliminate [5], K-Nearest Output Profiles (KNOP) [11] and Multiple Classifier Behavior (MCB) [12].

For the definitions below, let $\theta_j = \{x_1, \dots, x_K\}$ be the region of competence of the test sample x_j (K is the size of the region of competence), defined on the validation data, c_i a base classifier from the pool $C = \{c_1, \dots, c_M\}$ (M is the size of the pool), w_l the correct label of x_j and $\delta_{i,j}$ the level of competence of c_i for the classification of the input instance x_j.

Overall Local Accuracy (OLA)

In this method, the level of competence $\delta_{i,j}$ of a base classifier c_i is simply computed as the local accuracy achieved by c_i for the region of competence θ_j. (Equation 1). The classifier with the highest level of competence $\delta_{i,j}$ is selected.

$$\delta_{i,j} = \sum_{k=1}^{K} P(w_l \mid x_k \in w_l, c_i) \tag{1}$$

Local Classifier Accuracy (LCA)

This rule is similar to the OLA, with the only difference being that the local accuracy of c_i is estimated with respect to the output classes; w_l (w_l is the class assigned for x_j by c_i) for the whole region of competence, θ_j (Equation 2). The classifier with the highest level of competence $\delta_{i,j}$ is selected.

$$\delta_{i,j} = \frac{\sum_{x_k \in w_l} P(w_l \mid x_k, c_i)}{\sum_{k=1}^{K} P(w_l \mid x_k, c_i)} \tag{2}$$

Modified Local Accuracy (MLA)

The MLA technique works similarly to the LCA. The only difference is that each instance x_k belonging to the region of competence θ_j is weighted by its Euclidean distance to the query sample x_j. The classifier with the highest level of competence $\delta_{i,j}$ is selected.

KNORA-Eliminate (KNORA-E)

Given the region of competence θ_j, only the classifiers that achieved a perfect score, considering the whole region of competence, are considered competent for the classification of x_j. Thus, the level of competence $\delta_{i,j}$ is either "competent", $\delta_{i,j} = 1$ or "incompetent", $\delta_{i,j} = 0$. All classifiers considered competent are selected.

Multiple Classifier Behavior (MCB)

Given the query pattern x_j, the first step is to compute its K-Nearest-Neighbors $x_k, k = 1, \ldots, K$. Then, the output profiles of each neighbor $\tilde{x}_k$ are computed and compared to the output profile of the test instance $\tilde{x}_j$ according to a similarity metric $D_{OutProf}$. If $D_{OutProf} > threshold$, the pattern is removed from the region of competence. The level of competence $\delta_{i,j}$ is measured by the recognition performance of the base classifier c_i over the filtered region of competence. The classifier with the highest level of competence $\delta_{i,j}$ is selected.

K-Nearest Output Profiles (KNOP)

This rule is similar to the KNORA technique, with the only difference being that KNORA works in the feature space while KNOP works in the decision space using output profiles. First, the output profiles' transformation is applied over the input x_j, giving $\tilde{x}_j$. Next, the similarity between $\tilde{x}_j$ and the output profiles from the validation set is computed and stored in the set ϕ_j. The level of competence $\delta_{i,j}$ of a base classifier c_i for the classification of x_j is defined by the number of samples in ϕ_j that are correctly classified by c_i.

Oracle

The Oracle is classically defined in the literature as a strategy that correctly classifies each query instance x_j if any classifier c_i from the pool of classifiers C predicts the correct label for x_j. In this paper, we formalize the Oracle as the ideal DES technique which always selects the classifier that predicts the correct label x_j and rejects otherwise. The Oracle as a DES technique is defined in Equation 3:

$$\begin{cases} \delta_{i,j} = 1, & \text{if } c_i \text{ correctly classifies } x_j \\ \delta_{i,j} = 0, & \text{otherwise} \end{cases} \quad (3)$$

In other words, the level of competence $\delta_{i,j}$ of a base classifier c_i is 1 if it predicts the correct label for x_j, or 0 otherwise.

3   Dynamic Ensemble Selection Using Meta-Learning

A general overview of the proposed meta-learning framework is depicted in Figure 1. It is divided into three phases: Overproduction, Meta-training and Generalization. Each phase is detailed in the following sections.

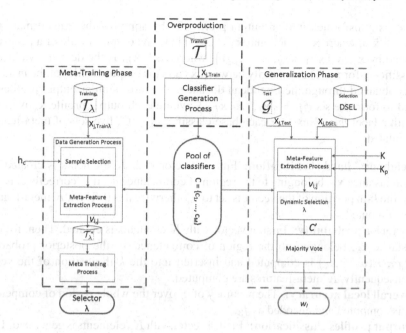

Fig. 1. Overview of the proposed framework. It is divided into three steps 1) Overproduction 2) Meta-training and 3) Generalization. [Adapted from [9]]

3.1 Overproduction

In this step, the pool of classifiers $C = \{c_1, \ldots, c_M\}$, where M is the pool size, is generated using the training dataset $\mathcal{T}$. The Bagging technique [13] is used in this work in order to build a diverse pool of classifiers.

3.2 Meta-Training

In this phase, the meta-features are computed and used to train the meta-classifier λ. As shown in Figure 1, the meta-training stage consists of three steps: sample selection, the meta-features extraction process and meta-training. A different dataset $\mathcal{T}_\lambda$ is used in this phase to prevent overfitting.

Sample Selection. We focus the training of λ on cases in which the extent of consensus of the pool is low. Thus, we employ a sample selection mechanism based on a threshold h_C, called the consensus threshold. For each $\mathbf{x}_{j,train_\lambda} \in \mathcal{T}_\lambda$, the degree of consensus of the pool, denoted by $H(\mathbf{x}_{j,train_\lambda}, C)$, is computed. If $H(\mathbf{x}_{j,train_\lambda}, C)$ falls below the threshold h_C, $\mathbf{x}_{j,train_\lambda}$ is passed down to the meta-features extraction process.

Meta-Features Extraction. In order to extract the meta-features, the region of competence of $\mathbf{x}_{j,train_\lambda}$, denoted by $\theta_j = \{\mathbf{x}_1, \ldots, \mathbf{x}_K\}$ must be first computed. The region of competence is defined in the $\mathcal{T}_\lambda$ set using the K-Nearest Neighbor algorithm.

Then, $\mathbf{x}_j$ is transformed into an output profile, $\tilde{\mathbf{x}}_j$ by applying the transformation T, $(T : \mathbf{x}_j \Rightarrow \tilde{\mathbf{x}}_j)$, where $\mathbf{x}_j \in \Re^D$ and $\tilde{\mathbf{x}}_j \in Z^M$ [11]. The output profile of a pattern $\mathbf{x}_j$ is denoted by $\tilde{\mathbf{x}}_j = \{\tilde{\mathbf{x}}_{j,1}, \tilde{\mathbf{x}}_{j,2}, \ldots, \tilde{\mathbf{x}}_{j,M}\}$, where each $\tilde{\mathbf{x}}_{j,i}$ is the decision yielded by the classifier c_i for $\mathbf{x}_j$. The similarity between $\tilde{\mathbf{x}}_j$ and the output profiles of the instances in $\mathcal{T}_\lambda$ is obtained through the Euclidean distance. The most similar output profiles are selected to form the set $\phi_j = \{\tilde{\mathbf{x}}_1, \ldots, \tilde{\mathbf{x}}_{K_p}\}$, where each output profile $\tilde{\mathbf{x}}_k$ is associated with a label $w_{l,k}$. Next, for each base classifier $c_i \in C$, five sets of meta-features are calculated:

f_1 - **Neighbors' hard classification:** First, a vector with K elements is created. For each instance $\mathbf{x}_k$, belonging to the region of competence θ_j, if c_i correctly classifies $\mathbf{x}_k$, the k-th position of the vector is set to 1, otherwise it is 0. Thus, K meta-features are computed.

f_2 - **Posterior probability:** First, a vector with K elements is created. Then, for each instance $\mathbf{x}_k$, belonging to the region of competence θ_j, the posterior probability of c_i, $P(w_l \mid \mathbf{x}_k)$ is computed and inserted into the k-th position of the vector. Consequently, K meta-features are computed.

f_3 - **Overall local accuracy:** The accuracy of c_i over the whole region of competence θ_j is computed and encoded as f_3.

f_4 - **Output profiles classification:** First, a vector with K_p elements is generated. Then, for each member $\tilde{\mathbf{x}}_k$, belonging to the set of output profiles ϕ_j, if the label produced by c_i for $\mathbf{x}_k$ is equal to the label $w_{l,k}$ of $\tilde{\mathbf{x}}_k$, the k-th position of the vector is set to 1, otherwise it is 0. A total of K_p meta-features are extracted using output profiles.

f_5 - **Classifier's Confidence:** The perpendicular distance between the input sample $\mathbf{x}_{j,train_\lambda}$ and the decision boundary of the base classifier c_i is calculated and encoded as f_5. f_5 is normalized to a $[0 - 1]$ range using the Min-max normalization.

A vector $v_{i,j} = \{f_1 \cup f_2 \cup f_3 \cup f_4 \cup f_5\}$ is obtained at the end of the process. It is important to mention that a different vector $v_{i,j}$ is generated for each base classifier c_i. If c_i correctly classifies $\mathbf{x}_{j,train_\lambda}$, the class attribute of $v_{i,j}$, $\alpha_{i,j} = 1$ (i.e., $v_{i,j}$ corresponds to the behavior of a competent classifier), otherwise $\alpha_{i,j} = 0$. $v_{i,j}$ is stored in the meta-features dataset (Figure 1).

Training. With the meta-features dataset, $\mathcal{T}_\lambda^*$, on hand, the last step of the meta-training phase is the training of the meta-classifier λ. The dataset $\mathcal{T}_\lambda^*$ is divided on the basis of 75% for training and 25% for validation. A Multi-Layer Perceptron (MLP) neural network with 10 neurons in the hidden layer is considered as the selector λ. The training process for λ is performed using the Levenberg-Marquadt algorithm, and is stopped if its performance on the validation set decreases or fails to improve for five consecutive epochs.

3.3 Generalization

Given an input test sample $\mathbf{x}_{j,test}$ from the generalization dataset $\mathcal{G}$, first, the region of competence θ_j and the set of output profiles ϕ_j, are calculated using the samples from the dynamic selection dataset D_{SEL} (Figure 1). For each classifier $c_i \in C$, the five subsets of meta-features are extracted, returning the meta-features vector $v_{i,j}$. Next, $v_{i,j}$ is passed down as input to the meta-classifier λ, which decides whether c_i is competent enough to classify $x_{j,test}$. In this case, the posterior probability obtained by the meta-classifier λ is considered as the estimation of the level of competence $\delta_{i,j}$ of the base classifier c_i in relation to $x_{j,test}$.

After each classifier of the pool is evaluated, the majority vote rule [2] is applied over the ensemble C', giving the label w_l of $x_{j,test}$. Tie-breaking is handled by choosing the class with the highest a posteriori probability.

4 Experiments

We evaluated the generalization performance of the proposed technique using eleven classification datasets, nine from the UCI machine learning repository, and two artificially generated using the Matlab PRTOOLS toolbox[2]. The experiment was conducted using 20 replications. For each replication, the datasets were randomly divided on the basis of 25% for training ($\mathcal{T}$), 25% for meta-training $\mathcal{T}_{\lambda}$, 25% for the dynamic selection dataset (D_{SEL}) and 25% for generalization ($\mathcal{G}$). The divisions were performed while maintaining the prior probability of each class. The pool of classifiers was composed of 10 Perceptrons. The values of the hyper-parameters K, K_p and h_c were set as 7, 5 and 70%, respectively. They were selected empirically based on previous publications [7,9].

4.1 Results

Table 1. Mean and standard deviation results of the accuracy obtained for the proposed meta-learning framework and the DES systems in the literature. The best results are in bold. Results that are significantly better ($p < 0.05$) are underlined.

Database	Proposed	KNORA-E	MCB	LCA	OLA	MLA	KNOP	Oracle
Pima	77.74(2.34)	73.16(1.86)	73.05(2.21)	72.86(2.98)	73.14(2.56)	73.96(2.31)	73.42(2.11)	95.10(1.19)
Liver Disorders	68.83 (5.57)	63.86(3.28)	63.19(2.39)	62.24(4.01)	62.05(3.27)	57.10(3.29)	65.23(2.29)	90.07(2.41)
Breast Cancer	97.41(1.07)	96.93(1.10)	96.83(1.35)	97.15(1.58)	96.85(1.32)	96.66(1.34)	95.42(0.89)	99.13(0.52)
Blood Transfusion	79.14(1.88)	74.59(2.62)	72.59(3.20)	72.20(2.87)	72.33(2.36)	70.17(3.05)	77.54(2.03)	94.20(2.08)
Banana	90.16(2.09)	88.83(1.67)	88.17(3.37)	89.28(1.89)	89.40(2.15)	80.83(6.15)	85.73(10.65)	94.75(2.09)
Vehicle	82.50(2.07)	81.19(1.54)	80.20(4.05)	80.33(1.84)	81.50(3.24)	71.15(3.50)	80.09(1.47)	96.80(0.94)
Lithuanian Classes	90.26(2.78)	88.83(2.50)	89.17(2.30)	88.10(2.20)	87.95(1.85)	77.67(3.20)	89.33(2.29)	98.35 (0.57)
Sonar	79.72(1.86)	74.95(2.79)	75.20(3.35)	76.51(2.06)	74.52(1.54)	74.85(1.34)	75.72(2.82)	94.46(1.63)
Ionosphere	89.31(0.95)	87.37(3.07)	85.71(2.12)	86.56(1.98)	86.56(1.98)	87.35(1.34)	85.71(5.52)	96.20(1.72)
Wine	96.94(4.08)	95.00(1.53)	95.55(2.30)	95.85(2.25)	96.16(3.02)	96.66(3.36)	95.00(4.14)	100.00(0.21)
Haberman	76.71(3.52)	71.23(4.16)	72.86(3.65)	70.16(3.56)	72.26(4.17)	65.01(3.20)	75.00(3.40)	97.36(3.34)

In Table 1, we compare the recognition rates obtained by the proposed meta-learning framework against dynamic selection techniques explained in this paper: Overall Local Accuracy (OLA) [6], Local Classifier Accuracy (LCA) [6], Modified Local Accuracy (MLA) [8], KNORA-Eliminate [5], K-Nearest Output Profiles (KNOP) [11] and the Multiple Classifier Behavior (MCB) [12]. We compare each pair of results using the

---

[2] www.prtools.org

Kruskal-Wallis non-parametric statistical test with a 95% confidence interval. The results of the proposed framework over the Pima, Liver Disorders, Blood Transfusion, Vehicle, Sonar and Ionosphere datasets are statistically superior to the result of the best DES from the literature. For the other datasets, Breast, Banana and Lithuanian, the results are statistically equivalent.

4.2 Dissimilarity Analysis

In this section, we conduct a dissimilarity analysis between distinct DES techniques. The analysis is performed based on the difference between the level of competence $\delta_{i,j}$ estimated by each DES technique for a given base classifier c_i, for each query sample x_j (Section 2). The goal of the dissimilarity analysis is twofold: to understand the behavior of different DES techniques (i.e., whether or not the criterion used by DES techniques present a similar behavior), and in order to see which DES criterion is closer to the behavior of the criterion used by the ideal DES scheme (Oracle) for the estimation of the competence level of a base classifier.

Given 8 dynamic selection techniques, the first step of the dissimilarity analysis is to compute the dissimilarity matrix D. This matrix D is an 8×8 symmetrical matrix, where each element $d_{A,B}$ represents the dissimilarity between two different DES techniques, A and B. Given that $\delta_{i,j}^{A}$ and $\delta_{i,j}^{B}$ are the levels of competence of c_i in relation to x_j for the techniques A and B, respectively, the dissimilarity $d_{A,B}$ is calculated by the difference between $\delta_{i,j}^{A}$ and $\delta_{i,j}^{B}$ (Equation 4).

$$d_{A,B} = \frac{1}{NM} \sum_{j=1}^{N} \sum_{i=1}^{M} \left(\delta_{i,j}^{A} - \delta_{i,j}^{B} \right)^2 \tag{4}$$

where N and M are the size of the validation dataset and the pool of classifiers, respectively.

For each dataset considered in this work, a dissimilarity matrix (e.g., D_{Pima}, D_{Liver}) is computed, with the mean dissimilarity values over 20 replications. Then, the average dissimilarity matrix $\bar{D}$ is obtained by computing the mean and standard deviation of the eleven dissimilarity matrices. Table 2 shows the average dissimilarity matrix $\bar{D}$. Both the average and the standard deviation values are presented. Each line or column of the dissimilarity matrix can be seen as one axe in the $8th$ dimensional space. Each axe in this space represents the distance to a specific DES technique, for instance, the first axe represents the distance to the proposed meta-learning framework; the second represents the distance to the KNORA technique and so forth.

Classifier Projection Space. The next step is to project the dissimilarity matrix $\bar{D}$ onto the Classifier Projection Space (CPS) for a better visualization of the relationship between all techniques. The CPS is an $\mathbb{R}^n$ space where each technique is represented as a point and the Euclidean distance between two techniques is equal to their dissimilarities [10]. Techniques that are similar to one another appear closer in the CPS while those with a higher dissimilarity are more distant. Thus, it is possible to obtain a spatial representation of the dissimilarity between all techniques. A two-dimensional

Table 2. The average dissimilarity matrix $\bar{D}$. The values are the mean and standard deviation computed over the eleven dissimilarity matrix.

	Meta-Learning	KNORA	MCB	LCA	OLA	MLA	KNOP	Oracle
Meta-Learning	0	0.36(0.06)	0.46(0.15)	0.40(0.07)	0.36(0.06)	0.40(0.04)	0.53(0.08)	0.54(0.03)
KNORA	0.36(0.06)	0	0.89(0.06)	0.42(0.01)	0.44(0.01)	0.71(0.04)	0.74(0.11)	0.68(0.01)
MCB	0.46(0.15)	0.89(0.06)	0	0.58(0.01)	0.89(0.06)	1.06(0.07)	0.75(0.03)	0.72(0.08)
LCA	0.40(0.07)	0.42(0.01)	0.58(0.01)	0	0.42(0.01)	0.45(0.02)	0.31(0.04)	0.60(0.06)
OLA	0.36(0.06)	0.44(0.01)	0.89(0.06)	0.42(0.01)	0	0.71(0.04)	0.74(0.11)	0.68(0.11)
MLA	0.40(0.04)	0.71(0.04)	1.06(0.07)	0.45(0.02)	0.71(0.04)	0	0.54(0.01)	0.63(0.07)
KNOP	0.53(0.08)	0.74(0.11)	0.75(0.03)	0.31(0.04)	0.74(0.11)	0.54(0.01)	0	0.86(0.12)
Oracle	0.54(0.03)	0.68(0.01)	0.72(0.08)	0.60(0.06)	0.68(0.11)	0.63(0.07)	0.86(0.12)	0

CPS is used for better visualization. To obtain a two-dimensional CPS, a dimensionality reduction of the dissimilarity matrix $\bar{D}$ in the $\mathbb{R}^8$ to $\tilde{D}$ in the $\mathbb{R}^2$ is required. This reduction is performed using Sammon mapping [14]; that is, a non-linear Multidimensional Scaling (MDS) projection onto a lower dimensional space such that the distances are preserved [10,14].

Given the dissimilarity matrix $\bar{D}$, a configuration X of m points in $\mathbb{R}^k, (k \leq m)$ is computed using a linear mapping, called classical scaling [14]. The process is performed through rotation and translation, such that the distances after dimensionality reduction are preserved. The projection X is computed as follows: first, a matrix of the inner products is obtained by the square distances $B = -\frac{1}{2}JD^2J$, where $J = I - \frac{1}{m}UU^T$, and I and U are the identity matrix and unit matrix, respectively. J is used as a normalization matrix such that the mean of the data is zero. The eigendecomposition of B is then obtained as, $B = Q\Lambda Q^T$, where Λ is a diagonal matrix containing the eigenvalues (in decreasing order) and Q is the matrix of the corresponding eigenvectors. The configuration of points in the reduced space is determined by the k largest eigenvalues. Therefore, X is uncorrelated in the $\mathbb{R}^k$, $X = Q_k\sqrt{\Lambda_k}$ space. In our case, $k = 2$.

The CPS projection is obtained by applying Sammon mapping over the matrix X. The mapping is performed by defining a function, called stress function $\mathcal{S}$ (Equation 5), which measures the difference between the original dissimilarity matrix $\bar{D}$ and the distance matrix of the projected configuration, $\tilde{D}$, where $\tilde{d}(i, j)$ is the distance between the classifiers i and j in the projection X.

$$\mathcal{S} = \frac{1}{\sum_{i=1}^{m-1}\sum_{j=i+1}^{m} d(i,j)^2} \sum_{i=1}^{m-1} \sum_{j=i+1}^{m} (d(i,j) - \tilde{d}(i,j)) \tag{5}$$

The two-dimensional CPS plot is shown in Figure 2. Figure 2(a) shows the average CPS plot obtained considering the average dissimilarity matrix $\bar{D}$, while Figure 2(b) shows an example of the CPS plot obtained for the Liver Disorders dataset D_{Liver}.

An important observation that can be drawn from Figure 2(a) is that the LCA, OLA and MLA appear close together in the dissimilarity space. Which means, that the criteria used by these three techniques to estimate the level of competence of a base classifiers present similar behaviors when averaged over several classification problems. Thus, they are very likely to achieve the same results [15]. This can be explained by

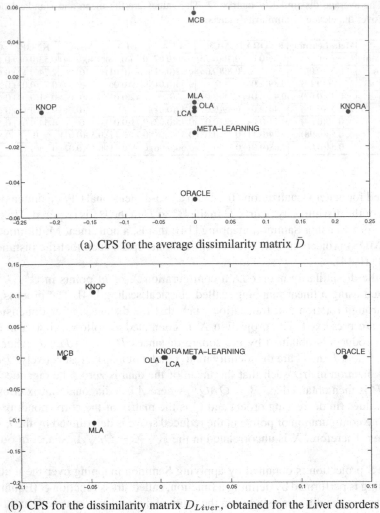

(a) CPS for the average dissimilarity matrix $\bar{D}$

(b) CPS for the dissimilarity matrix D_{Liver}, obtained for the Liver disorders dataset

Fig. 2. Two-dimensional CPS plot for the average dissimilarity matrix $\bar{D}$ and for the dissimilarity matrix obtained for the Liver disorders dataset D_{Liver}. It is important to mention that the axes of the CPS plot cannot be interpreted alone. Only the Euclidean distances between the points count.

the fact that these three techniques are based on the same information (the classification accuracy over a defined local region in the feature space), with little difference regarding the use of a posteriori information by the LCA technique or weights for the MLA technique.

The meta-learning framework appears closer to the Oracle in the two-dimensional CPS (Figures 2(a) and (b)). In addition, the meta-learning framework is also closer to the techniques from the local accuracy paradigm (LCA, OLA and MLA) than to any other DES technique, which can be explained by the fact that three out of the five meta-features comes from estimations of the local regions (f_1, f_2 and f_3).

Table 3 presents the dissimilarity measure for each DES technique in relation to the Oracle. Results show that the proposed meta-learning framework is closer to the behavior of the Oracle as it presents the lowest dissimilarity value on average, 0.54. The LCA technique comes closer, with an average dissimilarity value of 0.60. Thus, we suggest that the use of multiple criteria to estimate the level of competence of a base classifier results in a DES technique that obtains a estimation of the level of competence of a base classifier closer to that provided by an ideal DES scheme (Oracle).

Table 3. Mean and standard deviation of the dissimilarity between each DES technique from the Oracle for each classification problem. The smallest dissimilarity values are highlighted.

Database	Meta-Learning	KNORA-E	MCB	LCA	OLA	MLA	KNOP
Pima	**0.32(0.04)**	0.43(0.01)	0.47(0.08)	0.36(0.06)	0.43(0.01)	0.44(0.07)	0.41(0.02)
Liver Disorders	**0.50(0.04)**	0.61(0.01)	0.67(0.008)	0.56(0.06)	0.61(0.01)	0.60(0.07)	0.51(0.02)
Breast Cancer	**0.59(0.35)**	1.22(0.10)	1.20(0.10)	0.69(0.01)	1.20(0.10)	0.77(0.03)	1.20(0.10)
Blood Transfusion	**0.33(0.03)**	0.40(0.01)	0.46(0.01)	0.36(.003)	0.40(0.01)	0.44(0.08)	0.4(0.01)
Banana	0.33(0.10)	0.29(0.01)	0.36(0.01)	**0.24(0.01)**	0.29(0.01)	0.36(0.01)	0.34(0.01)
Vehicle	**0.36(0.07)**	0.49(0.01)	0.48(0.02)	**0.36(0.04)**	0.49(0.01)	0.37(0.05)	0.47(0.02)
Lithuanian Classes	0.47(0.14)	0.49(0.02)	0.56(0.02)	**0.39(0.04)**	0.49(0.02)	0.54(0.01)	0.51(0.03)
Sonar	**0.58(0.10)**	0.91(0.04)	0.88(0.01)	0.70(0.01)	0.91(0.04)	0.85(0.02)	0.84(0.06)
Ionosphere	**0.62(0.22)**	0.89(0.05)	0.88(0.06)	0.70(0.07)	0.89(0.05)	0.68(0.02)	0.88(0.06)
Wine	1.03(0.20)	0.88(0.11)	0.98(0.11)	**0.73(0.02)**	0.88(0.11)	0.93(0.06)	0.82(0.14)
Haberman	**0.79(0.04)**	0.89(0.05)	1.01(0.05)	0.82(0.02)	0.89(0.05)	0.92(0.04)	0.86(0.06)
Mean	**0.54(0.05)**	0.68(0.01)	0.72(0.08)	0.60(0.06)	0.68(0.11)	0.63(0.07)	0.86(0.12)

5   Conclusion

In this paper, we conducted a study about the dissimilarity between different DES techniques. These dissimilarities are computed in order to generate a dissimilarity matrix. Through Sammon Mapping, the dissimilarity matrix is embedded in a two-dimensional space, called the Classifier Projection Space (CPS), where the Euclidean distance between two feature representations reflects their dissimilarity.

Based on the visual representation provided by the CPS, we can draw two conclusions:

- The proposed technique is closer to the Oracle in the dissimilarity space, which indicates that the use of different types of information about the behavior of base classifiers is indeed necessary in order to achieve a DES technique that is closer to the Oracle.
- Techniques that use the same kind of information to compute the level of competence of the base classifiers, such as LCA, OLA and MLA, are more likely to present the same results when their performance is averaged over several problems.

Future works in this topic include: i) The design of new sets of meta-features; ii) Carrying out a comparison of different meta-features vectors in order to achieve a set of features that can better address the behavior of the Oracle; and, iii) Increasing the number of classification problems in the analysis.

References

1. Kittler, J., Hatef, M., Duin, R.P.W., Matas, J.: On combining classifiers. IEEE Transactions on Pattern Analysis and Machine Intelligence 20, 226–239 (1998)
2. Kuncheva, L.I.: Combining Pattern Classifiers: Methods and Algorithms. Wiley-Interscience (2004)
3. Cruz, R.M.O., Cavalcanti, G.D.C., Ren, T.I.: Handwritten digit recognition using multiple feature extraction techniques and classifier ensemble. In: 17th International Conference on Systems, Signals and Image Processing, pp. 215–218 (2010)
4. de Souza Britto Jr., A., Sabourin, R., Oliveira, L.: Dynamic selection of classifiers a comprehensive review. Pattern Recognition (in press, 2014), http://dx.doi.org/10.1016/j.patcog.2014.05.003
5. Ko, A.H.R., Sabourin, R., Britto Jr., A.S.: From dynamic classifier selection to dynamic ensemble selection. Pattern Recognition 41, 1735–1748 (2008)
6. Woods, K., Kegelmeyer Jr., W.P., Bowyer, K.: Combination of multiple classifiers using local accuracy estimates. IEEE Transactions on Pattern Analysis Machine Intelligence 19, 405–410 (1997)
7. Cruz, R.M.O., Cavalcanti, G.D.C., Ren, T.I.: A method for dynamic ensemble selection based on a filter and an adaptive distance to improve the quality of the regions of competence. In: International Joint Conference on Neural Networks, pp. 1126–1133 (2011)
8. Smits, P.C.: Multiple classifier systems for supervised remote sensing image classification based on dynamic classifier selection. IEEE Transactions on Geoscience and Remote Sensing 40(4), 801–813 (2002)
9. Cruz, R.M.O., Sabourin, R., Cavalcanti, G.D.: On meta-learning for dynamic ensemble selection. In: International Conference on Pattern Recognition (in press, 2014)
10. Pękalska, E., Duin, R.P.W., Skurichina, M.: A discussion on the classifier projection space for classifier combining. In: Roli, F., Kittler, J. (eds.) MCS 2002. LNCS, vol. 2364, pp. 137–148. Springer, Heidelberg (2002)
11. Cavalin, P.R., Sabourin, R., Suen, C.Y.: Dynamic selection approaches for multiple classifier systems. Neural Computing and Applications 22(3-4), 673–688 (2013)
12. Giacinto, G., Roli, F.: Dynamic classifier selection based on multiple classifier behaviour. Pattern Recognition 34, 1879–1881 (2001)
13. Breiman, L.: Bagging predictors. Machine Learning 24, 123–140 (1996)
14. Cox, T.F., Cox, M.A.A.: Multidimensional Scaling, 2nd edn. Chapman and Hall (2000)
15. Cruz, R.M.O., Cavalcanti, G.D., Tsang, I.R., Sabourin, R.: Feature representation selection based on classifier projection space and oracle analysis. Expert Systems with Applications 40(9), 3813–3827 (2013)

Hidden Markov Models Based on Generalized Dirichlet Mixtures for Proportional Data Modeling

Elise Epaillard[1] and Nizar Bouguila[2]

[1] Department of Electrical and Computer Engineering
[2] Concordia Institute for Information Systems Engineering
Concordia University, Montreal, Quebec, Canada
e_epail@encs.concordia.ca
nizar.bouguila@concordia.ca

Abstract. Hidden Markov models (HMMs) are known for their ability to well model and easily handle variable length time-series. Their use in the case of proportional data modeling has been seldom mentioned in the literature. However, proportional data are a common way of representing large data in a compact fashion and often arise in pattern recognition applications frameworks. HMMs have been first developed for discrete and Gaussian data and their extension to proportional data through the use of Dirichlet distributions is quite recent. The Dirichlet distribution has its limitations and is a special case of the more general generalized Dirichlet (GD) distribution that suffers from less restrictions on the modeled data. We propose here to derive the equations and the methodology of a GD-based HMM and to assess its superiority over a Dirichlet-based HMM (HMMD) through experiments conducted on both synthetic and real data.

Keywords: Hidden Markov models, generalized Dirichlet, mixtures, machine learning, EM-algorithm.

1 Introduction

HMMs are probabilistic generative models used in various fields such as speech processing [18], object and gesture classification [3,9] or anomaly detection [2,14]. Their use has been popularized by [18], and numerous extensions and adaptations to specific applications have been developed along the years.

Among the extensions developed for HMMs, the study of time-series generated from multiple processes and/or involving dynamics at different scales led to the development of factorial HMMs [11]. In this framework, each state is decomposed into a collection of sub-states, often assumed independent at each time step in order to reduce algorithmic complexity.

Classic HMMs naturally embed a geometric distribution as for state duration, i.e. state self-transitioning, with parameter depending on the state transition matrix [10]. Variable Duration HMMs have been a first attempt to modify the state duration probability distribution [17]. At each state transition, the duration of the new state is drawn from a probability mass function and the corresponding number of observations is generated before drawing a new state accordingly to the state transition matrix. An alternate approach that explicitly introduces the time variable into the state transition matrix is

N. El Gayar et al. (Eds.): ANNPR 2014, LNAI 8774, pp. 71–82, 2014.
© Springer International Publishing Switzerland 2014

proposed in [10]. Known as Non-Stationary HMMs, they have been shown equivalent to Variable Duration HMMs, though allowing an easier and computationally more efficient parameter estimation [10].

The most widely used estimation algorithm for HMMs is the so-called Baum-Welch algorithm, though its iterative nature can be prohibitive in some applications. [12] proposed a non-iterative method for parameters estimation. Based on subspace estimation, the idea has been theoretically derived in [1] and provides, under few conditions, a computationally fast method to estimate HMMs with finite alphabet output.

HMMs have been initially developed for discrete and Gaussian data [18]. The multiplication of applications in domains such as weather forecast or medical studies raised the need to modify the original HMM algorithm so it can efficiently work with new data types [13,15]. Longitudinal or panel data are time-series collected from multiple entities. Example of these data in the context of a medical study could be the evolution of some disease characteristics evaluated every day for a given period of time on a number of patients (see [16] for concrete example). At the entity level, data heterogeneity is involved by the presence of multiple data sources. HMMs have been shown to be able to model this heterogeneity by introducing a random variable in the model, known as the *random effect*, that follows a predefined probability distribution. Doing so, the conditional independence of the observed data given the latent states assumption is relaxed. [15] provides a review of the use of these HMMs that are known in the literature as Mixed HMMs. [13] discusses circular data processing, i.e data taking cyclic values (e.g. directions, angles,...). Von Mises, Wrapped Normal and Wrapped Cauchy are proposed as state emission probability distributions to handle such data. A Maximum-Likelihood estimation algorithm is derived and applied to circular time-series.

Proportional data (i.e. positive data that sum up to 1) results from numerous pattern recognition pre-processing procedures, the most common being histograms. Their use in an HMM framework has been first studied in [8] where Dirichlet mixtures are used as emission probability functions, involving a deep modification of the M-step of the Expectation-Maximization algorithm (EM) for Dirichlet parameters estimation. The limitation of the Dirichlet distribution has been brought to light by [5] and we propose here to derive the equations of an HMM based on mixtures of GD distributions (HM-MGD). This model is expected to be more general and versatile as the GD distribution embeds the Dirichlet distribution as a special case.

Section 2 fully develops the HMMGD derivation, Section 3 presents experimental work on synthetic data, and Section 4, on real data. We conclude and explain our future work in Section 5.

2   HMM Based on Generalized Dirichlet Mixtures

Based on [18], a first-order HMM is a probabilistic model assuming an ordered observation sequence $O = \{O_1, ..., O_T\}$ to be generated by some hidden states, each of them associated with a probability distribution governing the emission of the observed data. The hidden states $H = \{h_1, ..., h_T\}$, $h_j \in [1, K]$, with K the number of states, are assumed to form a Markov chain.

At each time t, a new state is entered based on a transition matrix $B = \{B_{ij} = P(h_t = j | h_{t-1} = i)\}$ that specifies the transition probabilities between states. Once in the new state, an observation is made following its associated probability distribution. For discrete observation symbols taken from a vocabulary $V = \{v_1, ..., v_S\}$, the emission matrix is defined as $D = \{D_j(k) = P(O_t = v_k | h_t = j)\}, [t, k, j] \in [1, S] \times [1, M] \times [1, K]$. For continuous observation vectors, emission probability distributions are usually taken as Gaussian mixtures [2,3,18] defined by their mean and covariance matrices, denoted θ. In the latter case, a matrix $C = \{C_{i,j} = P(m_t = i | h_t = j)\}$, $i \in [1, M]$, is defined with M the number of mixture components associated with state j (which can be assumed to be the same for all states without loss of generality). An initial probability distribution π controls the initial state. We denote an HMM as $\lambda = \{B, D, \pi\}$ or $\{B, C, \theta, \pi\}$.

HMMs are well fit for classification tasks that rely on the probability of an observation sequence given a model λ, computed using a forward-backward procedure [18]. Model training consists in the estimation of the parameters that maximize the probability of a given set of observations and is addressed with the Baum-Welch algorithm, an Expectation-Maximization process [18]. Finally finding the most probable sequence of states and mixture components that generated a series of observations can be solved with the Viterbi algorithm [18].

The number of hidden states and the parameters initial values have to be a priori set. Both are strongly linked to model's performance. Indeed, the former is a trade-off between performance and complexity [9], while the latter leads the Baum-Welch procedure to converge towards the closest local maximum of the likelihood function, not guaranteed to be the global one given its high modality [3].

In this paper we propose to develop HMMs with mixtures of GD as emission probability distributions. [8] derived the equations for HMMs with Dirichlet mixtures, yet these distributions have one main limitation residing in the fact that data covariance is always negative. Therefore, they might not be adapted to model all types of proportional data. The GD distribution overcomes this limitation and embeds the Dirichlet distribution as a special case.

2.1 Expected Complete-Data Log-Likelihood Equation Setting

An N-dimensional generalized Dirichlet distribution is defined as

$$GD(\mathbf{x}|\alpha, \beta) = \prod_{n=1}^{N} \frac{\Gamma(\alpha_n + \beta_n)}{\Gamma(\alpha_n)\Gamma(\beta_n)} x_n^{\alpha_n - 1} \left(1 - \sum_{r=1}^{n} x_r\right)^{\nu_n}, \quad (1)$$

where Γ denotes the Gamma function and $\alpha = [\alpha_1, ..., \alpha_N]$ and $\beta = [\beta_1, ..., \beta_N]$ the GD parameters, with $\alpha \in \mathbb{R}_+^N$, $\beta \in \mathbb{R}_+^N$, $\mathbf{x} \in \mathbb{R}_+^N$, and $\sum_{n=1}^{N} x_n < 1$. For $n \in [1, N-1]$, $\nu_n = \beta_n - \alpha_{n+1} - \beta_{n+1}$, and $\nu_N = \beta_N - 1$.

This change of probability distribution involves modifications in the EM parameters estimation process. The rest of the HMM algorithm is unchanged. We set notations for the quantities $\gamma_{h_t, m_t}^t \triangleq p(h_t, m_t | x_0, ..., x_T)$ and $\xi_{h_t, h_{t+1}}^t \triangleq p(h_t, h_{t+1} | x_0, ..., x_T)$, that represent the estimates of the states and mixture components, and of the local states sequence given the whole observation set, respectively. The E-step leads to γ_{h_t, m_t}^t and $\xi_{h_t, h_{t+1}}^t$ estimates for all $t \in [1, T]$. These two quantities are obtained using the initial parameters at step 1 and the result of the last M-step then. They are computed using a forward-backward procedure (not detailed here) as in HMM with mixtures of Gaussian.

The M-step aims at maximizing the data log-likelihood by maximizing its lower bound. If Z represents the hidden variables and X the data, the data likelihood $\mathcal{L}(\theta|X) = p(X|\theta)$ can be expressed as

$$
\begin{aligned}
E(X, \theta) - R(Z) &= \sum_Z p(Z|X) \ln(p(X, Z)) - \sum_Z p(Z|X) \ln(p(Z|X)) \\
&= \sum_Z p(Z|X) \ln(p(X)) \quad \text{(Bayes' rule)} \\
&= \ln(p(X)) \sum_Z p(Z|X) = \ln(p(X)) = \mathcal{L}(\theta|X) ,
\end{aligned}
\tag{2}
$$

with θ, representing all the HMM parameters, omitted on the given variables side of all the quantities involved. $E(X, \theta)$ is the value of the complete-data log-likelihood with the true/maximized parameters θ. $R(Z)$ is the log-likelihood of the hidden data given the observations and has the form of an entropy representing the amount of information brought by the hidden data itself (see eq. (12) for the detailed form of $R(Z)$). As we estimate the complete-data log-likelihood using non-optimized parameters, we have $E(X, \theta, \theta^{old}) \leq E(X, \theta)$, and hence $E(X, \theta, \theta^{old}) - R(Z)$ is a lower bound of the data likelihood.

The key quantity for data likelihood maximization is the expected complete-data log-likelihood which directly depends on the data and is written as

$$
E(X, \theta, \theta^{old}) = \sum_Z p(Z|X, \theta^{old}) \ln(p(X, Z|\theta)) .
\tag{3}
$$

The complete-data likelihood of an observation (the case of multiple observation sequences is addressed later) can be expanded as (eq. 4) that leads by identification to eq. (5).

$$
p(X, Z|\theta) = p(h_0) \prod_{t=0}^{T-1} p(h_{t+1}|h_t) \times \prod_{t=0}^{T} p(m_t|h_t) p(x_t|h_t, m_t) ,
\tag{4}
$$

$$
p(X, Z|\theta) = \pi_{h_0} \prod_{t=0}^{T-1} B_{h_t, h_{t+1}} \prod_{t=0}^{T} C_{h_t, m_t} GD(x_t|h_t, m_t) .
\tag{5}
$$

We substitute eq. (1) into eq. (5) and take the logarithm of the expression. Using the logarithm sum-product property the complete-data log-likelihood is split up into eight terms:

$$\ln(p(X, Z|\theta)) = \ln(\pi_{h_0}) + \sum_{t=0}^{T} \ln(C_{h_t, m_t}) + \sum_{t=0}^{T-1} \ln(B_{h_t, h_{t+1}})$$

$$+ \sum_{t=0}^{T} \sum_{n=1}^{N} \left\{ \ln(\Gamma(\alpha_{h_t, m_t, n} + \beta_{h_t, m_t, n})) + (\alpha_{h_t, m_t, n} - 1) \ln(x_n^t) \right.$$

$$\left. + \nu_{h_t, m_t, n} \ln(1 - \sum_{r=1}^{n} x_r^t) - \ln(\Gamma(\alpha_{h_t, m_t, n})) - \ln(\Gamma(\beta_{h_t, m_t, n})) \right\}.$$

$$(6)$$

Using eq. (6) into eq. (3), the expected complete-data log-likelihood can then be written:

$$E(X, \theta, \theta^{old}) = \sum_{k=1}^{K} \sum_{m=1}^{M} \gamma_{k,m}^0 \ln(\pi_k) + \sum_{t=0}^{T} \sum_{k=1}^{K} \sum_{m=1}^{M} \gamma_{k,m}^t \ln(C_{k,m})$$

$$+ \sum_{t=0}^{T-1} \sum_{i=1}^{K} \sum_{j=1}^{K} \xi_{i,j}^t \ln(B_{i,j}) + L(\alpha, \beta), \qquad (7)$$

with,

$$L(\alpha, \beta) = \sum_{t=0}^{T} \sum_{n=1}^{N} \sum_{k=1}^{K} \sum_{m=1}^{M} \left\{ \gamma_{k,m}^t \ln(\Gamma(\alpha_{k,m,n} + \beta_{k,m,n})) \right.$$

$$+ \gamma_{k,m}^t (\alpha_{k,m,n} - 1) \ln(x_n^t) + \gamma_{k,m}^t (\nu_{k,m,n} \ln(1 - \sum_{r=1}^{n} x_r^t))$$

$$\left. - \gamma_{k,m}^t \ln(\Gamma(\alpha_{k,m,n})) - \gamma_{k,m}^t \ln(\Gamma(\beta_{k,m,n})) \right\}. \qquad (8)$$

To set eq. (7) we make use of the two following properties, in which we omit the mention θ^{old} in the given variables side of the probabilities involved. Using the independence of h_t and m_t from h_{t+1}, we get $p(Z|X) = p(h_t = k, m_t = m|X)p(h_{t+1} = k')$ with $\sum_{k'=1}^{K} p(h_{t+1} = k') = 1$. Similar steps bring $p(Z|X) = p(h_t = k, h_{t+1} = k'|X, m_t = m)p(m_t = m)$, with $\sum_{m=1}^{M} p(m_t = m) = 1$.

Furthermore, if $D \geq 1$ observations are available, all can be used to avoid overfitting. In (7), a sum over $d \in [1, D]$ has to be added in front of the entire formula. The sum over time goes then from 0 to T_d, the length of the d-th observation sequence.

2.2   Update Equations of HMM and GD Parameters

Maximization of the expectation of the complete-data log-likelihood with respect to π, B, and C is solved introducing Lagrange multipliers in order to take into account the constraints due to the stochastic nature of these parameters. The resulting update equations are:

$$\pi_k^{new} \propto \sum_{d=1}^{D} \sum_{m=1}^{M} \gamma_{k,m}^{0,d}, \quad B_{i,j}^{new} \propto \sum_{d=1}^{D} \sum_{t=0}^{T_d-1} \xi_{k,k'}^{t,d}, \quad C_{k,m}^{new} \propto \sum_{d=1}^{D} \sum_{t=0}^{T_d} \gamma_{k,m}^{t,d}, \quad (9)$$

where k and k' are in the range $[1, K]$, and m, in the range $[1, M]$.

GD distributions parameters update is less straightforward. Indeed, a direct method would lead to maximize $L(\alpha, \beta)$. Instead of going through heavy computations, we propose to use a practical property of the GD distribution that reduces the estimation of a N-dimensional GD to the estimation of N Dirichlet distributions. The latter is a known problem and can be solved using a Newton method [8,19]. Using this property calls the need for the problem to be expressed in a transformed space that we refer to as the W-space. The data is transformed from its original space into its W-space by [5,20]:

$$W_l = \begin{cases} x_l & \text{for } l = 1 \\ x_l / \left(1 - \sum_{i=1}^{l-1} x_i\right) & \text{for } l \in [2, N]. \end{cases} \quad (10)$$

In the transformed space, each W_l follows a Beta distribution with parameters (α_l, β_l), which is a 2-dimensional Dirichlet distribution. The estimation of the N Beta distributions governing the N W_l clearly leads to the complete characterization of the GD distribution governing the observation vector $\mathbf{x}$. In the M-step of the HMMGD algorithm, the update of the GD distribution parameters can thus be done using N times a process similar to the one used in [8], considering the transformed data instead of the original one. Other parameters (B, C, π, γ, ξ) are estimated from the original data.

The initialization of the HMM parameters has been shown in [8] to be intractable as soon as the product KM grows up, if computed accurately. Following their framework, KM single Generalized Dirichlet distributions are initialized with a method of moments that uses the transformed data (detailed in [6]) and are then assigned to the HMM states. The parameters π, C, and B, are randomly initialized. Any EM-algorithm is iterative and thus needs a stop parameter. As the data log-likelihood is maximized by the means of its lower bound, convergence of this bound can be used as such. This lower bound is given by $E(X, \theta, \theta^{old}) - R(Z)$ (see eqs. (2) and (7)) and $R(Z)$ is derived using Bayes' rule:

$$p(Z|X) = p(h_0)p(m_0|h_0) \prod_{t=1}^{T} p(h_t|h_{t-1})p(m_t|h_t)$$

$$= p(h_0) \frac{p(m_0, h_0)}{p(h_0)} \prod_{t=1}^{T} \frac{p(h_t, h_{t-1})p(m_t, h_t)}{p(h_{t-1})p(h_t)}. \quad (11)$$

Denoting $\eta_t \triangleq p(h_t|X)$ and using the independence properties set earlier, the following expression is derived (see detail in [8], this expression is valid for any type of emission function):

$$R(Z) = \sum_{k=1}^{K} \left[\eta_k^0 \ln(\eta_k^0) + \eta_k^T \ln(\eta_k^T) - 2 \sum_{t=0}^{T} \eta_k^t \ln(\eta_k^t) \right]$$
$$+ \sum_{t=0}^{T} \sum_{m=1}^{M} \sum_{k=1}^{K} \gamma_{k,m}^t \ln(\gamma_{k,m}^t) + \sum_{t=0}^{T-1} \sum_{k=1}^{K} \sum_{k'=0}^{K} \xi_{i,j}^t \ln(\xi_{i,j}^t) . \tag{12}$$

This stands for a unique observation sample, if more are used, a summation over them has to be added in front of the whole expression and the index T has to be adapted to the length of each sequence. At each iteration, the difference between the former and current data likelihoods is computed. Once it goes below a predefined threshold, the algorithm stops and the current parameters values are kept to define the HMMGD. This threshold, empirically fixed to 10^{-6} in our experiments, is a trade-off between estimates precision and computational time.

3   Experiments on Synthetic Data

3.1   Process Description

We propose here to assess the superiority of HMMGD over HMMD with synthetic data. 1000 observations sequences of length randomly taken in the range $[10, 20]$ are generated from a known HMMGD with randomly chosen parameters. The generation of GD samples is described in [20]. The generative state and mixture component are recorded for each sample. As in [8], performance is computed as the proportion of states and mixture components correctly retrieved by an HMM trained on the generated data. Multiple experiments are run varying the number of states K, the number of mixture components M, and the data dimension N. The study of the influence of N is of particular importance as with proportional data, the greater N, the smaller the observation values. Too small values, through numerical processing, can lead to matrices invertibility issues which is not desirable for accurate estimation.

As stated earlier, the GD distribution relaxes the constraint on the sign of the data correlation coefficients. The proposed model is then expected to give a more accurate representation of the data in the case of data mostly positively correlated. On the other hand, with mostly negatively correlated data, HMMD should provide as good results with a reduced complexity. To verify this, we generate data from known HMMGDs and attempt to retrieve the state and mixture component that generated every sample using an HMMGD and an HMMD. We noticed that data generaed from HMMGDs with parameters randomly and uniformly drawn in the range $[1, 60]$, are quasi-automatically mostly positively correlated. To overcome this point we imposed some of the HMM parameters to follow a Dirichlet distribution expressed in the form of a GD distribution. We used the three following scenarii: 1- Data generated from HMMDs only, 2- Data generated from an hybrid HMM with on each state half of the components being Dirichlet and half GD distributions, 3- Data generated from HMMGDs only. Extensive testing confirmed our expectations. Results are illustrated in Figure 1 using a *correlation ratio* which is the number of positively correlated variables (minus the autocorrelations) over the number of negatively correlated ones. A ratio greater than 1 means the variables are mostly positively correlated and vice versa.

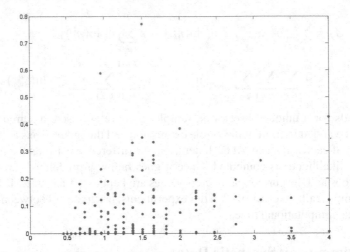

Fig. 1. Gain of accuracy using HMMGD compared to HMMD in function of the variables correlation ratio. The gain of accuracy is computed as the difference between the two models' performance.

Experiments have been led with $K = 3$ and $M = 2$. For scenario 1, HMMGD has a 85.3 % accuracy and HMMD 84.9%, confirming that both work equally well. For scenarii 2 and 3, HMMGD has an accuracy of 81.2% and 89.7%, respectively, and HMMD of 77.6% and 80.1%, respectively. As soon as some data are positively correlated, HMMGD outperforms HMMD. We observe that in scenario 2 (correlation ratio close to 1), for unclear reasons, it is more difficult for the HMMs to retrieve the correct state and component the sample comes from. Finally, the retrieval rate for data with a correlation ratio greater than 1 is of 86.1% for HMMGD and of 78.4% for HMMD, and of 84.8% and 83.2%, respectively, for correlation ratios smaller than 1. This shows HMMGDs overcome the weakness of HMMDs for positively correlated data.

Table 1 reports the results of experiments led fixing $N = 10$, generating 100 sequences only (because of time constraint), and letting K and M vary. According to the previous results, we only consider here mostly positively correlated data. For any combination (K, M), HMMGD achieves better results than HMMD showing the benefit of using HMMGD when proportional data is processed. As the product KM increases, the retrieval rate decreases which can be explained considering that the more distributions, the closer to each other they are, and the more difficult it is to clearly assign a sample to a distribution.

Table 1. HMMGD and HMMD retrieval rates with various (K, M) combinations

Parameters (K,M)	(2,2)	(2,3)	(3,2)	(2,4)	(4,2)	(3,3)	(3,4)	(4,3)	(4,4)	(5,5)	(10,5)
Product KM	4	6	6	8	8	9	12	12	16	25	50
HMMD retrieval rate (%)	84.2	75.9	82.0	82.0	86.2	81.2	72.9	73.3	61.9	66.0	52.4
HMMGD retrieval rate (%)	90.9	92.8	87.8	89.8	91.5	89.1	88.9	85.2	76.6	68.8	62.2

A bad initialization of the distribution parameters can give low retrieval rates. It can find its origin in the convergence of the clustering algorithm, used as the first step of the method of moments, towards local extrema. To overcome this issue, the initialization process can be run several times and the comparison of the lower bound of the data likelihood with these initial parameters be used to choose the best ones. However, this requires extra computations and does not guarantee a good convergence of the clustering procedure, even within several attempts. As we are only interested here in the relative performance of HMMGD compared to HMMD we did not use this option. Instead, in order not to introduce any bias from this issue, a unique clustering algorithm is used for both initializations.

Figure 2 reports the results of experiments in which we fixed $K = 3$ and $M = 2$ and let N increase until retrieval rates degrade dramatically. For scenario 1, equivalent results are obtained with both HMMs, HMMGD giving sometimes slightly better results at the cost of extra computations (not reported on Figure 2). In other cases, HMMGD systematically outperforms HMMD up to the point data dimension is too high to perform calculations accurately (intermediate matrices become singular). Fluctuations in the overall results are due to bad initializations that involve retrieval rates to dramatically drop on some isolated runs. The general shape of the curves and their relative distance clearly shows that, within an HMM framework, mixtures of GD distributions give the best results and allow working with data of higher dimension than Dirichlet ones. This performance improvement is obtained at the cost of a more complex model involving $(2N - 2)$ parameters to be estimated for every GD distribution compared to only N parameters for a Dirichlet one. These results are essential to target real applications for which HMMGD could be a potentially efficient tool.

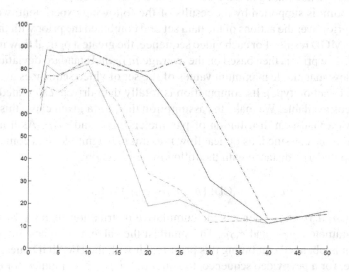

Fig. 2. Retrieval rate (%) of HMMGD (in black) and HMMD (in blue) against data dimension for scenarii 2 (dash lines) and 3 (solid lines)

4 Application to Real Data

We now compare the results of HMMGD and HMMD on real data. We base our experiments on the Weizmann Action Recognition data set [4] which is composed of video sequences representing 10 different actions (such as walk, run, jump,...) performed by 9 subjects. The features we use are Histograms of Oriented Optical Flow [7] and 10-bin histograms are built, with each bin representing a range of optical flow angles with respect to the horizontal axis. The optical flow magnitude weights the contribution of each pixel to the histogram. [7] showed that good classification results could be obtained with features of dimension higher than 30 however, we choose to use features of dimension 10 as, within our HMM-based framework, we did not find any improvement when using more bins. Finally, time savings, we divided the frame rate of the video sequences by 2.

Experiments are led using a Leave-One-Out cross validation, the results are averaged over 10 runs, and analyzed in terms of rank statistics. We empirically determined the optimal values $K = M = 4$ for both HMMs. With these parameters, the HMMD method achieves a 44.0% accuracy while the HMMGD achieves 54.8%. Though these results are low [7], they show the out-performance of HMMGD over HMMD. The rank statistics of order 2 are 71.3% and 82.0% for HMMD and HMMGD, respectively. Here again it is clear that the use of the GD model leads to higher likelihood than the Dirichlet one and is thus much more adapted for real proportional data modeling. Given the small size of the feature vectors (dimension 10) and the huge gap between the rank statistics of order 1 and 2, HMMGD seems to have the potential to perform accurate classification with a parameters fine tuning and the addition of a well-chosen prior.

This last point is supported by the results of the following experiment: we added a very simple prior over the actions of the data set and combined the prior with the already obtained HMMGD results. For each video sequence, the greatest optical flow magnitude is computed. The prior is then based on the average μ_{OF} and standard deviation σ_{OF} of the optical flow magnitude maximum values of the set of video sequences available for each class (i.e. action type). Its computation is totally data-driven, calculated from the training videos available. We make the assumption that, for a given class, this maximal value follows a Gaussian distribution of parameters μ_{OF} and σ_{OF}. As a new video sequence has to be classified, its optical flow maximum magnitude m is computed. The prior is computed as a distance with the following expression:

$$d_{prior} = |\text{CDF}(m, \mu_{OF}, \sigma_{OF}) - 0.5|, \qquad (13)$$

where $\text{CDF}(m, \mu_{OF}, \sigma_{OF})$ denotes the cumulative distribution function of the Gaussian with parameters μ_{OF} and σ_{OF}. The smallest the value, the highest the prior. The classification is obtained combining this prior result with the HMMGD ones.

Therefore, for a new video sequence, the quantity d_{prior} is computed for each class and a first classification result is obtained and stored. Then, a second classification result is obtained from the HMMGD method described in Section 2. For each class, we add up its rank in the HMMGD and prior results. We then assign the video sequence to the class with the lowest score (i.e. best cumulative rank). This simple prior used alone leads to a classification accuracy less than 50% however, combined with HMMGD results, the algorithm ends up with a 72.6% accuracy. The rank statistics of order 2 shows an even greater potential as it reaches 91.9%. Better results could be undoubtedly obtained with a more complex prior. However, the study of the best tuning and prior choice is out of the scope of this work that strives at showing the superior performance of HMMGD over HMMD. Figure 3 reports the rank statistics for the three studied methods.

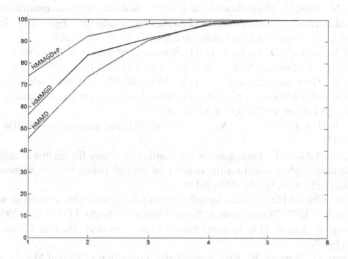

Fig. 3. Rank statistics of HMMD, HMMGD, and of the combination of HMMGD with a prior

5   Conclusion and Future Work

In this paper we theoretically derived a new HMM model for proportional data modeling based on mixtures of GD distributions. We then illustrated how this new model overcomes the limitations of Dirichlet-based HMMs in the case of positively correlated data using synthetic data. An extensive study of the impact of a number of parameters on the model's performance have been presented. Finally, we attempted to use this model on real data for action recognition. Though the first rank classification results are quite low, the study of rank statistics show a certain potential if a fine tuning is found and an appropriate prior used. The dramatic increase in classification accuracy observed when adding a very simple data-driven prior to the HMMGD framework reinforces this assessment. The HMMGD constitutes a new promising alternative when working with proportional data and has definitely to be used over HMMD methods for optimal results. Future work includes the study of HMMGD tuning for better performance and its application to other real-world tasks such as anomaly detection in crowded environment or texture classification.

References

1. Andersson, S., Ryden, T.: Subspace estimation and prediction methods for hidden Markov models. The Annals of Statistics 37(6B), 4131–4152 (2009)

2. Andrade, E.L., Blunsden, S., Fisher, R.B.: Hidden Markov models for optical flow analysis in crowds. In: ICPR (1), pp. 460–463. IEEE Computer Society (2006)
3. Bicego, M., Castellani, U., Murino, V.: A hidden Markov model approach for appearance-based 3d object recognition. Pattern Recogn. Lett. 26(16), 2588–2599 (2005)
4. Blank, M., Gorelick, L., Shechtman, E., Irani, M., Basri, R.: Actions as space-time shapes. In: ICCV, pp. 1395–1402. IEEE Computer Society (2005)
5. Bouguila, N., Ziou, D.: High-dimensional unsupervised selection and estimation of a finite generalized dirichlet mixture model based on minimum message length. IEEE Transactions on Pattern Analysis and Machine Intelligence 29(10), 1716–1731 (2007)
6. Chang, W.Y., Gupta, R.D., Richards, D.S.P.: Structural properties of the generalized dirichlet distributions. Contemporary Mathematics 516, 109–124 (2010)
7. Chaudhry, R., Ravichandran, A., Hager, G.D., Vidal, R.: Histograms of oriented optical flow and Binet-Cauchy kernels on nonlinear dynamical systems for the recognition of human actions. In: CVPR, pp. 1932–1939. IEEE (2009)
8. Chen, L., Barber, D., Odobez, J.M.: Dynamical dirichlet mixture model. IDIAP-RR 02, IDIAP (2007)
9. Cholewa, M., Glomb, P.: Estimation of the number of states for gesture recognition with hidden Markov models based on the number of critical points in time sequence. Pattern Recognition Letters 34(5), 574–579 (2013)
10. Djuric, P.M., Chun, J.H.: An mcmc sampling approach to estimation of nonstationary hidden Markov models. IEEE Transactions on Signal Processing 50(5), 1113–1123 (2002)
11. Ghahramani, Z., Jordan, M.I.: Factorial hidden Markov models. Machine Learning 29(2-3), 245–273 (1997)
12. Hjalmarsson, H., Ninness, B.: Fast, non-iterative estimation of hidden Markov models. In: Proc. IEEE Conf. Acoustics, Speech and Signal Process, vol. 4, pp. 2253–2256. IEEE (1998)
13. Holzmann, H., Munk, A., Suster, M., Zucchini, W.: Hidden Markov models for circular and linear-circular time series. Environmental and Ecological Statistics 13(3), 325–347 (2006)
14. Jiang, F., Wu, Y., Katsaggelos, A.K.: Abnormal event detection from surveillance video by dynamic hierarchical clustering. In: ICIP (5), pp. 145–148. IEEE (2007)
15. Maruotti, A.: Mixed hidden Markov models for longitudinal data: An overview. International Statistical Review 79(3), 427–454 (2011)
16. Maruotti, A., Rocci, R.: A mixed nonhomogeneous hidden Markov model for categorical data, with application to alcohol consumption. Statist. Med. (31), 871–886 (2012)
17. Rabiner, L.R.: A tutorial on hidden Markov models and selected applications in speech recognition. Proceedings of the IEEE 77(2), 257–286 (1989)
18. Rabiner, L.R., Juang, B.H.: An introduction to hidden Markov models. IEEE ASSP Magazine 3(1), 4–16 (1986)
19. Ronning, G.: Maximum-likelihood estimation of dirichlet distribution. Journal of Statistical Computation and Simulation 32, 215–221 (1989)
20. Wong, T.T.: Parameter estimation for generalized dirichlet distributions from the sample estimates of the first and the second moments of random variables. Computational Statistics and Data Analysis 54(7), 1756–1765 (2010)

Majority-Class Aware Support Vector Domain Oversampling for Imbalanced Classification Problems

Markus Kächele, Patrick Thiam, Günther Palm, and Friedhelm Schwenker

Institute of Neural Information Processing, Ulm University, James-Franck-Ring,
89081 Ulm, Germany
{markus.kaechele, patrick.thiam,
guenther.palm, friedhelm.schwenker}@uni-ulm.de

Abstract. In this work, a method is presented to overcome the difficulties posed by imbalanced classification problems. The proposed algorithm fits a data description to the minority class but in contrast to many other algorithms, awareness of samples of the majority class is used to improve the estimation process. The majority samples are incorporated in the optimization procedure and the resulting domain descriptions are generally superior to those without knowledge about the majority class. Extensive experimental results support the validity of this approach.

Keywords: Imbalanced classification, One-class SVM, Kernel methods.

1 Introduction and Related Work

Real world machine learning tasks can exhibit several problems that render solving them a severe challenge. Such problems include the unreliability of labels (such as incorrect or missing ones), degraded input data (e.g. by noise or unreliable preprocessing), a very high number of feature dimensions and only a few training samples, and imbalanced training sets, where there are much more samples of one class than the other. In this work, a possible solution for the classification of imbalanced training sets is proposed. Imbalanced datasets are a common problem in machine learning because for many applications, the ease of collecting data from different classes is not equal for each class. For example in medical tasks such as segmentation of cells, the process of collecting samples from healthy patients can be much easier than from patients with a more or less common illness.

The problem with unbalanced datasets is that classifiers that are trained with them usually only learn the larger class (the majority class) because the a-priori probability of a sample belonging to it is much higher than to the other class. A reason for that is that the measure that is usually optimized is classification accuracy, which can already be relatively high by only recognizing the majority class. To overcome this problem, many different solutions have been proposed. The solutions can be grouped into sampling methods, cost-sensitive methods and one-class methods. Sampling methods rebalance the training set by either

N. El Gayar et al. (Eds.): ANNPR 2014, LNAI 8774, pp. 83–92, 2014.

oversampling the minority class [5] or subsampling the majority class [18] (or combinations of both [7]). Subsampling has the advantages that the original data is not changed and that training is faster because of the reduced training set. However the disadvantage that information is thrown away (i.e. samples from the majority class). This issue can be resolved by training more than one classifier and applying ensemble methods on the results [6,12] however with the cost of training additional classifiers.

Oversampling, on the other hand, has the advantage that no information is lost and every sample is used for training. However, in order to enlarge the minority class, synthetic samples have to be generated. This process is critical and care has to be taken when choosing a hypothesis (or data model) to generate novel samples from. If the model is too close to the original data, the possibility of overtraining arises, however if the model is too general, the underlying distribution is lost. A well known algorithm of this category is Chawla et. al's synthetic minority over-sampling technique (SMOTE) [5], in which artificial samples are generated along the connecting lines between neighbouring samples. Many extensions and modifications to the original algorithm have since been proposed. BorderlineSMOTE [9] for example focuses on oversampling of samples that are suspected to be near the decision border. KernelSMOTE (KSMOTE) [20] is an extension that works by finding neighbouring samples in kernel space and then computing new samples using the pre-images in input space.

Other methods exist, that do not alter the training set in any way and rather change the algorithmic treatment of the different classes. Class specific weights/penalty factors can for example be used to instruct the optimization procedure to compensate for classes of different sizes. One way to do this is to introduce class specific boxconstraints for the SVM [15,14,1]. Another possibility is to directly encode the imbalance of the dataset into the creation of the classifier using different loss-metrics [10].

Another possibility are one-class methods, which are used to estimate the support of the minority class and to then generate samples from the inferred model. Popular model choices are Gaussian mixture models or One-class SVMs [17,16]. In [14], this approach has been successfully employed.

A closely related methodology can be found in the field of support vector candidate selection, in which samples are found that will most likely become support vectors in a later classification task (examples that lie near the class boundary for example). This way, the dataset is reduced by discarding uninformative samples and leaving only (potentially) informative ones (for further information, the reader is referred to [13,8]).

In this work, an oversampling method is presented that is based on estimating the support of the minority class using support vector domain description (SVDD). However, the original formulation is modified such that the model is aware of nearby majority class samples by incorporating them in the optimization function using negative weights. In the resulting domain description, regions with a large of number majority class samples but also isolated samples that lie near minority class samples are avoided by adapting the hyperplane to position them outside of the estimated domain. The domain description is then used to generate new samples to balance the classification problem.

The remainder of this work is organized as follows. In the next section the modified SVDD description is introduced. The sampling algorithm is explained in Section 3. Experimental results are presented in Section 4 together with a discussion, before the work is concluded in the last section.

2 Majority Class Aware Support Vector Domain Description

As in Tax and Duin's original SVDD formulation [17], the task is to find the minimum enclosing ball of radius R of the training samples $x_i \in \mathbb{R}^d$ to an unknown center a. In order to be insensitive to outliers, analogously to the definition of the SVM by Vapnik [19], so called *slack variables* ξ_i are introduced. The parameter C controls the trade-off between accuracy of the model (amount of samples inside the sphere) and generalization (tight fit of underlying distribution; outliers should be identified as such). The original objective was to minimize

$$F(R, a, \xi_i) = R^2 + C \sum_i \xi_i \tag{1}$$

under the constraints $(x - a)^T (x - a) \leq R^2 + \xi_i$ and $\forall i, \xi_i \geq 0$. Since the task here does not only consist of learning a data distribution but also the generation of new samples of a given class with a later classification experiment in mind, the material at hand (the samples of the minority class) is extended by *negative* examples (samples of the majority class), that should be avoided in the model learning task. In order to prevent problems that arise when the much larger majority class is included, an individual weight w_i for each sample is introduced. This way, they can either be switched on or off when needed, or weighted down to prevent domination of the minimization process. The constraints therefore change to:

$$w_i(R^2 - (x_i - a)^T (x_i - a)) + \xi_i \geq 0 \qquad \forall i, \xi_i \geq 0 \tag{2}$$

where $w_i \in \mathbb{R}$ are the sample weights. The constraint is built such that a weight $w_i < 0$ indicates that a sample should be outside the sphere and analogously $w_i > 0$ enforces the placement of the sample inside the sphere. Combining eq. 1 with the constraints and Lagrange multipliers α_i and γ_i leads to

$$L(R, a, \alpha_i, \xi_i) = R^2 + C \sum_i \xi_i - \sum_i \gamma_i \xi_i$$
$$- \sum_i \alpha_i \left[w_i(R^2 - \{x_i^2 - 2\langle a, x_i \rangle + a^2\}) + \xi_i \right] \tag{3}$$

Determining the partial derivatives with respect to R, a and ξ_i and setting them to 0 yields:

$$\frac{\partial L}{\partial R} = 2R - \sum_i \alpha_i w_i 2R \overset{!}{=} 0 \qquad \Rightarrow \sum_i \alpha_i w_i = 1 \tag{4}$$

and

$$\frac{\partial L}{\partial a} = -\left[\sum_i 2\alpha_i w_i x_i - 2\alpha_i w_i a\right] \overset{!}{=} 0$$

$$\Rightarrow a = \frac{\sum_i \alpha_i w_i x_i}{\sum_i \alpha_i w_i} \overset{(4)}{\Rightarrow} a = \sum_i \alpha_i w_i x_i \tag{5}$$

and

$$\frac{\partial L}{\partial \xi_i} = C - \alpha_i - \gamma_i \overset{!}{=} 0 \qquad \Rightarrow 0 \le \alpha_i \le C \tag{6}$$

Substitution of Equations 4, 5 and 6 into Equation 3 and rearrangement yields

$$L(R, a, \alpha_i, \xi_i) = \sum_i \alpha_i w_i x_i^2 - 2\sum_i \alpha_i w_i x_i (\sum_j \alpha_j w_j x_j) + (\sum_j \alpha_j w_j x_j)^2 \tag{7}$$

which leads to the dual form of the original problem:

$$L(R, a, \alpha_i, \xi_i) = \sum_i \alpha_i w_i \langle x_i, x_i \rangle - \sum_{ij} \alpha_i \alpha_j w_i w_j \langle x_i, x_j \rangle \tag{8}$$

The dual form has to be maximized under the constraints $0 \le \alpha_i \le C$ and $\sum_i \alpha_i w_i = 1$. This is a convex function and can be optimized using quadratic programming.

By incorporating a mapping function $\phi : \mathcal{S} \to \mathcal{F}$ from the domain of the samples to a high dimensional feature space $\mathcal{F}$, the dot products in Equation 8 can be replaced by $\langle \phi(x_i), \phi(x_j) \rangle$, which in turn can be substituted for a *kernel function* $K(x_i, x_j)$ using the kernel trick [3] to achieve non-linear models.

In Figure 1 the effect of the weights on the hyperplane is illustrated. One sample is selected and its weight is gradually decreased until it becomes negative so that the hyperplane starts to bend around the sample to exclude it.

3   Oversampling Using Modified Support Vector Domain Description

In contrast to SMOTE or KernelSMOTE, the proposed algorithm consists of the two phases (1) model building and (2) sample generation:

Phase 1: Model Building

The sample distribution is estimated using support vector domain description. The difference to the original one is that samples of the majority class are weighted negatively in order to keep them outside of the hypersphere. This is done to prevent that samples are generated near negative examples and therefore rendering them contradicting to the original training set. The weights should be determined based on the task at hand either manually or using cross validation.

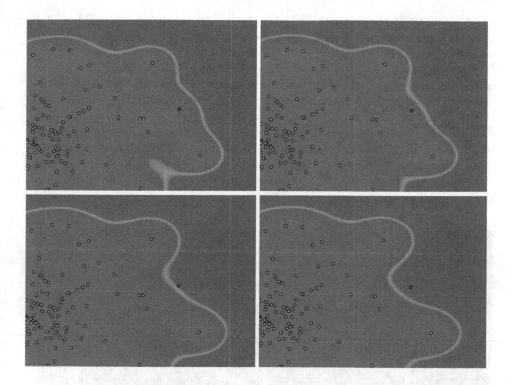

Fig. 1. Effect of sample weights on the hyperplane. By decreasing the weight for the red sample the hyperplane is bent such that the sample begins to traverse the border and resides outside the hypersphere in the end. In this manner, samples can be *forced* to be on either side of the hyperplane with a distance dependent on the magnitude of the weight.

The use of a suitable kernel function might be beneficial to create more complex domain boundaries and thus to allow a better fit of the underlying data distribution. In order to decrease training time, a preprocessing step can be applied to filter majority class samples that are nowhere near minority class samples.

Phase 2: Sample Generation

After the domain description is fit, it is used to infer novel samples. This step can be done using various methods. One possibility is to use rejection sampling to generate new samples by repeatedly drawing random numbers from the respective range of each feature dimension and then checking if the new sample is inside the hypersphere or not. This method has the advantage that it is simple to implement and can also be used to generate samples in regions inside the hypersphere where no minority sample directly resides (i.e. regions that are not directly connected to those regions that hold the original samples). A disadvantage is that the cost of producing new samples is tightly linked to the dimensionality of the features and also the complexity of the learned model. Possibly a large amount of random numbers will be used to generate the desired number of samples.

Another possibility is to perform a random walk starting from the samples of the minority class and to terminate randomly or when the path leaves the sphere. This algorithm has the advantage that it starts inside the boundary and therefore avoids sampling the empty space around the domain description. The disadvantage is that the area is not sampled uniformly and depending on the termination criterion of the random walk, densely populated regions will attract more samples than sparsely populated ones.

Using one of those methods, the minority class is resampled to approximately the same size as the majority class.

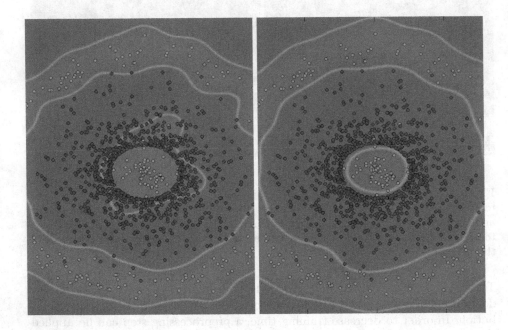

Fig. 2. Estimated sample distribution without and with negatively weighted majority class. In the centers of the figures, the differences are most dominant. On the left side, the model overlaps with the majority class to a great extent, while the center on the right side is constrained to the circle where only the positive samples reside. If the weight is continually increased, the optimization will try to exclude bordering points even more, narrowing the corridor on the outside and the circle in the inside.

4   Experimental Results

Experimental validation of the proposed algorithm has been carried out on a number of freely available, imbalanced datasets and is compared with four different algorithms. The algorithms were selected so, that they cover a broad spectrum of varieties from simple subsampling over cost-sensitive learning to oversampling. The experiments consisted of classifying imbalanced datasets of different size and with different imbalance ratios. For an overview, the reader is referred to Table 1. The comparison algorithms were:

- Support vector machine with randomly subsampled training sets
- Support vector machine with class specific boxconstraint
- SMOTE
- KernelSMOTE

Table 1. Overview of datasets with their characteristics. The datasets were selected so that a wide range of input dimensionalities, number of instances and imbalance ratios can be found. The datasets *diabetes*, *ecoli* and *glass* are part of the UCI machine learning repository (http://archive.ics.uci.edu/ml/).

Name	Dimensionality	Number of instances	Imbalance ratio
diabetes	8	768	1.9
ecoli	7	336	3.4
glass	9	214	4.6
ring	2	1170	3.3

As classification algorithm for SMOTE, KSMOTE and the algorithm proposed here, SVMs were chosen. The experimental setup consisted of stratified k-fold cross validation to obtain different imbalanced subsets (an exception to this was the random subsampling for the first comparison algorithm). Based on those subsets, the methods were trained on $k - 1$ subsets by oversampling the minority class to the same size as the majority class and then validated using the remaining subset. Parameter selection involved the *boxconstraint* of the SVM and the kernel parameter γ of the RBF kernel and was carried out using a grid search with a cross validation on a randomly selected subset of the whole dataset of approximately half of the original size. Each experiment was repeated 10 times. To evaluate the performances, the gmean measure as defined in Equation 9 was used

$$gmean = \sqrt{acc^+ * acc^-} \qquad (9)$$

where acc^+ and acc^- stand for the rates of true positives and true negatives, respectively. In Table 2 the results are summarized.

As can be seen, the proposed algorithm achieves competitive results and ranks first, together with KSMOTE. The experiments were conducted once with negatively weighted majority samples and once without. The variant with the weights clearly outperforms the one without. Only for the ring dataset the results without weights were better (and only slightly worse than KSMOTE in this case). To see the effects of the weights on the generated hyperplanes, the reader is referred to Figure 2.

Table 2. Summary of experimental results. The values denote the averaged gmean of the classification results from a 3-fold cross-validation for every dataset. To minimize statistical outliers, each experiment was repeated 10 times. As can be seen, the proposed algorithm exhibits superior performance over the remaining algorithms, except for KSMOTE, which performs approximately equally well. The weighted version of the SVDD sampling outperforms the unweighted version in almost every case.

Name	diabetes	ecoli	glass	ring
SVM (bagging)	0.705	0.791	0.913	0.940
SVM (cost)	0.641	0.823	0.918	0.978
SMOTE	0.691	0.882	0.850	0.940
KSMOTE	**0.731**	0.891	0.913	**0.957**
Proposed (w/o weights)	0.696	0.851	0.841	0.956
Proposed (w/ weights)	0.723	**0.905**	**0.945**	0.922

5 Discussion

In comparison to other algorithms such as SMOTE or KSMOTE, the proposed algorithm has the advantage that not only a local neighbourhood is used to infer the new samples. As with other one-class mechanisms, the underlying distribution is estimated and as a result, hypotheses are constructed that can be used to generate new samples. A major advantage of the algorithm as proposed here is that samples of the majority class are also considered in the model fitting phase. Overgeneral minority distributions can be avoided as well as regions where positive and negative samples overlap. If (a subset of) the negative samples are weighted strongly enough, the optimization procedure seeks to put such regions outside the sphere (thereby potentially cutting holes into the domain). The argument that SVM classifiers are able to deal with overlapping regions on their own (even in the same way, since both techniques are very similar) can be extenuated because samples can also be generated for other classification algorithms such as random forests [4] or multi-layer perceptrons. Another advantage is that the samples can effortlessly be generated with accompanying confidence values that indicate how certain a sample belongs to the class. This can be achieved using the distance to the hyperplane. A drawback of the approach is that the support vector domain description is parametrized by (usually at least) two parameters, namely a kernel parameter such as γ and the boxconstraint C. However this poses only a minor problem that can be solved using a grid search in the parameter space. In the experimentation process, the search for proper weights was not critical (i.e. there was no need for an extra cross validation for the weights). It mostly made a difference whether weights were used or not, but their exact value was less important (the reader is again referred to Figure 2).

6 Conclusion and Future Work

In this paper, a method was presented to solve the classification of imbalanced data by sampling the minority class using majority class aware support vector domain description. First, the data distribution of the class samples is estimated by the algorithm. Then, novel samples are generated using the proposed random sampling techniques. Sampling of new data points in overlapping or bordering regions can be avoided using individual weights that can for example be set negatively for majority class samples. Contrary, important regions can also be highlighted by giving them higher positive weights. Experimental validation was presented to emphasize the feasibility of the proposed mechanism. In future experiments, the applicability of the weighted SVDD for uncertainly labeled data, such as affect in human-computer interaction scenarios [11] will be investigated. The idea is that by incorporating uncertainty in the form of confidence values or fuzzy labels, more reliable models will emerge from the training process. Another idea could be to use the modified SVDD in Co-training like scenarios [2] to iteratively learn different classes and then reweight them using the gained knowledge to extract compact class descriptions.

Acknowledgements. This paper is based on work done within the Transregional Collaborative Research Centre SFB/TRR 62 *Companion-Technology for Cognitive Technical Systems* funded by the German Research Foundation (DFG). Markus Kächele is supported by a scholarship of the Landesgraduiertenförderung Baden-Württemberg at Ulm University.

References

1. Akbani, R., Kwek, S.S., Japkowicz, N.: Applying support vector machines to imbalanced datasets. In: Boulicaut, J.-F., Esposito, F., Giannotti, F., Pedreschi, D. (eds.) ECML 2004. LNCS (LNAI), vol. 3201, pp. 39–50. Springer, Heidelberg (2004)
2. Blum, A., Mitchell, T.: Combining Labeled and Unlabeled Data with Co-training. In: COLT: Proceedings of the Workshop on Computational Learning Theory. Morgan Kaufmann Publishers (1998)
3. Boser, B.E., Guyon, I.M., Vapnik, V.N.: A training algorithm for optimal margin classifiers. In: Proceedings of the 5th Annual Workshop on Computational Learning Theory, COLT 1992, pp. 144–152. ACM (1992)
4. Breiman, L.: Random forests. Machine Learning 45(1), 5–32 (2001)
5. Chawla, N.V., Bowyer, K.W., Hall, L.O., Kegelmeyer, W.P.: Smote: Synthetic minority over-sampling technique. Journal of Artificial Intelligence Research 16, 321–357 (2002)
6. Chawla, N.V., Lazarevic, A., Hall, L.O., Bowyer, K.W.: SMOTEBoost: Improving prediction of the minority class in boosting. Proceedings of the Principles of Knowledge Discovery in Databases (PKDD), 107–119 (2003)
7. Cohen, G., Hilario, M., Sax, H., Hugonnet, S.: Data imbalance in surveillance of nosocomial infections. In: Perner, P., Brause, R., Holzhütter, H.-G. (eds.) ISMDA 2003. LNCS, vol. 2868, pp. 109–117. Springer, Heidelberg (2003)
8. Guo, L., Boukir, S., Chehata, N.: Support vectors selection for supervised learning using an ensemble approach. In: Proceedings of the International Conference on Pattern Recognition (ICPR), pp. 37–40 (August 2010)
9. Han, H., Wang, W.-Y., Mao, B.-H.: Borderline-SMOTE: A new over-sampling method in imbalanced data sets learning. In: Huang, D.-S., Zhang, X.-P., Huang, G.-B. (eds.) ICIC 2005. LNCS, vol. 3644, pp. 878–887. Springer, Heidelberg (2005)

10. Hong, X., Chen, S., Harris, C.: A kernel-based two-class classifier for imbalanced data sets. IEEE Transactions on Neural Networks 18(1), 28–41 (2007)
11. Kächele, M., Glodek, M., Zharkov, D., Meudt, S., Schwenker, F.: Fusion of audio-visual features using hierarchical classifier systems for the recognition of affective states and the state of depression. In: De Marsico, M., Tabbone, A., Fred, A. (eds.) Proceedings of the International Conference on Pattern Recognition Applications and Methods (ICPRAM), pp. 671–678. SciTePress (2014)
12. Kächele, M., Schwenker, F.: Cascaded fusion of dynamic, spatial, and textural feature sets for person-independent facial emotion recognition. In: Proceedings of the International Conference on Pattern Recognition (ICPR) (to appear, 2014)
13. Li, M., Chen, F., Kou, J.: Candidate vectors selection for training support vector machines. In: Third International Conference on Natural Computation, ICNC 2007, vol. 1, pp. 538–542 (August 2007)
14. Raskutti, B., Kowalczyk, A.: Extreme re-balancing for svms: A case study. SIGKDD Explor. Newsl. 6(1), 60–69 (2004)
15. Schels, M., Scherer, S., Glodek, M., Kestler, H., Palm, G., Schwenker, F.: On the discovery of events in EEG data utilizing information fusion. Computational Statistics 28(1), 5–18 (2013)
16. Schölkopf, B., Platt, J.C., Shawe-Taylor, J.C., Smola, A.J., Williamson, R.C.: Estimating the support of a high-dimensional distribution. Neural Computation 13(7), 1443–1471 (2001)
17. Tax, D.M.J., Duin, R.P.W.: Support vector domain description. Pattern Recognition Letters 20, 1191–1199 (1999)
18. Van Hulse, J., Khoshgoftaar, T.M., Napolitano, A.: Experimental perspectives on learning from imbalanced data. In: Proceedings of the International Conference on Machine Learning, ICML 2007, pp. 935–942. ACM, New York (2007)
19. Vapnik, V.N.: Statistical Learning Theory, vol. 2. Wiley (1998)
20. Zeng, Z.-Q., Gao, J.: Improving SVM classification with imbalance data set. In: Leung, C.S., Lee, M., Chan, J.H. (eds.) ICONIP 2009, Part I. LNCS, vol. 5863, pp. 389–398. Springer, Heidelberg (2009)

Forward and Backward Forecasting Ensembles for the Estimation of Time Series Missing Data

Tawfik A. Moahmed[1], Neamat El Gayar[1], and Amir F. Atiya[2]

[1] Faculty of Computers and Information, Cairo University, Giza, Egypt
[2] Department of Computer Engineering, Faculty of Engineering, Cairo University, Giza, Egypt
t.ahmed@fci-cu.edu.eg,
elgayar.neamat@gmail.com,
amir@alumni.caltech.edu

Abstract. The presence of missing data in time series is big impediment to the successful performance of forecasting models, as it leads to a significant reduction of useful data. In this work we propose a multiple-imputation-type framework for estimating the missing values of a time series. This framework is based on iterative and successive forward and backward forecasting of the missing values, and constructing ensembles of these forecasts. The iterative nature of the algorithm allows progressive improvement of the forecast accuracy. In addition, the different forward and backward dynamics of the time series provide beneficial diversity for the ensemble. The developed framework is general, and can make use of any underlying machine learning or conventional forecasting model. We have tested the proposed approach on large data sets using linear, as well as nonlinear underlying forecasting models, and show its success.

Keywords: Time series prediction, missing data, ensemble prediction.

1   Introduction

Time series forecasting has become an important decision making tool [1]. Its application to many domains such as weather prediction [2], stock market forecasting [3], electric load estimation [4], river flow forecasting [5], economic forecasting [6], and sales prediction [7] has had a big impact on the profitability, utilization, risk mitigation, and efficiency of these processes. Time series forecasting can essentially be considered as a modeling problem. Some of the dominant approaches in the literature include linear models, such as autoregressive (AR), ARMA, [8] and exponential smoothing [9]. Recognizing that in many real-world situations the data generation process (DGP) may not follow a simple linear model, nonlinear models such as neural networks [10], support vector machines [10]), vector quantization [11], and nonlinear basis selection [44] have become another prominent group of models.

In ideal circumstances the time series is sampled with constant frequency and all samples are present. However, in real situations some samples are missing

N. El Gayar et al. (Eds.): ANNPR 2014, LNAI 8774, pp. 93–104, 2014.
© Springer International Publishing Switzerland 2014

due to human error or lapse in data collection. The missing data problem affects the performance of forecasting models in ways far beyond the amount of missing data.

In this paper we present a framework for missing value estimation for time series, based on the concept of ensembles. The framework that we propose is summarized as follows. We train the underlying forecasting model to forecast forward in time using the available or existing data as training data. In the second step we train the underlying forecasting model to forecast backward in time "backcast" using the available data as training set. Naturally, the training data in these two steps may not be sufficient to produce good forecasts in case there are many portions in the data missing. In the third step we apply the trained forward forecasting model and backward forecasting model to predict all missing values in the time series. Subsequently, we average both forecasts, to obtain the first ensemble forecast. We continue in this manner for a number of iterations, until no further improvements are expected. The advantage of the proposed model is that it utilizes the power of ensembles, and makes use of an iterative self-improving process, whereby each iteration is expected to improve its estimates over the previous iteration. It works akin to a consecutive sequence of "reflecting waves" of forward forecasting and backward forecasting. The proposed approach applies to the "missing completely at random (MCAR)" situation. This means that the fact that a record is missing does not depend on the underlying value of any of the data, it is purely by chance.

The paper is organized as follows. Next section presents a literature review. Section 3 presents the proposed procedure to estimate missing data; Section 4 presents the details of the simulation set-up. In section 5, the results and discussion are presented and section 6 includes conclusion and future work.

2 Related Work

The problem of missing data has been studied in the conventional statistics literature [12]. The most straightforward method, but at the same time least effective one, is the case-wise deletion method (CWD). It is based on simply removing the training patterns that have some missing values. Of course this leads to an unjustified loss of precious data. A more effective group of methods is the single imputation strategy. In this approach, instead of completely deleting the pattern, we impute (compute) a value for it, and use the resulting completed data set for training the prediction model.

A more involved group of methods is the so-called multiple imputation strategy [14] [15]. A more modern group of approaches uses some probabilistic concepts. The main approaches in this group are the maximum likelihood approach, and the expectation maximization (EM) approach [16].

Denk and Weber [17] describe the specific nature of missing value estimation for time series, and the differences from missing value for regular regression. They also investigate missing data patterns and categories, according to the type of the time series, whether cross-sectional, uni-variate, or multi-variate.

In addition, they provide an informative review concerning the dependency structure of the missingness of the variables (i.e. about the well-known concepts of missing completely at random MCAR, missing at random MAR, and missing not at random MNAR). The mean substitution method for conventional data sets (i.e. replacing a missing value by the mean of that variable) becomes infeasible for the time series case. In its place the "Carry Last Value Forward" method is practical and feasible. In this strategy a missing value is replaced by the most recent available value. On the other hand, imputation methods based on K-nearest neighbor matching are applicable for the time series case. In that approach, a missing portion of the time series is filled according to the K patterns that are closest to the considered pattern, based on the non-missing variable portion comparison [18]. Other pattern matching approaches have been proposed by Chiewchanwattana et al. [19]. Deterministic imputation approaches have also been considered in the literature. For example, spline smoothing or other curve fitting approaches could be used to fill in a missing portion of the time series [20] [21]. Vellcer and Colby [22] provide a review of a number of methods, including CWD, mean substitution, mean of adjacent time series values, and maximum likelihood estimation. They provide a comparison between these methods using synthetic time series.

Some of the approaches intertwine the time series modeling with the missing value imputation. Bermudez [23] present a new approach for the prediction of time series with missing data based on an alternative formulation for Holt's exponential smoothing model. Also Huo et al [24] develop the so-called two-directional exponential smoothing. Ferreiro [25] apply a similar approach for the case of autoregressive processes, where optimal values of the missing data points are derived in the context of the AR process. Durbin and Koopman [26] consider another such approach for state space time series models. There has also been work based on machine learning models. Uysal [27] proposes the use of radial basis functions for the missing value imputation. Eltoft and Kristiansen [28] use independent component analysis (ICA) and the dynamical functional artificial neural network (D-FANN) for filling the gaps in multivariate time series. Gupta and Lam [29] compare neural network prediction of missing values in the context of regression, and show that it consistently beat traditional approaches such as moving averages and regression. Kihoro et al. [30] apply neural networks for missing values in time series, and compare it with seasonal ARIMA forecasting. Pearl [31] uses Bayesian belief networks, as a probabilistic mechanism to estimate missing value distribution.

Some researchers considered neural network ensembles for the missing value problem, but in the context of regression, not time series prediction [33] [34]. The only studies that we have found in the literature on using ensembles for time series missing values are the following. Chiewchanwattana et al. [35] use an ensemble of FIR neural networks, where each one is trained to predict the missing values using a different target (obtained by other missing value methods). Sorjamaa and Lendasse [36] use an ensemble of self-organizing maps to predict missing values. They obtain the combination weights of the ensemble predictions using the nonnegative least squares algorithm. Sorjamaa and Lendasse [37] also consider the combination of self-organizing maps and a linear model for missing value imputation. Alternatively, Ahmed et al. [45] use the concept of semi-supervised co-training for the time series missing value problem. A similar approach was previously developed for classification but is adapted here for prediction whereby different networks in an ensemble boost each other's training performance. This work also examines the use of different base learners in the ensemble, and different confidence measures to accurately combine the predictions.

Viewing the approach proposed in this work in the context of the existing approaches, one can observe that it is of a multiple imputation type. It generalizes the multiple imputation and the iterative concepts of imputation to the case of time series, and at the same time combines these approaches with the concept of ensembles.

3   The Proposed Framework

For the problem of missing data in time series prediction, we propose two forecasting models that act on different dynamics of the time series, therefore possessing the beneficial diversity, and to combine them in an ensemble. There is a certain dynamic governing the forward flow of information, and allowing time series values to relate to previous time series values. There is also the backward dynamic that allows time series values to be expressed in terms of the future values. Both dynamics are different, and will therefore add diversity of the ensemble. Assume that the time series has a missing portion from $t = M + 1$ to $t = M + J$. We consider the forward forecasting model that forecasts the missing values in terms of the lagged previous values, i.e to estimate missing value at time t=M+1 we use $x_{M-L-1}, ..., x_M$ as input to our model. We also design the backward forecasting model "backcaster", that forecasts the missing values in terms of the subsequent time series values, i.e to estimate missing value at time t=M+J we use $x_{M+J+1}, ..., x_{M+J+L}$ as inpit to our models. Then, an ensemble is constructed, consisting of the forecaster and the backcaster. But this is only the first step. Once we have filled the missing values we create an augmented training set (consisting of the original training patterns, and new patterns that are available after estimating the missing values of time series). So we use this more complete training set to retrain the forward forecasting system, and the backcasting system, to obtain better models. We continue in this iterative manner, until the extra improvement tapers off. Typically three or four steps of this iteration should be sufficient. Note that a deseasonalization step may be needed if the time series is deemed to possess seasonality. See Figure 1 for a detailed structure of the proposed framework.

We have a number of observations concerning the proposed method:

– There is some aspect of co-training in this approach. Co-training is a methodology, developed in the pattern classification field, whereby two or more models' predictions are incorporated in each other's training data [38]. Subsequent retraining should lead to improved performance.

- The proposed approach is a general framework, and could apply to any underlying forecasting model, whether conventional, linear, neural networks, or other.
- It is well-known that the success of ensemble approaches hinges on the diversity of the constituent models [13] [39]. We can make the point that the forward forecasting process and the backward forecasting process possess fairly different and complementary dynamics. This is because how the past affects the future is generally different from how one can infer the past from the future or present. Both forecasters also use different and generally widely spaced input variables.

1. Given time series with missing data
   $\mathbf{T} = x_1, x_2, x_3, ?, ?, x_6, \cdots \cdots, x_{30}, x_{31}, ?, ?, ?, x_{35} \cdots \cdots x_n$
   - Apply a seasonality test. If seasonality exists, apply a deseasonalization step.
   - For iteration i=1 to N
     (a) Use $\mathbf{T}$ to tune the model parameters by K-fold validation
     (b) Train the model with the tuned parameters using T
     (c) Use the trained model to estimate the missing values "Forward Missing Estimation" (FME_i)
        $FME_i = x_1, x_2, x_3, x_4', x_5', x_6 \cdots \cdots, x_{30}, x_{31}, x_{32}', x_{33}', x_{34}', x_{35} \cdots \cdots, x_n$
     (d) Reverse the order of time series values T
        $\mathbf{RT} = x_n, x_{35}, ?, ?, ?, x_{31}, x_{30}, x_6, ?, ?, x_3, x_2, x_1$
     (e) Use $\mathbf{RT}$ to tune the model parameters by K-fold validation
     (f) Train model with tuned parameters using RT
     (g) Use the trained model to estimate the missing values "Backward Missing Estimation" (BME_i)
        $BME_i = x_n, \ldots\ldots x_{35}, x_{34}', x_{33}', x_{32}', x_{31}, x_{30}\ldots x_6, x_5', x_4', x_3, x_2, x_1$
     (h) Calculate the Average of the FME_i and BME_i to get AME_i
     (i) Replace each missing value the calculated average one to get $\mathbf{NT}$
        $\mathbf{NT}$ = time series with estimated missing values
     (j) T=NT
   - End
2. Restore the seasonality if the time series was deseasonalized.
3. End

Fig. 1. Ensemble of Forward and Backward missing estimation

4   Experimental Setup

To test the proposed approach, we have applied it to the M3 and the NN3 competition data sets. The M3 competition is the latest in a sequel of M forecasting competitions [40]. It consists of 3003 business-type time series, covering the types of economy, industry, finance, demographic, and others. It consists of yearly series (645 series), quarterly series (756 series) , monthly series (1428series) and others (174series). In this study we consider the ones that have at least 80 points. We ended up with 1020 time series. The NN3 competition [41] is a similar competition, geared towards computational intelligence forecasting models. The type of data is also mostly economics and business related. It consists of 111 monthly time series. We excluded the short ones, and therefore ended up with 65 time series. Both the M3 and the NN3 have become important benchmarks for testing and comparing forecasting models. Having that many diverse time series gives confidence into comparison results. All time series in these data sets are complete, with no missing values. So we removed some values to artificially create time series with missing values. We have considered missing ratios of 10%, 20%, 30%, 40%, and 50%. We have tested the proposed framework using three different underlying forecasting models. This is needed in order to verify the generality of the proposed approach. We considered the following three forecasting models, **Feed Forward Neural Network** as an example of a nonlinear model, **Holt's Exponential Smoothing** as an example of a linear model and **Moving Average** as a simple model. We use the symmetric mean absolute percentage error "SMAPE". It is a normalized error measure, and this feature is beneficial in our study, since each data set has different time series from various sectors, and they have therefore different value ranges. The SMAPE is defined as:

$$SMAPE = \frac{1}{M} \sum \frac{\left| y'_m - y_m \right|}{\frac{|y'_m| + |y_m|}{2}} \tag{1}$$

Time series often possess seasonal and trend components. An effective strategy has been to deseasonalize the time series before applying the forecasting model (e.g. a neural network). The study in [42] reports that deseasonalization is beneficial, because it relieves the forecasting model from an undue burden of predicting the seasonal cycle, in addition to its main job of forecasting the time series. Another work [10] also shows that deseasonalization is useful. However, unlike [42], they mention that detrending was detrimental. Based on these previous works' consistently favoring deseasonalization, and the conflicting and questionable benefit of detrending, we decided to apply deseasonalization, but not detrending.

4.1 Parameter Setting

A critical step in any machine learning or forecasting model is the parameter estimation process. In every model there are typically one or two key parameters that affect the performance greatly, and must therefore be set with care. In this work we use the K-fold validation approach for parameter selection. For single hidden layer neural network, there are two critical parameters that have to be tuned using K-fold validation: the number of input variables (the number of lagged values) and the number of hidden neurons. For the number of lagged values, we consider the range [1,2,3,4,5] whereas for the number of hidden neurons,

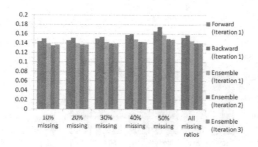

Fig. 2. MLP-NN3-SMAPE

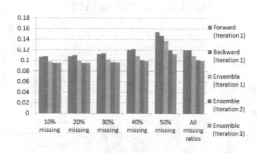

Fig. 3. MLP-M3-SMAPE

we specify the range to be [0,1,3,5,7]. Note that 0 hidden neurons means that the network is in effect a simple linear network. For Holt's exponential smoothing model the main parameters are α and γ. We search the following range for these parameters: α , $\gamma = [0, 0.1, 0.2, 0.3, 0.4, 0.5, 0.6, 0.7, 0.8, 0.9, 1]$. For each α and γ pair, we generate smoothed "estimated" values of all the training time series values, and calculate the mean square error (MSE) between the smoothed time series values and true ones. We select the α and γ pair with minimum MSE. The model parameter of the moving average model is the window size. We simply test different window sizes to estimate missing values and select the window size that results in the minimum MSE. Once we fix the window size for the moving average model, more iterations will not change the accuracy of the missing value estimation. So we just apply a single iteration when using the moving average as a forecasting model.

5   Results and Discussion

Figures 2 and 3 show the SMAPE for the M3 data set and the NN3 data set for the neural network forecasting model. Figures 4 and 5 show the corresponding results for the Holt's exponential smoothing model, while Figures 6

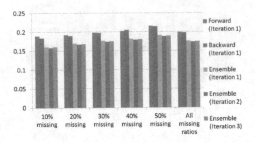

Fig. 4. Expo-NN3-SMAPE

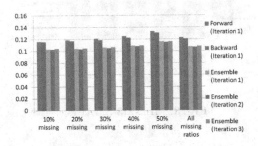

Fig. 5. Expo-M3-SMAPE

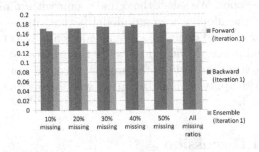

Fig. 6. MA-NN3-SMAPE

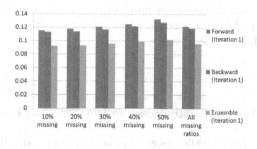

Fig. 7. MA-M3-SMAPE

and 7 show the corresponding results of the moving average model. The figures compare between the performance in predicting the missing values for the forward forecasting, the backward forecasting, and the ensemble of forward and backward forecasting for each of the different iterations. The goal here is to check if the ensembles of backward and forward forecasts, and if the successive iterations show some gain. One can observe that the neural network exhibits fairly consistent improvement up to the third iteration. For the Holt's exponential smoothing model there is an improvement up to the second iteration. For the moving average model the ensemble provides considerable improvement. Also, one can observe that in most cases the improvement is more significant for larger percentages of missing values. This is understandable because for these cases the impact will be larger, because more training patterns will be more accurate. Overall, one can conclude that the use of successive iterations of forward and backward forecasting, and their ensembles are a beneficial strategy. Of course the largest gain comes from the first iteration. The reason is that it leads to a large increase in the training set size, as missing value estimates are now included. In subsequent iterations the gain stems mainly from having more accurate estimates of the missing values in the training set, which leads to a better training, and hence more accurate models. With successive iterations, this added value diminishes, and there is not much more improvement to offer. A prudent strategy is to use two or three iterations.

6   Conclusion and Future Work

In this work we have introduced a new approach to handle and estimate missing data for time series forecasting. We used successive ensembles of forward and backward forecasting models. The forward and the backward forecasting provide useful and complementary diversity, and therefore their inclusion in an ensemble is beneficial and adds value. In addition, successive application of this ensemble provides some refinement for the estimates. Experiments have been conducted on two large data sets each of which contains many time series with different characteristics. The results show the beneficial effect of the proposed ensembles, and their successive iterations, for a number of underlying forecasting models. We therefore believe that the proposed framework should be one of the useful contenders for handling missing values.

References

1. Brockwell, P.J., Davis, R.A.: Introduction to Time Series and Forecasting. Springer, New York (1996)
2. Hill, H.S., Mjelde, J.W.: Challenges and opportunities provided by seasonal climate forecasts: a literature review. Journal of Agricultural and Applied Economics 34, 603–632 (2002)
3. Abu-Mostafa, Y., Atiya, A., Magdon-Ismail, M., White, H. (eds.): Special Issue on Neural Networks in Financial Engineering. IEEE Transactions on Neural Networks 12, 653–656 (2001)
4. Parlos, A., Oufi, E., Muthusami, J., Patton, A., Atiya, A.: Development of an intelligent long-term electric load forecasting system. In: Proc. Intelligent Systems Applications to Power Systems Conference, ISAP 1996, pp. 288–292 (1996)
5. Atiya, A., El-Shoura, S., Shaheen, S., El-Sherif, M.: A comparison between neural network forecasting techniques - case study: river flow forecasting. IEEE Transactions Neural Networks 10, 402–409 (1999)
6. Clements, M.P., Hendry, D.F.: Macro-economic forecasting and modelling. The Economic Journal 105, 1001–1013 (1995)
7. Chu, C.-W., Zhang, G.P.: A comparative study of linear and nonlinear models for aggregate retail sales forecasting. International Journal of Production Economics, 217–231 (2003)
8. Box, G., Jenkins, G.: Time Series Analysis, Forecasting and Control. Holden-Day Inc., San Francisco (1976)
9. Andrawis, R., Atiya, A.F.: A new Bayesian formulation for Holt's exponential smoothing. Journal of Forecasting 28, 218–234 (2009)
10. Ahmed, N.K., Atiya, A.F., El Gayar, N., El-Shishiny, H.: An empirical comparison of machine learning models for time series forecasting. Econometric Reviews 29(5-6), 594–621 (2010)
11. Lendasse, A., Francois, D., Wertz, V., Verleysen, M.: Vector quantization: a weighted version for time-series forecasting. Future Generation Computer Systems 21, 1056–1067 (2005)
12. Allison, P.D.: Missing Data. Sage Publications, Inc. (2001)
13. Andrawis, R., Atiya, A.F., El-Shishiny, H.: Forecast combination model using computational intelligence/linear models for the NN5 time series forecasting competition. International Journal of Forecasting 27, 672–688 (2011)
14. Rubin, D.B.: An overview of multiple imputation, in Survey Research Section. American Statistical Association (1988)
15. Honaker, J., King, G.: What to do about missing values in time series cross-section data. American Journal of Political Science 54, 561–581 (2010)
16. Dempster, A.P., Laird, N.M., Rubin, D.B.: Maximum likelihood from incomplete data via the EM algorithm. Journal of the Royal Statistical Society. Series B 39, 1–38 (1977)
17. Denk, M., Weber, M.: Avoid filling Swiss cheese with whipped cream: imputation techniques and evaluation procedures for cross-country time series. IMF Working Paper WP/11/151 (2011)

18. Tarsitano, A., Falcone, M.: Missing-Values Adjustment For Mixed-Type Data. Working Paper WP15-2010, Department of Economics and Statistics. University of Calabria (2010), http://www.ecostat.unical.it/RePEc/WorkingPapers/WP15_2010.pdf

19. Chiewchanwattana, S., Lursinsap, C., Chu, C.-H.H.: Imputing incomplete time-series data based on varied-window similarity measure of data sequences. Pattern Recognition Letters 28, 1091–1103 (2007)

20. de Jong, P.: The simulation smoother for time series models. Biometrika 82, 339–350 (1995)

21. He, Y., Yucel, R., Raghunathan, T.E.: A functional multiple imputation approach to incomplete longitudinal data. Statistics in Medicine Early View. Wiley Online Library (2011), http://wileyonlinelibrary.com/

22. Velicer, W.F., Colby, S.M.: A comparison of missing-data procedures for ARIMA time-series analysis. Educational and Psychological Measurement, 596–615 (2005)

23. Bermúdez, J.D.: Forecasting time series with missing data using Holt's model. Journal of Statistical Planning and Inference 139, 2791–2799 (2009)

24. Huo, J., Cox, C., Seaver, W., Robinson, R., Jiang, Y.: Application of two-directional time series models to replace missing data. Journal of Environmental Engineering 136, 435–443 (2010)

25. Ferreio, O.: Methodologies for the estimation of missing observations in time series. Statistical Probability Letters 5, 65–69 (1987)

26. Durbin, J., Koopman, S.J.: Time Series Analysis by State Space Methods. Oxford Univ. Press (2004)

27. Uysal, M.: Reconstruction of time series data with missing values. Journal of Applied Science 7, 922–925 (2007)

28. Eltoft, T., Kristiansen, Ø: ICA and nonlinear time series prediction for recovering missing data segments in multivariate signals. In: Proceedings of the Third International Conference on Independent Component Analysis and Blind Signal Separation (ICA 2001), pp. 716–721 (2001)

29. Gupta, A., Lam, M.S.: Estimating missing values using neural networks. The Journal of the Operational Research Society 47, 229–238 (1996)

30. Kihoro, J.K., Otieno, R.O., Wafula, C.: Seasonal time series data imputation: comparison between feed forward neural networks and parametric approaches. East African Journal of Statistics 1, 68–83 (2007)

31. Pearl, J.: Causality: Models, Reasoning, and Inference, 2nd edn. Cambridge University Press, New York (2009)

32. Tresp, V., Hofmann, R.: Missing and noisy data in nonlinear time-series prediction. Neural Computation 10, 731–747 (1998)

33. Hassan, M., Atiya, A., El Gayar, N., El-Fouly, R.: Novel ensemble techniques for regression with missing data. New Mathematics and Natural Computation 5, 635–652 (2009)

34. Twala, B., Cartwright, M., Shepperd, M.: Ensemble of missing data techniques to improve software prediction accuracy. In: ICSE 2006: Proceeding of the 28th International Conference on Software Engineering, pp. 909–912. ACM Press, New York (2006)

35. Chiewchanwattana, S., Lursinsap, C., Chu, C.-H.H.: Time-serises data prediction based on reconstruction of Missing samples and selective ensembling of FIR neural networks. In: Proceeding of the 9th International Conference on Neural Information Processing (ICONIP 2002), Singapore, p. 2156 (2002)

36. Sorjamaa, A., Lendasse, A.: Fast missing value imputation using ensemble of SOMs, Espoo, Finland: Aalto University School of Science and Technology, Report TKK-ICS-R33 (2010)
37. Sorjamaa, A., Lendasse, A.: Time series prediction as a problem of missing values: Application to ESTSP2007 and NN3 competition benchmarks. In: Proceedings of the 2007 International Joint Conference on Neural Networks (IJCNN 2007), Orlando, Florida (2007)
38. Salaheldin, A., El Gayar, N.F.: Complementary feature splits for co-training. In: Proceedings ISSPA, pp. 1303–1308 (2012)
39. Gavin Brown, T.H., Wyatt, J., Yao, X.: Diversity creation methods: A survey and categorization. Pattern Recognition Society (2004)
40. Makridakis, S., Hibon, M.: The M3-competition: results, conclusions and implications. International Journal of Forecasting 16, 451–476 (2000)
41. NN3 competition, http://www.neural-forecasting.com/nn3-competition.html
42. Zhang, G.P., Qi, M.: Neural network forecasting for seasonal and trend time series. European Journal of Operational Research 160, 501–514 (2005)
43. Ben Taieb, S., Bontempi, G., Atiya, A., Sorjamaa, A.: A review and comparison of strategies for multi-step ahead time series forecasting based on the NN5 forecasting competition. Expert Systems with Applications 39, 7067–7083 (2012)
44. Atiya, A.F., Aly, M., Parlos, A.G.: Sparse basis selection: new results and application to adaptive video traffic flow forecasting. IEEE Transactions on Neural Networks 16, 1136–1146 (2005)
45. Mohamed, T.A., El Gayar, N., Atiya, A.F.: A co-training approach for time series prediction with missing data. In: Haindl, M., Kittler, J., Roli, F. (eds.) MCS 2007. LNCS, vol. 4472, pp. 93–102. Springer, Heidelberg (2007)

Dynamic Weighted Fusion of Adaptive Classifier Ensembles Based on Changing Data Streams

Christophe Pagano[1], Eric Granger[1], Robert Sabourin[1], Gian Luca Marcialis[2], and Fabio Roli[2]

[1] Laboratoire d'imagerie, de vision et d'intelligence artificielle
École de technologie supérieure
Université du Québec, Montreal, Canada
cpagano@livia.etsmtl.ca, {eric.granger,robert.sabourin}@etsmtl.ca
[2] Pattern Recognition and Applications Group
Dept. of Electrical and Electronic Engineering
University of Cagliari, Cagliari, Italy
{marcialis,roli}@diee.unica.it

Abstract. Adapting classification systems according to new input data streams raises several challenges in changing environments. Although several adaptive ensemble-based strategies have been proposed to preserve previously-acquired knowledge and reduce knowledge corruption, the fusion of multiple classifiers trained to represent different concepts can increase the uncertainty in prediction level, since only a sub-set of all classifier may be relevant. In this paper, a new score-level fusion technique, called S_{wavg_h}, is proposed where each classifier is dynamically weighted according to the similarity between an input pattern and the histogram representation of each concept present in the ensemble. During operations, the Hellinger distance between an input and the histogram representation of every previously-learned concept is computed, and the score of every classifier is weighted dynamically according to the resemblance to the underlying concept distribution. Simulation produced with synthetic problems indicate that the proposed fusion technique is able to increase system performance when input data streams incorporate abrupt concept changes, yet maintains a level of performance that is comparable to the average fusion rule when the changes are more gradual.

Keywords: Pattern Classification, Multi-Classifier Systems, Adaptive Systems, Dynamic Weighting, Score-Level Fusion, Change Detection.

1 Introduction

A challenge in many real-world pattern classification problems is processing data sampled from underlying class distributions changing over time. In this case input data may reflect various different concepts[1] that re-occur during operations.

---

[1] A concept can be defined as the underlying class distribution of data captured under specific operating conditions.

N. El Gayar et al. (Eds.): ANNPR 2014, LNAI 8774, pp. 105–116, 2014.
© Springer International Publishing Switzerland 2014

For example, in face recognition for video surveillance, the facial model of an individual of interest should be designed using representative reference data from all the possible capture conditions under which it can be observed in the operational environments. These include variations in faces captured under different pose angles, illuminations, resolution, etc. However, fully representative video sequences or sets of reference stills are rarely available a priori for system design, although they may be provided over the time. In this paper, adaptive ensembles methods are considered to perform supervised incremental learning from new reference data that exhibit changes. In this context, different concepts are learned gradually by training incremental learning classifiers using blocks of reference data. It is assumed that classifiers are not selected from a pool, but gradually initiated as new data concepts emerge in the reference data, and combined into an ensemble. During operations, these ensembles process input data streams that incorporate those different concepts.

To adapt a pattern classifier based on new reference data, several techniques have been proposed in the literature, which can be characterized by their level of the adaptation. Incremental learning classifiers (like ARTMAP [4] neural networks) are designed to adapt their internal parameters in response to new data, while ensemble methods allow to adapt the generation (i.e. the internal parameters of base classifiers), selection and fusion of an ensemble of classifiers (EoC) [10]. Incremental classifiers can be updated efficiently, but learning over the time from reference data that represent significantly different concepts can corrupt their previously acquired knowledge [5,16]. In addition to being more robust to problems with a limited number of references, ensemble methods such as the Learn++ algorithm [16] are well suited to prevent knowledge corruption when adapted to new data. Indeed, previously acquired knowledge (i.e. concepts) can be preserved by training a new classifier on new reference data.

Active approaches have been proposed in the literature to adapt classification systems to data sampled from changing class distributions. They exploit a change detection mechanism, such as the Diversity for Dealing with Drifts [13], to drive incremental learning. Other methods such as the Just-in-Time architecture use change detection to regroup reference templates per concept [1]. Recently, Pagano et al. [15] proposed an active approach based the Hellinger drift detection method (HDDM) [7], where an ensemble of incremental learning classifiers is updated depending on the nature of the detected change. If newly available data exhibit small or gradual changes w.r.t. previously-learned concepts, the incremental classifier trained for the closest concept is updated. Otherwise, a new classifier trained on the newly available data is added to the ensemble. This method ensures that every classifier in the ensemble represents a different concept.

Such adaptive ensemble methods may allow to mitigate knowledge corruption, since every concept encountered in reference data may be represented in the ensemble. However, the fusion of classifiers during operations is a critical mechanism to exploit this information. A fusion function for an EoC in changing environments should be robust to a growing number of classifiers, and adapt dynamically to changing concepts that appear in new reference data. Given specific operating conditions, every member of the ensemble is susceptible to incorporate valuable knowledge to classify a given input pattern, but exploitation of this information relies on the way the outputs are combined. For example, to recognize faces captured in video feeds from a profile view, the knowledge of the classifiers trained with profile views is more valuable than the ones trained with frontal views. Depending on the operational input to the EoC, some classifiers may provide more reliable predictions, and considering too many unrelated classifiers can increase the uncertainty of the overall system [9].

In this paper, a concept-based dynamic weighting technique is proposed for score-level fusion that accounts for concepts related to each pattern from an input data stream. This weighted average fusion technique is referred to as S_{wavg_h}. Assuming an adaptive ensemble system (such as presented in [15]), composed of K classifiers $\{C_1, ..., C_K\}$ associated to histogram representations $\{\mathcal{H}_1, ..., \mathcal{H}_K\}$ of reference data captured from the concept they incorporate. The adaptation of the EoC to new reference data is guided by change detection using HDDM [7]. Several dynamic selection methods have been proposed in the literature, where a subset of classifiers is considered to be competent to classify a given operational input. Although regions of competence can be evaluated dynamically, additional training data is required, and their representativeness is crucial to avoid over-training [3,10]. As the availability of such data may not be guaranteed, S_{wavg_h} will rely on information available on the concepts represented in the EoC, the histogram representations $\{\mathcal{H}_1, ..., \mathcal{H}_K\}$, to weight the decisions of each classifiers.

To account for their possible limited representativeness, S_{wavg_h} proposes to benefit from the diversity of opinion of the EoC, through a dynamic weighting of every classifier of the ensemble instead of a subset selection. More precisely, during operations, the proposed score-level fusion strategy will dynamically weight classification scores $s_{i,k}(\mathbf{q})$ (for $i = 1, ..., N$ classes, and $k = 1, ..., K$ classifiers in the ensemble) for input pattern $\mathbf{q}$, according on classifier relevance or knowledge. The Hellinger distance between $\mathbf{q}$ and every concept representation $\mathcal{H}_k$ of the data learned by the classifier C_k is computed, and used to weight its decision. This allows to dynamically favour the decision of classifiers trained with reference data close to every input.

To evaluate the performance of the proposed fusion, an ensemble of incremental learning Probabilistic Fuzzy ARTMAP (PFAM) [12] classifiers is trained on synthetic problems, providing reference data blocks over the time. The two problems generate reference input streams exhibiting gradual and abrupt patterns of change, and testing input streams incorporating every possible concept.

2 Adaptive Ensembles Methods

Fig. 1 represents a block diagram of a generic active adaptive EoC system based on change detection. New reference data for design and update is provided as data blocks $D[t]$ at time steps $t = 1, ..., T$. During the design of the classifier

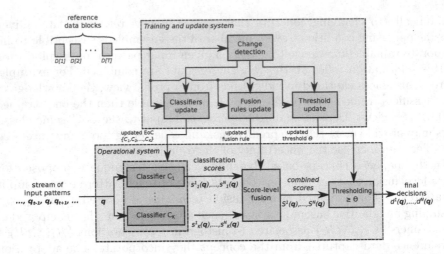

Fig. 1. An active adaptive ensemble system learning from a stream of data blocks

system, at each time step t, a new block of reference data $D[t]$ is processed by the training and update system. Depending on the nature of the change (which can be detected using distribution models or classification rules [15]), the classifiers of the ensemble, the selection and fusion rules and the decision threshold can be updated. During operations, an input pattern $\mathbf{q}$ from an input data stream is first processed by every classifier or the EoC to generate the scores $s_k^i(\mathbf{q})$ $(i = 1, ..., N, k = 1, ..., K)$. The overall scores $S^i(\mathbf{q})$ are then computed using the fusion rules, and compared to the decision threshold Θ to generate the final decisions $d^i(\mathbf{q})$.

Several passive approaches to ensemble adaptation, with varied ensemble generation, selection and fusion strategies have been proposed in the literature, adapting the fusion rule [2,17], the classifiers [8,6] or both at the same time [16]. Active approaches differ from the passive ones in their use of a change detection mechanism to drive ensemble adaptation. For example Minku et al. [13] proposed the Diversity for Dealing with Drifts algorithm, which maintains two ensembles with different diversity levels, one low and one high, in order to assimilate a new concept emerging in the observed data. When a significant change is detected though the monitoring of the system's error rate, the high diversity ensemble is used to assimilate new data and converge to a low diversity ensemble, and a new high diversity one is generated and maintained through bagging. Alippi et al. [1] also proposed a Just-in-Time classification algorithm, using a density-based change detection to regroup reference patterns per detected concept, and update an on-line classifier using this knowledge when the observed data drifts toward a known concept. More recently, Pagano et al. [15] proposed a ensemble generation strategy relying on a Hellinger drift detection method (HDDM) [7] change detection process. It is composed of an EoC of K 2-class classifiers $\{C_1, ..., C_K\}$, associated to set of histogram representations $\{\mathcal{H}_1, ..., \mathcal{H}_K\}$ of the data blocks used to train them, as well as a decision threshold Θ. During training, at each time step t, the HDDM is used to detect whether the histogram representation $\mathcal{H}$ of the reference data block $D[t]$ represents an abrupt change w.r.t. all the previously-learned data, represented by $\{\mathcal{H}_1, ..., \mathcal{H}_K\}$. If an abrupt change is detected, a new classifier C_{K+1} is trained with the data from $D[t]$ and added to the ensemble, the corresponding histogram representation $\mathcal{H}_{K+1} = \mathcal{H}$ is stored, and the decision threshold Θ is updated. If no change is detected, the reference patterns in $D[t]$ are used to update to classifier C_{k*} corresponding to the closest concept representation $\mathcal{H}_{k*}$ to $\mathcal{H}$. This methodology ensures the generation of an ensemble composed by classifiers representing different concepts.

These adaptive ensembles are generated to provide a diverse representation of all the previously-encountered concepts. However, a fusion rule for these ensembles should also adapt to changing concepts that appear during operations.

3   Dynamic Weighting Based on Concepts

This paper focuses on active adaptive ensembles based on HDDM for change detection [15]. It is assumed that, during design phases, each classifier has been trained or adapted using reference data that corresponds to one concept. These ensembles are composed of classifiers producing continuous scores $s \in [0, 1]$. During operations, a input pattern $\mathbf{q}$ is presented to the system, and the scores $\{s_1^i(\mathbf{q}), ..., s_K^i(\mathbf{q})\}$ are produced by each classifier. Those are combined to generate the overall score $S^i(\mathbf{q})$. Then, the final decision $d^i(\mathbf{q})$ is true for class i if $S^i(\mathbf{q}) \geq \Theta$.

In this paper, a the score-level fusion technique S_{wavg_h} is proposed, where each score is dynamically weighted based on a input pattern's resemblance to the corresponding classifier concept. Weights are defined by the Hellinger distance $\delta_h(\mathbf{q}, \mathcal{H}_k)$ between the pattern $\mathbf{q}$ and the histogram concept representation $\mathcal{H}_k$. Although representing different concepts, the classifiers of the ensemble are somewhat correlated as they model a part of the underlying distribution. Following recommendations in [9], average score-level fusion rule should be used for fusion. In order to account for the concept represented by each classifier, S_{wavg_h} increases the influence of the classifiers representing the closest concept to $\mathbf{q}$ in a weighted average following:

$$S_{wavg_h}^i(\mathbf{q}) = \sum_{k=1}^{K} \frac{(1 - \delta_h(\mathbf{q}, \mathcal{H}_k)) \cdot s_k^i(\mathbf{q})}{\sum_{k=1}^{K}(1 - \delta_h(\mathbf{q}, \mathcal{H}_k))} \tag{1}$$

4   Experimental Methodology

In this paper, a class-modular architecture, with an adaptive ensemble per class, is considered for proof of concept validation [14]. Two-class classifiers are trained on patterns sampled from the target class, versus a random mixture of non-target samples (from other and unknown classes). Following those considerations, only one target class is considered in experiments, and the class subscript i is omitted.

4.1 Simulation Scenario

Labelled blocks of data $D[t]$ become available to the system for design and update, at time $t = 1, 2..., T$. These blocks are comprised of reference patterns from the target and the non-target classes, generated from two 2-class synthetic classification problems represented in 2D space. For each problem, the introduction of the new data block $D[t], t = 2, ..., T$, represents a gradual or an abrupt change w.r.t. to the patterns from the previous ones. These are sampled from a different concept than the ones previously modelled by the system. Each block has been completed with an equal number of non-target patterns to design the classifiers. At each time step t, the systems are tested with the same database $dTest_g$ containing a mixture of patterns from all the possible concepts.

4.2 Synthetic Problems

In order to evaluate the proposed fusion, two synthetic problems are considered: (1) the rotating Gaussians problem, where new data blocks represents gradual changes w.r.t. the previous ones, and (2), the checkerboard Gaussians, introducing more abrupt changes.

The Rotating Gaussians. Fig. 2 presents the rotating Gaussians problem, inspired by the drifting Gaussians problem presented in [7]. Patterns are sampled from two spherical Gaussian distributions originally centred on $(0.25, 0.5)$ and

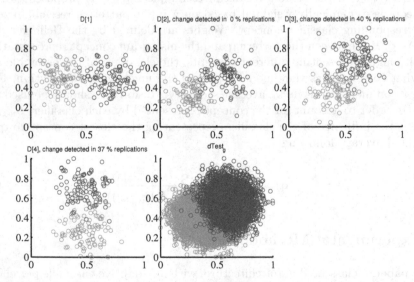

Fig. 2. Rotating Gaussians problem. Green (light) points correspond to target patterns, and red (dark) points to non-target ones. Abrupt changes between each block and the previous ones have been detected using the Hellinger drift detection methods for an histogram size of 15 bins, for 30 replications of random data generation.

$(0.75, 0.5)$ at $t = 1$, with a standard deviation of 0.15 in each direction, for $T = 4$ blocks. From $t = 2$ to $t = 4$, the centers gradually rotate of $\pi/2$ around $(0.5, 0.5)$. This problem provides gradual changes between the data blocks in order to evaluate the proposed fusion when classifiers are trained with close concepts. The gradual nature of the changes is confirmed by the results of a change detection performed for 30 replications (see Fig. 2), which didn't detect a new concept (abrupt change) in more than 50% of the replications for the data blocks.

The Checkerboard Gaussian. Fig. 3 presents the checkerboard Gaussian problem, providing patterns sampled from 16 Gaussian distributions with a standard deviation of 0.07, arranged in a 4x4 checkerboard. A concept is defined as a group of two adjacent Gaussian distributions, which generates $T = 8$ different training blocks. The presentation order of the 8 concepts has been randomized, as shown on Fig. 3. This problem enables to evaluate the proposed fusion rule when classifiers are trained with distinct concepts. Change detection performed for 30 replications (see Fig. 3) detected a new concept (abrupt change) in more than 50% of the replications for 4 data blocks.

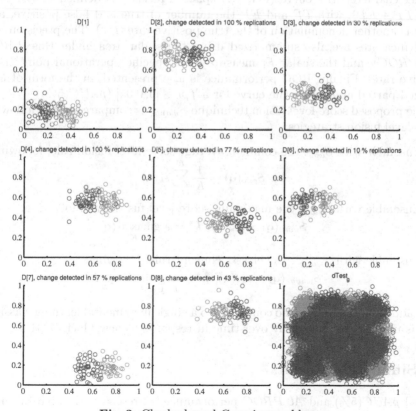

Fig. 3. Checkerboard Gaussian problem

4.3   Protocol for Simulations

Adaptive ensembles are composed of Probabilistic Fuzzy ARTMAP (PFAM) [12] classifiers. The DNPSO training strategy presented in [5] is used to train and optimize the PFAM classifiers, using a training dataset $dTrain[t]$ and validation dataset $dVal_1[t]$, and the decision threshold Θ is selected based on ROC curve produced by the full system over a second validation dataset $dVal_2[t]$, respecting the operational constraint $fpr \leq 5\%$.

At each time step t, $D[t]$ is thus composed by $dTrain[t]$, $dVal_1[t]$ and $dVal_2[t]$, with 25 patterns per class each. $dTest_g$ is composed of 50 patterns per class and per concept, and is fixed for each time step. As the dataset is randomly generated from the sources, the simulations have been repeated for 30 replications, and the performance is presented as average and standard deviations values computed using a Student distribution and a confidence interval of 10%. Histogram representation of concepts are computed using 15 bins.

4.4   Performance Measures and Reference Systems

To measure performance, systems are characterized in the precision-recall operating characteristics curve (P-ROC) space. Precision is defined as the ratio $TP/(TP + FP)$, with TP and FP the number of true and false positive, and recall is another denomination of the true positive rate (tpr). The precision and recall measures are also summarized using the scalar area under this P-ROC ($AUPROC$), and the scalar F_1 measure for a specific operational point (false positive rate). Finally, ROC performance is also presented, in the form of normalized partial area under the curve for a $fpr \in [0, 0.05]$ ($pAUC(5\%)$)

The proposed score-level fusion technique S_{wavg_h} is compared to the following score-level fusion strategies:

— Ensemble with S_{avg}, the common average score-level fusion rule, following:

$$S_{avg}(\mathbf{q}) = \frac{1}{K} \sum_{k=1}^{K} s_k(\mathbf{q}) \tag{2}$$

— Ensemble with S_{max}, the maximum score-level fusion rule, following:

$$S_{max}(\mathbf{q}) = s_{k*}(\mathbf{q}), \ k^* = \underset{k=1,\dots,K}{\operatorname{argmax}} \ s_k(\mathbf{q}) \tag{3}$$

— Ensemble with S_{wavg_c}, the closest concept rule, following:

$$S_{wavg_c}(\mathbf{q}) = s_{k*}(\mathbf{q}), \ k^* = \underset{k=1,\dots,K}{\operatorname{argmin}} \ \delta_h(\mathbf{q}, \mathcal{H}_k) \tag{4}$$

In addition, the EoC is also compared to a single incremental learning classifier that is updated incrementally over time in response to new blocks $D[t]$.

5   Simulation Results

Global $pAUC(5\%)$ and $AUPROC$ performance is presented in Table 1a and b for the Rotating Gaussians problem. In terms of $pAUC(5\%)$, it can be observed

Table 1. ROC and PR performance of adaptive classification systems for the Rotating Gaussians problem

(a) $pAUC(5\%)$ measure

Adaptive classifiers	Time Step			
	1	2	3	4
Incr. classifier	0.23 ± 0.08	0.40 ± 0.06	0.47 ± 0.05	0.35 ± 0.05
Ensemble with S_{max}	0.30 ± 0.07	0.35 ± 0.08	0.40 ± 0.08	0.37 ± 0.08
Ensemble with S_{avg}	0.30 ± 0.07	0.48 ± 0.06	0.52 ± 0.06	0.52 ± 0.04
Ensemble with S_{avg_h}	0.30 ± 0.07	0.33 ± 0.10	0.43 ± 0.09	0.43 ± 0.09
Ensemble with S_{avg_c}	0.30 ± 0.07	0.23 ± 0.10	0.14 ± 0.05	0.08 ± 0.04

(b) $AUPROC$ measure

Adaptive classifiers	Time Step			
	1	2	3	4
Incr. classifier	0.77 ± 0.04	0.83 ± 0.05	0.89 ± 0.02	0.78 ± 0.04
Ensemble with S_{max}	0.80 ± 0.04	0.80 ± 0.05	0.81 ± 0.04	0.77 ± 0.06
Ensemble with S_{avg}	0.80 ± 0.04	0.92 ± 0.01	0.94 ± 0.01	0.93 ± 0.01
Ensemble with S_{avg_h}	0.79 ± 0.04	0.87 ± 0.03	0.91 ± 0.02	0.91 ± 0.03
Ensemble with S_{avg_c}	0.79 ± 0.04	0.78 ± 0.06	0.78 ± 0.05	0.73 ± 0.04

(c) F_1 measure for a $fpr \leq 5\%$ on validation data

Adaptive classifiers	Time Step			
	1	2	3	4
Incr. classifier	0.74 ± 0.04	0.75 ± 0.06	0.83 ± 0.02	0.66 ± 0.07
Ensemble with S_{max}	0.80 ± 0.03	0.78 ± 0.06	0.76 ± 0.08	0.64 ± 0.12
Ensemble with S_{avg}	0.80 ± 0.03	0.82 ± 0.02	0.84 ± 0.03	0.80 ± 0.06
Ensemble with S_{avg_h}	0.80 ± 0.03	0.74 ± 0.07	0.77 ± 0.06	0.75 ± 0.10
Ensemble with S_{avg_c}	0.80 ± 0.03	0.71 ± 0.10	0.55 ± 0.14	0.22 ± 0.10

that the single incremental classifier and adaptive ensembles with S_{max}, S_{avg} and S_{wavg_h} have statistically similar performances at $t = 2$ and $t = 3$, significantly higher than S_{wavg_c}. However, at $t = 4$, S_{avg} and S_{wavg_h} remain stable, while the single incremental classifier and S_{max} significantly decline. In terms of $AUPROC$, S_{avg} and S_{wavg_h} remain statistically higher than all the other systems, but cannot be differentiated. The same overall observations can be made with the F_1 performance at the selected operating point of $fpr \leq 5\%$ (Table 1c). S_{avg} and S_{wavg_h} performance remains significantly higher than the other systems.

Results indicate a correlation between the classifiers. Although trained with slightly different concepts, several classifiers still provide valuable information to classify the input patterns, and relying only on the closest classifier (S_{wavg_c}) or the maximum score (S_{max}) underperforms other approaches. In addition, it can be observed that when dealing with small (or gradual) changes, S_{wavg_h} is equivalent in performance to an average fusion rule.

Table 2. ROC and PR performance of adaptive classification systems for the Checkerboard Gaussians problem

(a) $pAUC(5\%)$ measure

Adaptive classifiers		Time Step							
		1	2	3	4	5	6	7	8
Incr. classifier		0.02 ± 0.00	**0.04 ±** 0.01	**0.05 ±** 0.01	0.03 ± 0.00	0.03 ± 0.01	0.03 ± 0.01	**0.06 ±** 0.01	0.09 ± 0.01
Ensemble with	S_{max}	0.03 ± 0.00	0.03 ± 0.00	0.03 ± 0.00	0.02 ± 0.00	0.02 ± 0.00	0.02 ± 0.00	0.02 ± 0.00	0.03 ± 0.00
	S_{avg}	0.03 ± 0.00	**0.04 ±** 0.01	**0.05 ±** 0.01	**0.04 ±** 0.01	0.04 ± 0.01	0.04 ± 0.01	0.04 ± 0.01	0.04 ± 0.01
	S_{avg_h}	**0.04 ±** 0.01	**0.04 ±** 0.00	**0.04 ±** 0.01	**0.05 ±** 0.01	**0.06 ±** 0.01	**0.07 ±** 0.01	**0.08 ±** 0.01	**0.14 ±** 0.01
	S_{avg_c}	0.03 ± 0.00	0.02 ± 0.00	0.02 ± 0.00	0.02 ± 0.00	0.03 ± 0.00	0.03 ± 0.01	0.02 ± 0.00	0.03 ± 0.00

(b) $AUPROC$ measure

Adaptive classifiers		Time Step							
		1	2	3	4	5	6	7	8
Incr. classifier		0.42 ± 0.02	**0.51 ±** 0.01	**0.54 ±** 0.01	0.50 ± 0.01	0.51 ± 0.01	0.52 ± 0.01	0.59 ± 0.02	0.64 ± 0.02
Ensemble with	S_{max}	0.43 ± 0.02	0.34 ± 0.02	0.31 ± 0.01	0.30 ± 0.01	0.28 ± 0.01	0.26 ± 0.01	0.26 ± 0.00	0.25 ± 0.00
	S_{avg}	0.43 ± 0.02	**0.51 ±** 0.01	**0.54 ±** 0.01	0.52 ± 0.01	0.51 ± 0.01	0.51 ± 0.01	0.51 ± 0.01	0.53 ± 0.01
	S_{avg_h}	0.47 ± 0.01	**0.51 ±** 0.01	**0.55 ±** 0.01	**0.58 ±** 0.01	**0.60 ±** 0.01	**0.61 ±** 0.01	**0.63 ±** 0.01	**0.67 ±** 0.01
	S_{avg_c}	0.43 ± 0.02	0.46 ± 0.01	0.48 ± 0.01	0.49 ± 0.01	0.52 ± 0.01	0.53 ± 0.01	0.51 ± 0.01	0.52 ± 0.01

(c) F_1 measure for a $fpr \leq 5\%$ on validation data

Adaptive classifiers		Time Step							
		1	2	3	4	5	6	7	8
Incr. classifier		**0.51 ±** 0.04	**0.44 ±** 0.02	0.22 ± 0.03	**0.40 ±** 0.02	0.30 ± 0.02	0.18 ± 0.02	**0.23 ±** 0.02	0.21 ± 0.02
Ensemble with	S_{max}	**0.53 ±** 0.03	0.21 ± 0.05	0.11 ± 0.01	0.11 ± 0.01	0.10 ± 0.01	0.09 ± 0.00	0.09 ± 0.00	0.09 ± 0.00
	S_{avg}	**0.53 ±** 0.03	**0.39 ±** 0.04	0.21 ± 0.04	0.18 ± 0.02	0.16 ± 0.02	0.12 ± 0.02	0.12 ± 0.02	0.13 ± 0.02
	S_{avg_h}	0.42 ± 0.03	**0.44 ±** 0.02	**0.34 ±** 0.02	**0.39 ±** 0.02	**0.42 ±** 0.02	**0.38 ±** 0.02	**0.23 ±** 0.02	**0.36 ±** 0.01
	S_{avg_c}	**0.53 ±** 0.03	0.31 ± 0.05	0.26 ± 0.04	0.15 ± 0.03	0.10 ± 0.01	0.10 ± 0.01	0.09 ± 0.01	0.08 ± 0.01

Global $pAUC(5\%)$ and $AUPROC$ performance is presented in Table 2a and b for the Checkerboard Gaussian problem. With both measures, it can be observed that the single incremental classifier and adaptive ensembles with S_{avg} and S_{wavg_h} present statistically similar performances above the other systems, up to $t = 3$. Then, from $t = 4$ until the end of the simulation, S_{wavg_h} significantly outperforms the others. Similar observations can be made from the F_1 performance at the selected operating point of $fpr \leq 5\%$ (Table 2c), where S_{wavg_h} exhibits significantly better performance from $t = 3$ until the end of the simulation.

As with the Rotating Gaussians problem, S_{wavg_c} exhibits considerably lower performance, which underlings the importance of combining all the classifiers of the EoC, even if they've been trained with different concepts. However, as the Checkerboard Gaussians problem provides more numerous and distinct concepts, the performance of S_{wavg_h} finally exceeded the average fusion rule, through a dynamic adaptation of the influence of each classifier depending on the concept which which they have been trained.

6 Conclusion

In this paper a new concept-based dynamically weighted score-level average fusion technique is proposed for adaptive ensemble-based classification. Assuming an adaptive ensemble system that grows to incrementally to learn different concepts from blocks of input data, and which update rule relies on histogram-based change detection, this fusion function dynamically update the weights of the classifiers based on the input pattern resemblance to each classifier concept. More precisely, these weights are defined by the Hellinger distance between the input pattern and the histogram representation of the concepts learned by the system.

Simulation results produced with synthetic problems indicate that the proposed fusion outperforms the average score-level fusion rule when learning from data streams incorporating abrupt concept changes. In addition, it maintains a comparable level of performance when the changes are more gradual.

In this paper, a classification scenario only presenting changes in the input stream of reference patterns is considered. In future work, pruning strategies should be evaluated to deal with real-world changing operational environments. Indeed, previously-learned concept might become irrelevant over time in those conditions. In addition, the proposed fusion strategy should be tested with larger problems incorporating more concepts, in order to evaluate its robustness when the number of classifiers in the ensemble increases.

Acknowledgements. This work was partially supported by the Natural Sciences and Engineering Research Council of Canada, and the Ministère du Développement Économique, de l'Innovation et de l'Exportation du Québec.

References

1. Alippi, C., Boracchi, G., Roveri, M.: Just-In-Time Classifiers for Recurrent Concepts. IEEE Trans. Neural Networks and Learning Systems 24(4), 620–634 (2013)
2. Blum, A.: Empirical Support for Winnow and Weighted-Majority Algorithms: Results on a Calendar Scheduling Domain. Machine Learning 26(1), 5–23 (1997)
3. Britto Jr., A.S., Sabourin, R., Oliveira, L.S.: Dynamic Selection of Classifiers - a Comprehensive Review. Pattern Recognition, (May 17, 2014), http://dx.doi.org/10.1016/j.patcog.2014.05.003, ISSN: 0031-3203
4. Carpenter, G.A., Grossberg, S., Markuzon, N., Reynolds, J.H., Rosen, D.B.: Fuzzy ARTMAP: A Neural Network Architecture for Incremental Supervised Learning of Analog Multidimensional Maps. IEEE Trans. Neural Networks 3(5), 698–713 (1992)
5. Connolly, J.F., Granger, E., Sabourin, R.: An Adaptive Classification System for Video-Based Face Recognition. Information Sciences 192, 50–70 (2012)
6. Connolly, J.F., Granger, E., Sabourin, R.: Dynamic Multi-Objective Evolution of Classifier Ensembles For Video-Based Face Recognition. Applied Soft Computing 13(6), 3149–3166 (2013)
7. Ditzler, G., Polikar, R.: Hellinger Distance Based Drift Detection for Nonstationary Environments. In: IEEE Sym. on Comp. Intelligence in Dynamic and Uncertain Environments, Paris, France, April 11-15, pp. 41–48 (2011)
8. Gama, J., Medas, P., Castillo, G., Rodrigues, P.: Learning with Drift Detection Advances in Artificial Intelligence. In: Proc. of 17th Brazilian Symp. on Artificial Intelligence, Sao Luis, Brazil, September 29-October 1, pp. 286–295 (2004)
9. Kittler, J., Alkoot, F.M.: Sum Versus Vote Fusion in Multiple Classifier Systems. Trans. Pat. Analysis and Machine Intelligence 25(1), 110–115 (2003)
10. Kuncheva, L.I.: Combining Pattern Classifiers: Methods and Algorithms. Wiley (2004)
11. Kuncheva, L.I.: Classifier Ensembles for Changing Environments. In: Proc. Int'l. Workshop on MCS, Cagliari, Italy, pp. 1–15 (2004)
12. Lim, C.P., Harrison, R.F.: Probabilistic Fuzzy ARTMAP: an Autonomous Neural Network Architecture for Bayesian Probability Estimation. Proc. Int'l. Conf. on Artificial Neural Networks, pp. 148–153 (1995)
13. Minku, L.L., Yao, X.: DDD: A New Ensemble Approach for Dealing With Concept Drift. IEEE Trans. on Knowledge and Data Engineering 24(4) (2012)
14. Pagano, C., Granger, E., Sabourin, R., Gorodnichy, D.: Detector Ensembles for Face Recognition in Video Surveillance. In: Int. J. Conf. on Neural Networks, Brisbane, pp. 1–8 (2012)
15. Pagano, C., Granger, E., Sabourin, R., Roli, F., Marcialis, G.L.: Adaptive Ensembles for Face Recognition in Changing Video Surveillance Environments. Information Sciences (in press since July 2014)
16. Polikar, R., Upda, L., Upda, S.S., Honavar, V.: Learn++: an Incremental Learning Algorithm for Supervised Neural Networks. IEEE Trans. Systems, Man and Cybernetics 31, 497–508 (2001)
17. Xingquan, Z., Xindong, W., Ying, Y.: Dynamic Classifier Selection for Effective Mining from Noisy Data Streams. In: Proc. IEEE Int'l Conf. on Data Mining, pp. 305–312 (2004)

Combining Bipartite Graph Matching and Beam Search for Graph Edit Distance Approximation

Kaspar Riesen[1], Andreas Fischer[2], and Horst Bunke[3]

[1] Institute for Information Systems, University of Applied Sciences and Arts
Northwestern Switzerland, Riggenbachstrasse 16, 4600 Olten, Switzerland
kaspar.riesen@fhnw.ch

[2] Biomedical Science and Technologies Research Centre, Polytechnique Montreal
2500 Chemin de Polytechnique, Montreal H3T 1J4, Canada
andreas.fischer@polymtl.ca

[3] Institute of Computer Science and Applied Mathematics, University of Bern,
Neubrückstrasse 10, 3012 Bern, Switzerland
bunke@iam.ch

Abstract. Graph edit distance (GED) is a powerful and flexible graph dissimilarity model. Yet, exact computation of GED is an instance of a quadratic assignment problem and can thus be solved in exponential time complexity only. A previously introduced approximation framework reduces the computation of GED to an instance of a linear sum assignment problem. Major benefit of this reduction is that an optimal assignment of nodes (including local structures) can be computed in polynomial time. Given this assignment an approximate value of GED can be immediately derived. Yet, the primary optimization process of this approximation framework is able to consider local edge structures only, and thus, the observed speed up is at the expense of approximative, rather than exact, distance values. In order to improve the overall approximation quality, the present paper combines the original approximation framework with a fast tree search procedure. More precisely, we regard the assignment from the original approximation as a starting point for a subsequent beam search. In an experimental evaluation on three real world data sets a substantial gain of assignment accuracy can be observed while the run time remains remarkable low.

1 Introduction

Graphs, which consist of a finite set of nodes connected by edges, are the most general data structure in computer science. Due to the ability of graphs to represent properties of entities and binary relations at the same time, a growing interest in graph-based object representation can be observed in various fields. In bio- and chemoinformatics, for instance, graph based representations are intensively used [1–3]. Another field of research where graphs have been studied with emerging interest is that of web content and data mining [4, 5]. Image classification [6, 7], graphical symbol and character recognition [8, 9], and computer network analysis [10] are further areas of research where graph based representations draw the attention.

N. El Gayar et al. (Eds.): ANNPR 2014, LNAI 8774, pp. 117–128, 2014.
© Springer International Publishing Switzerland 2014

Various procedures for evaluating the similarity or dissimilarity of graphs – known as *graph matching* – have been proposed in the literature [11]. The present paper addresses the issue of processing arbitrarily structured and arbitrarily labeled graphs. Hence, the graph matching method actually employed has to be able to cope with directed and undirected, as well as with labeled and unlabeled graphs. If there are labels on nodes, edges, or both, no constraints on the label alphabet should compromise the representational power of the employed graphs. Anyhow, the matching framework should in any case be flexible enough to be adopted and tailored to certain problem specifications. As it turns out, *graph edit distance* [12, 13] meets both requirements, viz. flexibility and expressiveness.

The major drawback of graph edit distance is its computational complexity which is is exponential in the number of nodes of the involved graphs. Consequently, exact edit distance can be computed for graphs of a rather small size only. In recent years, a number of methods addressing the high computational complexity of graph edit distance computation have been proposed (e.g. [14–17]). The authors of the present paper also introduced an algorithmic framework which allows the approximate computation of graph edit distance in a substantially faster way than traditional methods [18]. Yet, the substantial speed-up in computation time is at the expense of an overestimation of the actual graph edit distance.

The reason for this overestimation is that the core of our framework is able to consider only local, rather than global, edge structure. The main objective of the present paper is to significantly reduce the overestimation of edit distances. To this end, the distance approximation procedure of [18] is combined with a fast (but suboptimal) tree search algorithm, namely *beam search*. Beam search has been employed before as a stand-alone approximation scheme for graph edit distance computation [17]. The present paper adapts this search algorithm for the task of systemically improving the original node assignment and the corresponding edit distance approximation.

The remainder of this paper is organized as follows. Next, in Sect. 2 the concept and computation of graph edit distance as well as the original framework for graph edit distance approximation [18] are summarized. In Sect. 3 the combination of this framework with a beam search procedure is introduced. An experimental evaluation on diverse data sets is carried out in Sect. 4, and in Sect. 5 we draw some conclusions.

2   Graph Edit Distance Computation

2.1   Exact Computation Based on A*

Given two graphs, the source graph g_1 and the target graph g_2, the basic idea of graph edit distance is to transform g_1 into g_2 using some distortion operations. A standard set of distortion operations is given by *insertions*, *deletions*, and *substitutions* of both nodes and edges. We denote the substitution of two nodes u and v by $(u \to v)$, the deletion of node u by $(u \to \varepsilon)$, and the insertion of node

v by $(\varepsilon \to v)^1$. A sequence $v = (e_1, \ldots, e_k)$ of k edit operations that transform g_1 completely into g_2 is called an *edit path* between g_1 and g_2.

Let $\Upsilon(g_1, g_2)$ denote the set of all possible edit paths between two graphs g_1 and g_2. To find the most suitable edit path out of $\Upsilon(g_1, g_2)$, one introduces a cost $c(e_i)$ for each edit operation $e_i \in v$, measuring the strength of the corresponding operation. The idea of such a cost is to define whether or not an edit operation represents a strong modification of the graph. Clearly, between two similar graphs, there should exist an inexpensive edit path, representing low cost operations, while for dissimilar graphs an edit path with high cost is needed. Consequently, the *edit distance* of two graphs is defined by the minimum cost edit path between two graphs:

$$d(g_1, g_2) = \min_{(e_1, \ldots, e_k) \in \Upsilon(g_1, g_2)} \sum_{i=1}^{k} c(e_i)$$

The exact computation of graph edit distance is usually carried out by means of a tree search algorithm which explores the space of all possible mappings of the nodes and edges of the first graph to the nodes and edges of the second graph. A widely used method is based on the A* algorithm [19]. The basic idea is to organize the underlying search space as an ordered tree. The root node of the search tree represents the starting point of our search procedure, inner nodes of the search tree correspond to partial edit paths, and leaf nodes represent complete – not necessarily optimal – edit paths.

Such a search tree is constructed dynamically at runtime as follows. The nodes of the source graph g_1 are processed in a fixed order $u_1, u_2, \ldots, u_n$. The deletion $(u_i \to \varepsilon)$ and all available substitutions $\{(u_i \to v_{(1)}), \ldots, (u_i \to v_{(t)})\}$ of a node u_i are thereby considered simultaneously. This produces $(t + 1)$ successor nodes in the search tree. If all nodes of the first graph have been processed in an inner node of the tree, the remaining nodes of the second graph are inserted in a single step (which completes the edit path).

A set *open* of partial edit paths contains the search tree nodes to be processed in the next steps. The most promising partial edit path $v \in open$, i.e. the one with minimal cost so far, is always chosen first (best-first search algorithm). This procedure guarantees that the complete edit path found by the algorithm first is always optimal in the sense of providing minimal cost among all possible competing paths.

2.2   Bipartite Graph Edit Distance Approximation

A major drawback of the procedure described in the last section is its computational complexity. In fact, the problem of graph edit distance can be reformulated as an instance of a *Quadratic Assignment Problem (QAP)* [20]. QAPs have been

---

[1] For edges we use a similar notation.

introduced in [21] and belong to the most difficult combinatorial optimization problems for which only exponential run time algorithms are known to date[2].

The graph edit distance approximation framework introduced in [18] reduces the QAP of graph edit distance computation to an instance of a *Linear Sum Assignment Problem* (*LSAP*) which can be – in contrast with QAPs – efficiently solved.

In order to translate the problem of graph edit distance computation to an instance of an LSAP, the graphs to be matched are subdivided into individual nodes plus local structures in a first step. Next, these independent sets of nodes including local structures are optimally assigned to each other. Finally, an approximate graph edit distance value is derived from this optimal node assignment. In the next paragraphs of this section, these three major steps of our framework are discussed in greater detail.

Assume that the graphs to be matched consists of node sets $V_1 = \{u_1, \ldots, u_n\}$ and $V_2 = \{v_1, \ldots, v_m\}$, respectively. A cost matrix $\mathbf{C}$ is then defined as follows:

$$
\mathbf{C} =
\begin{bmatrix}
c_{11} & c_{12} & \cdots & c_{1m} & c_{1\varepsilon} & \infty & \cdots & \infty \\
c_{21} & c_{22} & \cdots & c_{2m} & \infty & c_{2\varepsilon} & \ddots & \vdots \\
\vdots & \vdots & \ddots & \vdots & \vdots & \ddots & \ddots & \infty \\
c_{n1} & c_{n2} & \cdots & c_{nm} & \infty & \cdots & \infty & c_{n\varepsilon} \\
c_{\varepsilon 1} & \infty & \cdots & \infty & 0 & 0 & \cdots & 0 \\
\infty & c_{\varepsilon 2} & \ddots & \vdots & 0 & 0 & \ddots & \vdots \\
\vdots & \ddots & \ddots & \infty & \vdots & \ddots & \ddots & 0 \\
\infty & \cdots & \infty & c_{\varepsilon m} & 0 & \cdots & 0 & 0
\end{bmatrix}
$$

Entry c_{ij} thereby denotes the cost of a node substitution $(u_i \to v_j)$, $c_{i\varepsilon}$ denotes the cost of a node deletion $(u_i \to \varepsilon)$, and $c_{\varepsilon j}$ denotes the cost of a node insertion $(\varepsilon \to v_j)$.

Obviously, the left upper corner of the cost matrix represents the costs of all possible node substitutions, the diagonal of the right upper corner the costs of all possible node deletions, and the diagonal of the bottom left corner the costs of all possible node insertions. Note that each node can be deleted or inserted at most once. Therefore any non-diagonal element of the right-upper and left-lower part is set to ∞. The bottom right corner of the cost matrix is set to zero since substitutions of the form $(\varepsilon \to \varepsilon)$ should not cause any costs.

Note that the described extension of cost matrix $\mathbf{C}$ to dimension $(n + m) \times (n + m)$ is necessary since assignment algorithms for LSAPs expect every entry of the first set to be assigned with exactly one entry of the second set (and vice versa), and we want the optimal matching to be able to possibly include several node deletions and/or insertions. Moreover, matrix $\mathbf{C}$ is by definition quadratic. Consequently, standard algorithms for LSAPs can be used to find the minimum cost assignment.

In order to integrate knowledge about the graph's edge structure, to each cost of a node edit operation c_{ij} the minimum sum of edge edit operation costs,

---

[2] QAPs belong to the class of $\mathcal{NP}$-*complete* problems. That is, an exact and efficient algorithm for the graph edit distance problem can not be developed unless $\mathcal{P} = \mathcal{NP}$.

implied by the corresponding node operation, is added. That is, we encode the matching cost arising from the local edge structure in the individual entries of matrix $\mathbf{C}$.

The second step of our framework consists in applying an assignment algorithm to the square cost matrix $\mathbf{C}$ in order to find the minimum cost assignment of the nodes and their local edge structure of g_1 to the nodes and their local edge structure of g_2. Note that this task exactly corresponds to an instance of an LSAP and can thus be solved in polynomial time by means of Munkres' algorithm [22], the algorithm of Volgenant-Jonker [23], or others [24][3].

Formally, LSAP optimization procedures operate on a cost matrix $\mathbf{C} = (c_{ij})$ and find a permutation $(\varphi_1, \ldots, \varphi_{n+m})$ of the integers $(1, 2, \ldots, (n + m))$ that minimizes the overall mapping cost $\sum_{i=1}^{(n+m)} c_{i\varphi_i}$. In our scenario, this permutation corresponds to a mapping

$$\psi = \{(u_1 \to v_{\varphi_1}), (u_2 \to v_{\varphi_2}), \ldots, (u_{m+n} \to v_{\varphi_{m+n}})\}$$

of nodes. Note that mapping ψ includes node assignments of the form $(u_i \to v_j)$, $(u_i \to \varepsilon)$, $(\varepsilon \to v_j)$, and $(\varepsilon \to \varepsilon)$ (the latter can be dismissed, of course). Mapping ψ can also be interpreted as partial edit path considering edit operations on nodes only.

In the third step of our framework the partial edit path ψ is completed according to the node edit operations. Note that edit operations on edges are implied by edit operations on their adjacent nodes, i.e. whether an edge is substituted, deleted, or inserted, depends on the edit operations performed on all of its adjacent nodes. Hence, given the set of node operations in ψ the global edge structures from g_1 and g_2 can be edited accordingly. The cost of the complete edit path is finally returned as an approximate graph edit distance. We denote the approximated distance value between graphs g_1 and g_2 according to mapping ψ with $d_{\langle\psi\rangle}(g_1, g_2)$ (or $d_{\langle\psi\rangle}$ for short).

Note that the edit path corresponding to $d_{\langle\psi\rangle}(g_1, g_2)$ considers the edge structure of g_1 and g_2 in a global and consistent way while the optimal node mapping ψ from step 2 is able to consider the structural information in an isolated way only (single nodes and their adjacent edges). Hence, the distances found by this approximation framework are – in the optimal case – equal to, or – in a suboptimal case – larger than the exact graph edit distance. Yet, the proposed reduction of graph edit distance to an LSAP allows the approximate graph edit distance computation in polynomial time complexity. For the remainder of this paper we denote this graph edit distance approximation algorithm with *BP* (*Bipartite*).

3   Improving the Node Assignment Using Beam Search

In an experimental evaluation in [18] we observed that the overestimation of BP is very often due to a few incorrectly assigned nodes in ψ. That is, only few node

---

[3] In [18] Munkres' algorithm is deployed, while in [25] also other algorithms have been tested for graph edit distance approximation.

assignments from the second step are responsible for additional (unnecessary) edge operations in the third step (and the resulting overestimation of the true edit distance). Our novel procedure ties in at this observation. That is, the node assignment ψ of our framework is used as a starting point for a subsequent search in order to improve the quality of the distance approximation (rather than using the assignment for graph edit distance approximation directly).

The basic idea of our search procedure is that the original node assignment ψ is systematically varied by swapping the target nodes v_{φ_i} and v_{φ_j} of two node assignments $(u_i \to v_{\varphi_i}) \in \psi$ and $(u_j \to v_{\varphi_j}) \in \psi$. For each swap it is verified whether (and to what extent) the derived distance approximation stagnates, increases or decreases. For a systematic variation of mapping ψ a tree search with ψ as the starting point is carried out.

The tree nodes in our search procedure correspond to triples $(\psi, q, d_{\langle\psi\rangle})$, where ψ is a certain node assignment, q denotes the depth of the tree node in the search tree and $d_{\langle\psi\rangle}$ is the approximate distance value corresponding to ψ. The root node of the search tree refers to the optimal node assignment

$$\psi = \{(u_1 \to v_{\varphi_1}), (u_2 \to v_{\varphi_2}), \ldots, (u_{m+n} \to v_{\varphi_{m+n}})\}$$

found by our former algorithm BP. Hence, the root node (with depth = 0) is given by the triple $(\psi, 0, d_{\langle\psi\rangle})$. Subsequent tree nodes $(\psi', q, d_{\langle\psi'\rangle})$ with depth $q = 1, \ldots, (m + n)$ contain node assignments ψ' with swapped element $(u_q \to v_{\varphi_q})$.

As usual in tree search based methods, a set *open* is employed that holds all of the unprocessed tree nodes. We keep the tree nodes in *open* sorted in ascending order according to their depth in the search tree (known as *breadth-first search*). Thus, at position 1 of *open* the tree node with smallest depth among all unprocessed tree nodes can be found. As a second order criterion the approximate edit distance $d_{\langle\psi\rangle}$ is used. That is, if two tree nodes have same depth in the search tree, they are queued in *open* according to ascending distance values.

Note that a best-first search algorithm, where *open* is sorted in ascending order according to the cost of the respective solution, would not be suitable for the present task. Best-first search algorithms expect that the cost of a solution increases monotonically with the increase of the depth in the search tree. Obviously, this is not the case in our scenario since for two tree nodes $(\psi', q', d_{\langle\psi'\rangle})$ and $(\psi'', q'', d_{\langle\psi''\rangle})$ with $q' < q''$, it must not necessarily hold that $d_{\langle\psi'\rangle} < d_{\langle\psi''\rangle}$. This is due to the fact that each tree node in the search tree represents a complete node mapping with the corresponding graph edit distance approximation value (in contrast with exact computations of graph edit distance, where inner tree nodes always refer to incomplete mappings).

The extended framework BP with the tree search based improvement is given in Alg. 1 (the first three lines correspond to the three major steps of the original approximation). Before the main loop of the search procedure starts, *open* is initialized with the root node (line 4). As long as *open* is not empty, we retrieve (and remove) the triple $(\psi, q, d_{\langle\psi\rangle})$ at the first position in *open* (the one with

Algorithm 1. BP-Beam(g_1, g_2) (Meta Parameter: b)

1. Build cost matrix $\mathbf{C} = (c_{ij})$ according to the input graphs g_1 and g_2
2. Compute optimal node assignment $\psi = \{u_1 \rightarrow v_{\varphi_1}, u_2 \rightarrow v_{\varphi_2}, \ldots, u_{m+n} \rightarrow v_{\varphi_{m+n}}\}$ on $\mathbf{C}$
3. $d_{best} = d_{\langle\psi\rangle}(g_1, g_2)$
4. Initialize $open = \{(\psi, 0, d_{\langle\psi\rangle}(g_1, g_2))\}$
5. **while** $open$ is not empty **do**
6.   Remove first tree node in $open$: $(\psi, q, d_{\langle\psi\rangle}(g_1, g_2))$
7.   **for** $j = (q+1), \ldots, (m+n)$ **do**
8. $\psi' = \psi \setminus \{u_{q+1} \rightarrow v_{\varphi_{q+1}}, u_j \rightarrow v_{\varphi_j}\} \cup \{u_{q+1} \rightarrow v_{\varphi_j}, u_j \rightarrow v_{\varphi_{q+1}}\}$
9.   Derive approximate edit distance $d_{\langle\psi'\rangle}(g_1, g_2)$
10. $open = open \cup \{(\psi', q+1, d_{\langle\psi'\rangle}(g_1, g_2))\}$
11.   **if** $d_{\langle\psi'\rangle}(g_1, g_2) < d_{best}$ **then**
12. $d_{best} = d_{\langle\psi'\rangle}(g_1, g_2)$
13.   **end if**
14.   **end for**
15.   **while** size of $open > b$ **do**
16.   Remove tree node with highest approximation value $d_{\langle\psi\rangle}$ from $open$
17.   **end while**
18. **end while**
19. **return** d_{best}

minimal depth and minimal distance value), generate the successors of this specific tree node and add them to $open$ (line 6 – 10). That is, similarly to exact computation of the graph edit distance the search tree is dynamically built at run time.

The successors of tree node $(\psi, q, d_{\langle\psi\rangle})$ are generated as follows. The assignments of our original node matching ψ are processed according to the depth q of the current search tree node. That is, at depth q the assignment $u_q \rightarrow v_{\varphi_q}$ is processed and swapped with other assignments. Formally, in order to build the set of successor of node $(\psi, q, d_{\langle\psi\rangle})$ all pairs of node assignments $(u_{q+1} \rightarrow v_{\varphi_{q+1}})$ and $(u_j \rightarrow v_{\varphi_j})$ with $j = (q+1), \ldots, (n+m)$ are individually regarded. For each of these pairs, the target nodes $v_{\varphi_{q+1}}$ and v_{φ_j} are swapped resulting in two new assignments $(u_{q+1} \rightarrow v_{\varphi_j})$ and $(u_j \rightarrow v_{\varphi_{q+1}})$. In order to derive node mapping ψ' from ψ, the original node assignment pair is removed from ψ and the swapped node assignment is added to ψ (see line 8). On line 9 the corresponding distance value $d_{\langle\psi'\rangle}$ is derived and finally, the triple $(\psi', q + 1, d_{\langle\psi'\rangle})$ is added to $open$ (line 10). Since index j starts at $(q + 1)$ we also allow that a certain assignment $u_{q+1} \rightarrow v_{\varphi_{q+1}}$ remains unaltered at depth $(q + 1)$ in the search tree.

Since every tree node in our search procedure corresponds to a complete solution and the cost of these solutions neither monotonically decrease nor increase with growing depth in the search tree, we need to buffer the best possible distance approximation found during the tree search (lines 11 – 13 take care of that by checking the distance value of every successor node that has been created).

Note that the algorithmic procedure described so far exactly corresponds to a breadth-first search. That is, the procedure described above explores the space of all possible variations of ψ through pairwise swaps and return the best possible approximation (which corresponds to the exact edit distance, of course). However, such an exhaustive search is both unreasonable and intractable.

In [17] a variant of an A*-algorithm, referred to as *beam search*, has been used in order to approximate graph edit distance from scratch. The basic idea of beam search is that only a fixed number b of nodes to be processed are kept in *open*. This idea can be easily integrated in our search procedure as outlined above. Whenever the **for**-loop on lines 7 – 14 has added altered assignments to *open*, only the b assignments with the lowest approximate distance values are kept, and the remaining tree nodes in *open* are removed. This means that not the full search space is explored, but only those nodes are expanded that belong to the most promising assignments (line 15 – 17). Note that parameter b can be used as trade-off parameter between run time and approximation quality. That is, it can be expected that larger values of b lead to both better approximations and increased run time (and vice versa).

From now on we refer to this variant of our framework as *BP-Beam* with parameter b.

4   Experimental Evaluation

For experimental evaluations three data sets from the IAM graph database repository[4] for graph based pattern recognition and machine learning are used. The first graph data set involves graphs that represent molecular compounds (AIDS), the second graph data set consists of graphs representing fingerprint images (FP), and the third data set consists of graphs representing symbols from architectural and electronic drawings (GREC). For details about the underlying data and/or the graph extraction processes on all data sets we refer to [26].

In Table 1 the achieved results are shown. On each data set and for each graph edit distance algorithm two characteristic numbers are computed, viz. the mean relative overestimation of the exact graph edit distance $(\varnothing o)$ and the mean run time to carry out one graph matching $(\varnothing t)$. The algorithms employed are A* and BP (reference systems) and six differently parametrized versions of our novel procedure BP-Beam $(b \in \{5, 10, 15, 20, 50, 100\})$.

First we focus on the degree of overestimation. The original framework (BP) overestimates the graph distance by 12.68% on average on the AIDS data, while on the Fingerprint and GREC data the overestimations of the true distances amount to 6.38% and 2.98%, respectively. These values can be reduced with our extended framework on all data sets. For instance on the AIDS data, the mean relative overestimation can be reduced to 1.93% with $b = 5$. With $b = 5$ also on the other data sets a substantial reduction of $\varnothing o$ can be reported (from 6.38% to 0.61% and from 2.98% to 0.49% on the FP and GREC data set, respectively). Increasing the values of parameter b allows to further decrease the relative overestimation. That is, with $b = 100$ the mean relative overestimation amounts to only 0.87% on the AIDS data set. On the Fingerprint data the overestimation can be heavily reduced from 6.38% to 0.32% with $b = 100$ and on the GREC data set the mean relative overestimation is reduced from 2.98% to 0.27% with this parametrization.

---

[4] www.iam.unibe.ch/fki/databases/iam-graph-database

The substantial improvement of the approximation accuracy can also be observed in the scatter plots in Fig. 1. These scatter plots give us a visual representation of the accuracy of the suboptimal methods on the AIDS data set[5]. We plot for each pair of graphs their exact (horizontal axis) and approximate (vertical axis) distance value. The reduction of the overestimation using our proposed extension is clearly observable and illustrates the power of our extended framework.

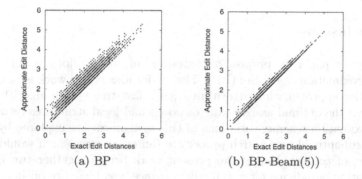

(a) BP   (b) BP-Beam(5))

Fig. 1. Exact (x-axis) vs. approximate (y-axis) graph edit distance

Table 1. The mean relative overestimation of the exact graph edit distance ($\varnothing o$) and the mean run time for one matching ($\varnothing t$ in ms) using a specific graph edit distance algorithm

Algorithm	Data Set					
	AIDS		FP		GREC	
	$\varnothing o$	$\varnothing t$	$\varnothing o$	$\varnothing t$	$\varnothing o$	$\varnothing t$
A* (Exact)	-	5629.53	-	5000.85	-	3103.76
BP	12.68	0.44	6.38	0.56	2.98	0.43
BP-Beam(5)	1.93	3.98	0.61	2.91	0.49	5.83
BP-Beam(10)	1.79	7.27	0.56	5.17	0.47	10.97
BP-Beam(15)	1.68	10.51	0.51	7.32	0.41	15.90
BP-Beam(20)	1.28	13.48	0.46	9.41	0.33	20.71
BP-Beam(50)	0.95	31.39	0.35	21.58	0.29	46.49
BP-Beam(100)	0.87	60.40	0.32	41.87	0.27	86.00

As expected, the run time of BP-Beam is clearly affected by parameter b. That is, doubling the values for parameter b (from 5 to 10, 10 to 20, or 50 to 100)

---

[5] On the other data sets similar results can be observed.

approximately doubles the run time of our procedure. Comparing the mean run time of BP-Beam(5) with the original framework, we observe that our extension increases run time approximately by factor 9, 6, and 13 on the three data sets. Yet, on all data sets the run time remains remarkable low (a few milliseconds per matching on average only). Furthermore, even with $b = 100$ the average run time lies below 0.1s per matching on every data set. Compared to the huge run time for exact computation (3 or more seconds per matching), the increase of the run time through our extension remains very small.

5 Conclusions

In the present paper we propose an extension of our previous graph edit distance approximation algorithm (BP). The major idea of our work is to combine the bipartite approximation algorithm with a fast tree search algorithm. Formally, given the optimal assignments of nodes and local structures returned by our approximation scheme, variations of this assignment are explored by means of a fast, suboptimal tree search procedure (an exact tree search would be unreasonable, of course). Hence, the present work brings together two different approximation paradigms for graph edit distance, viz. bipartite optimization of local structures and fast beam search. With several experimental results we show that this combination is clearly beneficial as it leads to a substantial reduction of the overestimations typical for BP. Though the run times are increased when compared to our former framework (as expected), they are still far below the run times of the exact algorithm.

In the current version of our extension the node assignment ($u_q \to v_{\varphi_q}$) to be swapped at search step q are selected in fixed order. In future work we plan, among other activities, to use heuristics for a more elaborated selection order of the node operations to be swapped.

Acknowledgements. This work has been supported by the *Hasler Foundation* Switzerland and the *Swiss National Science Foundation project P300P2-151279*.

References

1. Mahé, P., Ueda, N., Akutsu, T.: Graph kernels for molecular structures – activity relationship analysis with support vector machines. Journal of Chemical Information and Modeling 45(4), 939–951 (2005)
2. Borgwardt, K.: Graph Kernels. PhD thesis, Ludwig-Maximilians-University Munich (2007)
3. Ralaivola, L., Swamidass, S., Saigo, H., Baldi, P.: Graph kernels for chemical informatics. Neural Networks 18(8), 1093–1110 (2005)
4. Schenker, A., Bunke, H., Last, M., Kandel, A.: Graph-Theoretic Techniques for Web Content Mining. World Scientific (2005)
5. Cook, D., Holder, L. (eds.): Mining Graph Data. Wiley-Interscience (2007)

6. Harchaoui, Z., Bach, F.: Image classification with segmentation graph kernels. In: IEEE Conference on Computer Vision and Pattern Recognition, pp. 1–8 (2007)
7. Luo, B., Wilson, R., Hancock, E.: Spectral embedding of graphs. Pattern Recognition 36(10), 2213–2223 (2003)
8. Lladós, J., Sánchez, G.: Graph matching versus graph parsing in graphics recognition. Int. Journal of Pattern Recognition and Artificial Intelligence 18(3), 455–475 (2004)
9. Rocha, J., Pavlidis, T.: A shape analysis model with applications to a character recognition system. IEEE Transactions on Pattern Analysis and Machine Intelligence 16(4), 393–404 (1994)
10. Dickinson, P., Bunke, H., Dadej, A., Kraetzl, M.: Matching graphs with unique node labels. Pattern Analysis and Applications 7(3), 243–254 (2004)
11. Conte, D., Foggia, P., Sansone, C., Vento, M.: Thirty years of graph matching in pattern recognition. Int. Journal of Pattern Recognition and Artificial Intelligence 18(3), 265–298 (2004)
12. Sanfeliu, A., Fu, K.: A distance measure between attributed relational graphs for pattern recognition. IEEE Transactions on Systems, Man, and Cybernetics (Part B) 13(3), 353–363 (1983)
13. Bunke, H., Allermann, G.: Inexact graph matching for structural pattern recognition. Pattern Recognition Letters 1, 245–253 (1983)
14. Boeres, M.C., Ribeiro, C.C., Bloch, I.: A randomized heuristic for scene recognition by graph matching. In: Ribeiro, C.C., Martins, S.L. (eds.) WEA 2004. LNCS, vol. 3059, pp. 100–113. Springer, Heidelberg (2004)
15. Sorlin, S., Solnon, C.: Reactive tabu search for measuring graph similarity. In: Brun, L., Vento, M. (eds.) GbRPR 2005. LNCS, vol. 3434, pp. 172–182. Springer, Heidelberg (2005)
16. Justice, D., Hero, A.: A binary linear programming formulation of the graph edit distance. IEEE Trans. on Pattern Analysis ans Machine Intelligence 28(8), 1200–1214 (2006)
17. Neuhaus, M., Riesen, K., Bunke, H.: Fast suboptimal algorithms for the computation of graph edit distance. In: Yeung, D.-Y., Kwok, J.T., Fred, A., Roli, F., de Ridder, D. (eds.) SSPR 2006 and SPR 2006. LNCS, vol. 4109, pp. 163–172. Springer, Heidelberg (2006)
18. Riesen, K., Bunke, H.: Approximate graph edit distance computation by means of bipartite graph matching. Image and Vision Computing 27(4), 950–959 (2009)
19. Hart, P., Nilsson, N., Raphael, B.: A formal basis for the heuristic determination of minimum cost paths. IEEE Transactions of Systems, Science, and Cybernetics 4(2), 100–107 (1968)
20. Cortés, X., Serratosa, F., Solé-Ribalta, A.: Active graph matching based on pairwise probabilities between nodes. In: Gimel'farb, G., Hancock, E., Imiya, A., Kuijper, A., Kudo, M., Omachi, S., Windeatt, T., Yamada, K. (eds.) SSPR&SPR 2012. LNCS, vol. 7626, pp. 98–106. Springer, Heidelberg (2012)
21. Koopmans, T., Beckmann, M.: Assignment problems and the location of economic activities. Econometrica 25, 53–76 (1975)
22. Munkres, J.: Algorithms for the assignment and transportation problems. Journal of the Society for Industrial and Applied Mathematics 5, 32–38 (1957)
23. Jonker, R., Volgenant, T.: A shortest augmenting path algorithm for dense and sparse linear assignment problems. Computing 38, 325–340 (1987)
24. Burkard, R., Dell'Amico, M., Martello, S.: Assignment Problems. Society for Industrial and Applied Mathematics, Philadelphia (2009)

25. Fankhauser, S., Riesen, K., Bunke, H.: Speeding up graph edit distance computation through fast bipartite matching. In: Jiang, X., Ferrer, M., Torsello, A. (eds.) GbRPR 2011. LNCS, vol. 6658, pp. 102–111. Springer, Heidelberg (2011)
26. Riesen, K., Bunke, H.: IAM graph database repository for graph based pattern recognition and machine learning. In: da Vitoria Lobo, N., Kasparis, T., Roli, F., Kwok, J.T., Georgiopoulos, M., Anagnostopoulos, G.C., Loog, M. (eds.) S+SSPR 2008. LNCS, vol. 5342, pp. 287–297. Springer, Heidelberg (2008)

Computing Upper and Lower Bounds of Graph Edit Distance in Cubic Time

Kaspar Riesen[1], Andreas Fischer[2], and Horst Bunke[3]

[1] Institute for Information Systems, University of Applied Sciences and Arts
Northwestern Switzerland, Riggenbachstrasse 16, 4600 Olten, Switzerland
kaspar.riesen@fhnw.ch
[2] Biomedical Science and Technologies Research Centre, Polytechnique Montreal
2500 Chemin de Polytechnique, Montreal H3T 1J4, Canada
andreas.fischer@polymtl.ca
[3] Institute of Computer Science and Applied Mathematics, University of Bern,
Neubrückstrasse 10, 3012 Bern, Switzerland
bunke@iam.ch

Abstract. Exact computation of graph edit distance (GED) can be solved in exponential time complexity only. A previously introduced approximation framework reduces the computation of GED to an instance of a linear sum assignment problem. Major benefit of this reduction is that an optimal assignment of nodes (including local structures) can be computed in polynomial time. Given this assignment an approximate value of GED can be immediately derived. Yet, since this approach considers local – rather than the global – structural properties of the graphs only, the GED derived from the optimal assignment is suboptimal. The contribution of the present paper is twofold. First, we give a formal proof that this approximation builds an upper bound of the true graph edit distance. Second, we show how the existing approximation framework can be reformulated such that a lower bound of the edit distance can be additionally derived. Both bounds are simultaneously computed in cubic time.

1 Introduction

Graph-based representations, which are used in the field of structural pattern recognition, have found widespread applications in the last decades [1,2]. In fact, graphs offer two major advantages over feature vectors. First, in contrast with vectors graphs provide a direct possibility to describe structural relations in the patterns under consideration. Second, while the size of a graph can be adapted to the size and complexity of a given pattern, vectors are constrained to a predefined length, which has to be preserved for all patterns encountered in a particular application.

Graph matching refers to the task of evaluating the similarity of graphs. A huge amount of graph matching methodologies have been developed in the last four decades [1]. They include methods stemming from *spectral graph theory* [3], *relaxation labeling* [4], or *graph kernel theory* [5], to name just a few.

N. El Gayar et al. (Eds.): ANNPR 2014, LNAI 8774, pp. 129–140, 2014.
© Springer International Publishing Switzerland 2014

Among the vast number of graph matching methods available, the concept of *graph edit distance* [6] is in particular interesting because it is able to cope with directed and undirected, as well as with labeled and unlabeled graphs. If there are labels on nodes, edges, or both, no constraints on the respective label alphabets have to be considered. Moreover, through the use of a cost function graph edit distance can be adapted and tailored to various problem specifications.

A major drawback of graph edit distance is its computational complexity. In fact, the problem of graph edit distance can be reformulated as an instance of a *Quadratic Assignment Problem (QAP)*. QAPs belong to the most difficult combinatorial optimization problems for which only exponential run time algorithms are available to date.[1]

In recent years, a number of methods addressing the high complexity of graph edit distance computation have been proposed. In [7], for instance, an efficient algorithm for edit distance computation of planar graphs has been proposed. Another approach described in [8] formulates the graph edit distance problem as a binary linear programming problem. This reformulation is applicable to graphs with unlabeled and undirected edges only, and determines lower and upper bounds of graph edit distance in $O(n^7)$ and $O(n^3)$ time, respectively (n refers to the number of nodes in the graphs). The authors of [9] propose the use of continuous-time quantum walks for graph edit distance computation without explicitly determining the underlying node correspondences.

Most of the approximation methods for graph edit distance restrict their applicability to special classes of graphs. In [10] the authors of the present paper introduced an algorithmic framework for the approximation of graph edit distance which is applicable to any kind of graphs. The basic idea of this approach is to reduce the difficult QAP of graph edit distance computation to a *linear sum assignment problem* (LSAP) which can be efficiently solved. This approximation framework builds the basis for the present work. In [10] the result of an initial node assignment is used to derive a valid, yet suboptimal, edit path between the graphs. In the present paper we give a formal prove that this approximation builds an upper bound of the true edit distance. Moreover, we show how the same approximation framework can be exploited to instantly derive a lower bound of the graph edit distance. Both bounds can be simultaneously computed in $O((n+m)^3)$ time, where n and m refers to the number of nodes in the graphs under consideration.

2 Exact Graph Edit Distance Computation

2.1 Graph Edit Distance

Let L_V and L_E be finite or infinite label sets for nodes and edges, respectively. A *graph* g is a four-tuple $g = (V, E, \mu, \nu)$, where V is the finite set of nodes, $E \subseteq V \times V$ is the set of edges, $\mu : V \to L_V$ is the node labeling function, and $\nu : E \to L_E$ is the edge labeling function.

---

[1] Note that QAPs are known to be $\mathcal{NP}$-*complete*, and therefore, an exact and efficient algorithm for the graph edit distance problem can not be developed unless $\mathcal{P} = \mathcal{NP}$.

Given two graphs, the source graph $g_1 = (V_1, E_1, \mu_1, \nu_1)$ and the target graph $g_2 = (V_2, E_2, \mu_2, \nu_2)$, the basic idea of graph edit distance [6] is to transform g_1 into g_2 using some edit operations. A standard set of edit operations is given by *insertions*, *deletions*, and *substitutions* of both nodes and edges. We denote the substitution of two nodes $u \in V_1$ and $v \in V_2$ by $(u \to v)$, the deletion of node $u \in V_1$ by $(u \to \varepsilon)$, and the insertion of node $v \in V_2$ by $(\varepsilon \to v)$, where ε refers to the empty "node". For edge edit operations we use a similar notation.

Definition 1. *A sequence $(e_1, \ldots, e_k)$ of k edit operations e_i that transform g_1 completely into g_2 is called a (complete) edit path $\lambda(g_1, g_2)$ between g_1 and g_2. A partial edit path, i.e. a subsequence of $(e_1, \ldots, e_k)$, edits proper subsets of nodes and/or edges of the underlying graphs.*

Note that in an edit path $\lambda(g_1, g_2)$ each node of g_1 is either deleted or uniquely substituted with a node in g_2, and analogously, each node in g_2 is either inserted or matched with a unique node in g_1. The same applies for the edges. Yet, edit operations on edges are always defined by the edit operations on their adjacent nodes. That is, whether an edge (u, v) is substituted, deleted, or inserted, depends on the edit operations actually performed on both adjacent nodes u and v.

Since edge edit operations are uniquely defined via node edit operations, it is sufficient that edit operations $e_i \in \lambda(g_1, g_2)$ only cover the nodes from V_1 and V_2. That is, an edit path $\lambda(g_1, g_2)$ explicitly describes the correspondences found between the graphs' nodes V_1 and V_2, while the edge edit operations are implicitly given by these node correspondences.

Let $\Upsilon(g_1, g_2)$ denote the set of all admissible and complete edit paths between two graphs g_1 and g_2. To find the most suitable edit path out of $\Upsilon(g_1, g_2)$, one introduces a cost $c(e)$ for every edit operation e, measuring the strength of the corresponding operation. The idea of such a cost is to define whether or not an edit operation e represents a strong modification of the graph. By means of cost functions for elementary edit operations, graph edit distance allows the integration of domain specific knowledge about object similarity. Furthermore, if in a particular case prior knowledge about the labels and their meaning is not available, automatic procedures for learning the edit costs from a set of sample graphs are available as well [11].

Clearly, between two similar graphs, there should exist an inexpensive edit path, representing low cost operations, while for dissimilar graphs an edit path with high cost is needed. Consequently, the edit distance of two graphs is defined as follows.

Definition 2. *Let $g_1 = (V_1, E_1, \mu_1, \nu_1)$ be the source and $g_2 = (V_2, E_2, \mu_2, \nu_2)$ the target graph. The graph edit distance $d_{\lambda_{\min}}(g_1, g_2)$, or $d_{\lambda_{\min}}$ for short, between g_1 and g_2 is defined by*

$$d_{\lambda_{\min}}(g_1, g_2) = \min_{\lambda \in \Upsilon(g_1, g_2)} \sum_{e_i \in \lambda} c(e_i) \quad , \tag{1}$$

where $\Upsilon(g_1, g_2)$ denotes the set of all complete edit paths transforming g_1 into g_2, c denotes the cost function measuring the strength $c(e_i)$ of edit operation e_i

(including the cost of the implied edge edit operations), and $\lambda_{\min}$ *refers to the minimal cost edit path found in* $\Upsilon(g_1, g_2)$.

For our further investigations it will be necessary to subdivide any graph distance value $d_\lambda(g_1, g_2)$ corresponding to a (not necessarily minimal) edit path $\lambda \in \Upsilon(g_1, g_2)$ into the sum of costs $C_\lambda^{\langle V \rangle}$ for all node edit operations $e_i \in \lambda$ and the sum of costs $C_\lambda^{\langle E \rangle}$ for all edge edit operations implied by the node operations $e_j \in \lambda$. That is,

$$d_\lambda(g_1, g_2) = C_\lambda^{\langle V \rangle} + C_\lambda^{\langle E \rangle} \tag{2}$$

2.2 Exact Computation of Graph Edit Distance

Optimal algorithms for computing the edit distance $d_{\lambda_{\min}}(g_1, g_2)$ are typically based on combinatorial search procedures that explore the space of all possible mappings of the nodes and edges of g_1 to the nodes and edges of g_2 (i.e. the search space corresponds to $\Upsilon(g_1, g_2)$). Such an exploration is often conducted by means of A* based search techniques [12].

The basic idea of A* based search methods is to organize the underlying search space as an ordered tree. The root node of the search tree represents the starting point of our search procedure, inner nodes of the search tree correspond to partial edit paths, and leaf nodes represent complete – not necessarily optimal – edit paths. Such a search tree is dynamically constructed at runtime by iteratively creating successor nodes linked by edges to the currently considered node in the search tree.

The search tree nodes, i.e. (partial or complete) edit paths λ, to be processed in the next steps are typically contained in a set *OPEN*. In order to determine the most promising (partial) edit path $\lambda \in OPEN$, i.e. the edit path to be used for further expansion in the next iteration, an assessment function $f(\lambda) = g(\lambda) + h(\lambda)$ is usually used, which includes the accumulated cost $g(\lambda)$ of the edit operations $e_i \in \lambda$ plus a heuristic estimation $h(\lambda)$ of the future cost to complete λ. One can show that, given that the estimation of the future cost is lower than, or equal to, the real cost, the algorithm is admissible. Hence, this procedure guarantees that a complete edit path $\lambda_{\min}$ found by the algorithm first is always optimal in the sense of providing minimal cost among all possible competing paths.

Note that the edge operations implied by the node edit operations can be derived from every partial or complete edit path λ during the search procedure. The cost of these implied edge operations are dynamically added to the corresponding paths $\lambda \in OPEN$ and are thus considered in the edit path assessment $f(\lambda)$.

3   Bipartite Graph Matching

The computational complexity of exact graph edit distance is exponential in the number of nodes of the involved graphs. That is considering m nodes in g_1 and n nodes in g_2, $\Upsilon(g_1, g_2)$ contains $O(m^n)$ edit paths to be explored. This means that for large graphs the computation of edit distance is intractable. The graph edit distance approximation framework introduced in [10] reduces the difficult *Quadratic Assignment Problem (QAP)* of graph edit distance computation to an instance of a *Linear Sum Assignment Problem (LSAP)*. For solving LSAPs a large number of algorithms exist [13]. The time complexity of the best performing exact algorithms for LSAPs is cubic in the size of the problem. The LSAP is defined as follows.

Definition 3. *Given two disjoint sets $S = \{s_1, \ldots, s_n\}$ and $Q = \{q_1, \ldots, q_n\}$ and an $n \times n$ cost matrix $\mathbf{C} = (c_{ij})$, where c_{ij} measures the suitability of assigning the i-th element of the first set to the j-th element of the second set. The Linear Sum Assignment Problem (LSAP) is given by finding the minimum cost permutation*

$$(\varphi_1, \ldots, \varphi_n) = \underset{(\varphi_1, \ldots, \varphi_n) \in \mathcal{S}_n}{\arg\min} \sum_{i=1}^{n} c_{i\varphi_i} \quad ,$$

where $\mathcal{S}_n$ refers to the set of all n! possible permutations of n integers, and permutation $(\varphi_1, \ldots, \varphi_n)$ refers to the assignment where the first entity $s_1 \in S$ is mapped to entity $q_{\varphi_1} \in Q$, the second entity $s_2 \in S$ is assigned to entity $q_{\varphi_2} \in Q$, and so on.

By reformulating the graph edit distance problem to an instance of an LSAP, three major issues have to be resolved. First, LSAPs are generally stated on independent sets with equal cardinality. Yet, in our case the elements to be assigned to each other are given by the sets of nodes (and edges) with unequal cardinality in general. Second, solutions to LSAPs refer to assignments of elements in which every element of the first set is assigned to exactly one element of the second set and vice versa (i.e. a solution to an LSAP corresponds to a bijective assignment of the the underlying entities). Yet, graph edit distance is a more general assignment problem as it explicitly allows both deletions and insertions to occur on the basic entities (rather than only substitutions). Third, graphs do not only consist of independent sets of entities (i.e. nodes) but also of structural relationships between these entities (i.e. edges that connect pairs of nodes). LSAPs are not able to consider these relationships in a global and consistent way. The first two issues are perfectly – and the third issue partially – resolvable by means of the following definition of a square cost matrix whereon the LSAP is eventually solved.

Definition 4. *Based on the node sets $V_1 = \{u_1, \ldots, u_n\}$ and $V_2 = \{v_1, \ldots, v_m\}$ of g_1 and g_2, respectively, a cost matrix $\mathbf{C}$ is established as follows.*

$$\mathbf{C} = \begin{bmatrix} c_{11} & c_{12} & \cdots & c_{1m} & c_{1\varepsilon} & \infty & \cdots & \infty \\ c_{21} & c_{22} & \cdots & c_{2m} & \infty & c_{2\varepsilon} & \ddots & \vdots \\ \vdots & \vdots & \ddots & \vdots & \vdots & \ddots & \ddots & \infty \\ c_{n1} & c_{n2} & \cdots & c_{nm} & \infty & \cdots & \infty & c_{n\varepsilon} \\ c_{\varepsilon1} & \infty & \cdots & \infty & 0 & 0 & \cdots & 0 \\ \infty & c_{\varepsilon2} & \ddots & \vdots & 0 & 0 & \ddots & \vdots \\ \vdots & \ddots & \ddots & \infty & \vdots & \ddots & \ddots & 0 \\ \infty & \cdots & \infty & c_{\varepsilon m} & 0 & \cdots & 0 & 0 \end{bmatrix} \tag{3}$$

Entry c_{ij} thereby denotes the cost of a node substitution $(u_i \to v_j)$, $c_{i\varepsilon}$ denotes the cost of a node deletion $(u_i \to \varepsilon)$, and $c_{\varepsilon j}$ denotes the cost of a node insertion $(\varepsilon \to v_j)$.

Note that matrix $\mathbf{C} = (c_{ij})$ is by definition quadratic. Hence, the first issue (sets of unequal size) is instantly eliminated. Obviously, the left upper corner of the cost matrix $\mathbf{C} = (c_{ij})$ represents the costs of all possible node substitutions, the diagonal of the right upper corner the costs of all possible node deletions, and the diagonal of the bottom left corner the costs of all possible node insertions. Note that every node can be deleted or inserted at most once. Therefore any non-diagonal element of the right-upper and left-lower part is set to ∞. The bottom right corner of the cost matrix is set to zero since substitutions of the form $(\varepsilon \to \varepsilon)$ should not cause any cost.

Given the cost matrix $\mathbf{C} = (c_{ij})$, the LSAP optimization consists in finding a permutation $(\varphi_1, \ldots, \varphi_{n+m})$ of the integers $(1, 2, \ldots, (n + m))$ that minimizes the overall assignment cost $\sum_{i=1}^{(n+m)} c_{i\varphi_i}$. This permutation corresponds to the assignment

$$\psi = ((u_1 \to v_{\varphi_1}), (u_2 \to v_{\varphi_2}), \ldots, (u_{m+n} \to v_{\varphi_{m+n}}))$$

of the nodes of g_1 to the nodes of g_2. Note that assignment ψ includes node assignments of the form $(u_i \to v_j)$, $(u_i \to \varepsilon)$, $(\varepsilon \to v_j)$, and $(\varepsilon \to \varepsilon)$ (the latter can be dismissed, of course). Hence, the definition of the cost matrix in Eq. 3 also resolves the second issue stated above and allows insertions and/or deletions to occur in an optimal assignment.

The third issue is about the edge structure of both graphs which cannot be entirely considered by LSAPs. In fact, so far the the cost matrix $\mathbf{C} = (c_{ij})$ considers the nodes of both graphs only, and thus mapping ψ does not take any structural constraints into account. In order to integrate knowledge about the graph structure, to each entry c_{ij}, i.e. to each cost of a node edit operation $(u_i \to v_j)$, the minimum sum of edge edit operation costs, implied by the corresponding node operation, is added. That is, we encode the minimum matching cost arising from the local edge structure in the individual entries $c_{ij} \in \mathbf{C}$.

Formally, assume that node u_i has adjacent edges E_{u_i} and node v_j has adjacent edges E_{v_j}. With these two sets of edges, E_{u_i} and E_{v_j}, an individual cost matrix similarly to Eq. 3 can be established and an optimal assignment of the elements E_{u_i} to the elements E_{v_j} using an LSAP solving algorithm can be computed. Following this procedure, the assignment of adjacent edges is not constrained by an assignment of adjacent nodes other than u_i and v_j. Therefore, the estimated edge edit costs implied by $(u_i \to v_j)$ are less than, or equal to, the costs implied by a complete edit path. These minimum edge edit costs are eventually added to the entry c_{ij}. To entry $c_{i\varepsilon}$, which denotes the cost of a node deletion, the cost of the deletion of all adjacent edges of u_i is added, and to the entry $c_{\varepsilon j}$, which denotes the cost of a node insertion, the cost of all insertions of the adjacent edges of v_j is added. This particular encoding of the minimal edge edit operation cost enables the LSAP to consider information about the local, yet not global, edge structure of a graph. Hence, this heuristic procedure partially resolves the third issue.

4 Upper and Lower Bounds of Graph Edit Distance

4.1 Upper Bound d_ψ

Given the node assignment ψ two different distance values approximating the exact graph edit distance $d_{\lambda_{\min}}(g_1, g_2)$ can be inferred. As stated above, the LSAP optimization finds an assignment ψ in which every node of g_1 is either assigned to a unique node of g_2 or deleted. Likewise, every node of g_2 is either assigned to a unique node of g_1 or inserted. Hence, mapping ψ refers to an admissible and complete edit path between the graphs under consideration, i.e. $\psi \in \Upsilon(g_1, g_2)$. Therefore, the edge operations, which are implied by edit operations on their adjacent nodes, can be completely inferred from ψ. This gives us a first approximation value $d_\psi(g_1, g_2)$, or d_ψ for short, defined by (cf. Eq. 2)

$$d_\psi(g_1, g_2) = C_\psi^{\langle V \rangle} + C_\psi^{\langle E \rangle} . \tag{4}$$

Note that in case of $d_{\lambda_{\min}}$ the sum of edge edit cost $C_{\lambda_{\min}}^{\langle E \rangle}$ is dynamically built while the search tree is constructed and eventually added to every partial edit path $\lambda \in OPEN$. Yet, the sum of edge costs $C_\psi^{\langle E \rangle}$ is added to the cost of the complete edit path ψ only after the optimization process has been terminated. This is because LSAP solving algorithms are not able to take information about assignments of adjacent nodes into account during run time. In other words, for finding the edit path $\psi \in \Upsilon(g_1, g_2)$ based on the cost matrix $\mathbf{C} = (c_{ij})$ the structural information of the graphs is considered in an isolated way only (single nodes and their adjacent edges). This observation brings us to the following Lemma.

Lemma 1. *The distance $d_\psi(g_1, g_2)$ derived from the node assignment ψ constitutes an upper bound of the true graph edit distance $d_{\lambda_{\min}}(g_1, g_2)$. That is,*

$$d_\psi(g_1, g_2) \geq d_{\lambda_{\min}}(g_1, g_2)$$

holds for every pair of graphs g_1, g_2.

Proof. We distinguish two cases.

1. $\psi - \lambda_{\min}$: That is, the edit path ψ returned by our approximation framework is identical with the edit path $\lambda_{\min}$ computed by an exact algorithm. It follows that $d_\psi = d_{\lambda_{\min}}$.
2. $\psi \neq \lambda_{\min}$: In this case the approximate edit distance d_ψ cannot be smaller than $d_{\lambda_{\min}}$. Otherwise an exact algorithm for graph edit distance computation, which exhaustively explores $\Upsilon(g_1, g_2)$, would return ψ as edit path with minimal cost, i.e. $\psi = \lambda_{\min}$. Yet, this is a contradiction to our initial assumption that $\psi \neq \lambda_{\min}$.

4.2   Lower Bound d'_ψ

The distance value $d_\psi(g_1, g_2)$ is directly used as an approximate graph edit distance between graphs g_1 and g_2 in previous publications (e.g. in [10]). We now define another approximation of the true graph edit distance based on mapping ψ. As we will see below, this additional approximation builds a lower bound of the true graph edit distance $d_{\lambda_{\min}}(g_1, g_2)$.

First, we consider the the minimal sum of assignment costs $\sum_{i=1}^{(n+m)} c_{i\varphi_i}$ returned by our LSAP solving algorithm. Remember that every entry $c_{i\varphi_i}$ reflects the cost of the corresponding node edit operation $(u_i \to v_{\varphi_i})$ plus the minimal cost of editing the incident edges of u_i to the incident edges of v_{φ_i}. Hence, the sum $\sum_{i=1}^{(n+m)} c_{i\varphi_i}$ can be – similarly to Eq. 4 – subdivided into costs for node edit operations and costs for edge edit operations. That is,

$$\sum_{i=1}^{(n+m)} c_{i\varphi_i} = C_\psi^{\langle V \rangle} + C_\varphi^{\langle E \rangle} . \tag{5}$$

Analogously to Eq. 4, $C_\psi^{\langle V \rangle}$ corresponds to the sum of costs for node edit operations $e_i \in \psi$. Yet, note the difference between $C_\varphi^{\langle E \rangle}$ and $C_\psi^{\langle E \rangle}$. While $C_\psi^{\langle E \rangle}$ reflects the costs of editing the edge structure from g_1 to the edge structure of g_2 in a globally consistent way (with respect to all edit operations in ψ applied on both adjacent nodes of every edge), the sum $C_\varphi^{\langle E \rangle}$ is based on the optimal permutation $(\varphi_1, \ldots, \varphi_{(n+m)})$ and in particular on the limited, because local, information about the edge structure integrated in the cost matrix $\mathbf{C} = (c_{ij})$. Moreover, note that every edge (u, v) is adjacent with two individual nodes u, v and thus the sum of edge edit costs $C_\varphi^{\langle E \rangle}$ considers every edge twice in two independent edit operations. Therefore, we define our second approximation value $d'_\psi(g_1, g_2)$, or d'_ψ for short, by

$$d'_\psi(g_1, g_2) = C_\psi^{\langle V \rangle} + \frac{C_\varphi^{\langle E \rangle}}{2} \tag{6}$$

Clearly, Eq. 6 can be reformulated as

$$d'_\psi(g_1, g_2) = C_\psi^{\langle V \rangle} + \frac{\sum_{i=1}^{(n+m)} c_{i\varphi_i} - C_\psi^{\langle V \rangle}}{2} \tag{7}$$

and thus, $d'_\psi(g_1, g_2)$ only depends on quantities $C_\psi^{\langle V \rangle}$ and $\sum_{i=1}^{(n+m)} c_{i\varphi_i}$, which are already computed for d_ψ. Therefore, d'_ψ can be derived without any additional computations from the established approximation d_ψ.

Note that the approximation d_ψ corresponds to an admissible and complete edit path with respect to the nodes and edges of the underlying graphs. Yet, the second approximation d'_ψ is not related to a valid edit path since the edges of both graphs are not uniquely assigned to each other (or deleted/inserted at most once). The following Lemma shows an ordering relationship between d_ψ and d'_ψ.

Lemma 2. *For the graph edit distance approximations $d_\psi(g_1, g_2)$ (Eq. 4) and $d'_\psi(g_1, g_2)$ (Eq. 6) the inequality*

$$d'_\psi(g_1, g_2) \leq d_\psi(g_1, g_2)$$

holds for every pair of graphs g_1, g_2 and every complete node assignment ψ.

Proof. According to Eq. 4 and Eq. 6 we have to show that

$$\frac{C_\varphi^{\langle E \rangle}}{2} \leq C_\psi^{\langle E \rangle} \quad .$$

Assume that the node edit operation $(u_i \to v_j)$ is performed in ψ. Therefore, the edges E_{u_i} incident to node u_i are edited to the edges E_{v_j} of v_j in $C_\varphi^{\langle E \rangle}$ as well as in $C_\psi^{\langle E \rangle}$.

The sum of edge costs $C_\varphi^{\langle E \rangle}$ considers the minimal cost edit path between the edges E_{u_i} to the edges of E_{v_j} with respect to $(u_i \to v_j)$ only. In the case of $C_\psi^{\langle E \rangle}$, however, every edge in E_{u_i} and E_{v_j} is edited with respect to the node operations actually carried out on *both* adjacent nodes of every edge (rather than considering $(u_i \to v_j)$ only). Hence, $C_\varphi^{\langle E \rangle}$ is restricted to the best case, while $C_\psi^{\langle E \rangle}$ considers the consistent case of editing the edge sets.

Note that $C_\varphi^{\langle E \rangle}$ is built on the minimized sum $\sum_{i=1}^{(n+m)} c_{i\varphi_i}$. Yet, the cost sequence $c_{1\varphi_1}, \ldots, c_{(n+m)\varphi_{(n+m)}}$ considers every edge $(u_l, u_k) \in E_1$ twice, viz. once in entry $c_{l\varphi_l}$ and once in entry $c_{k\varphi_k}$. The same accounts for the edges in E_2. The cost of edge operations considered in $c_{l\varphi_l}$ as well as in $c_{k\varphi_k}$ refers to the best possible case of editing the respective edge sets. The sum of cost of these two best cases considered in $C_\varphi^{\langle E \rangle}$ are clearly smaller than, or equal to, twice the actual cost considered in $C_\psi^{\langle E \rangle}$.

We can now show that d'_ψ constitutes a lower bound for $d_{\lambda_{\min}}$.

Lemma 3. *The distance $d'_\psi(g_1, g_2)$ derived from the node assignment ψ constitutes a lower bound of the true graph edit distance $d_{\lambda_{\min}}(g_1, g_2)$. That is,*

$$d'_\psi(g_1, g_2) \leq d_{\lambda_{\min}}(g_1, g_2)$$

holds for every pair of graphs g_1, g_2.

Proof. We distinguish two cases.

1. $\psi = \lambda_{\min}$: An optimal algorithm would return ψ as optimal solution and thus $d_\psi = d_{\lambda_{\min}}$. From Lemma 2 we know that $d'_\psi \leq d_\psi$ and thus $d'_\psi \leq d_{\lambda_{\min}}$.
2. $\psi \neq \lambda_{\min}$: In this case ψ corresponds to a suboptimal edit path with cost d_ψ greater than (or possibly equal to) $d_{\lambda_{\min}}$. The question remains whether or not $d_{\lambda_{\min}} < d'_\psi$ might hold in this case. According to Lemma 2 we know that $d'_{\lambda_{\min}} \leq d_{\lambda_{\min}}$ and thus assuming that $d_{\lambda_{\min}} < d'_\psi$ holds, it follows that $d'_{\lambda_{\min}} < d'_\psi$. Yet, this is contradictory to the optimality of the LSAP solving algorithm that guarantees to find the assignment ψ with lowest cost d'_ψ.

We can now conclude this section with the following theorem.

Theorem 1.

$$d'_\psi(g_1, g_2) \leq d_{\lambda_{\min}}(g_1, g_2) \leq d_\psi(g_1, g_2) \quad \forall g_1, g_2$$

Proof. See Lemmas 1, 2, and 3.

5   Experimental Evaluation

In Table 1 the achieved results on three data sets from the IAM graph database repository[2] are shown. The graph data sets involve graphs that represent molecular compounds (AIDS), fingerprint images (FP), and symbols from architectural and electronic drawings (GREC). On each data set and for both bounds two characteristic numbers are computed, viz. the mean relative deviation of d_ψ and d'_ψ from the exact graph edit distance $d_{\lambda_{\min}}$ ($\varnothing e$) and the mean run time to carry out one graph matching ($\varnothing t$).

Table 1. The mean relative error of the exact graph edit distance ($\varnothing e$) in percentage and the mean run time for one matching ($\varnothing t$ in ms)

| Distance | Data Set | | | | | |
| | AIDS | | FP | | GREC | |
	$\varnothing e$	$\varnothing t$	$\varnothing e$	$\varnothing t$	$\varnothing e$	$\varnothing t$
$d_{\lambda_{\min}}$	-	5629.53	-	5000.85	-	3103.76
d_ψ	+12.68	0.44	+6.38	0.56	+2.98	0.43
d'_ψ	-7.01	0.44	-0.38	0.56	-3.67	0.43

First we focus on the exact distances $d_{\lambda_{\min}}$ provided by A*. As $d_{\lambda_{\min}}$ refers to the exact edit distance, the mean relative error $\varnothing e$ is zero on all data sets. We observe that the mean run time for the computation of $d_{\lambda_{\min}}$ lies between 3.1s and 5.6s per matching. Using the approximation framework, a massive speed-up of computation time can be observed. That is, on all data sets the the computation of both distance approximations d_ψ and d'_ψ is possible in less than or approximately 0.5ms on average (note that both distance measures are simultaneously computed and thus offer the same matching time).

Regarding the overestimation of d_ψ and the underestimation of d'_ψ we observe the following. The original framework, providing the upper bound d_ψ, overestimates the graph distance by 12.68% on average on the AIDS data, while on the Fingerprint and GREC data the overestimations of the true distances amount to 6.38% and 2.98%, respectively. On the GREC data, the upper bound d_ψ is a more accurate approximation than the lower bound d'_ψ, where the underestimation amounts to 3.67%. Yet, the deviations of d_ψ are substantially reduced on

---

[2] www.iam.unibe.ch/fki/databases/iam-graph-database

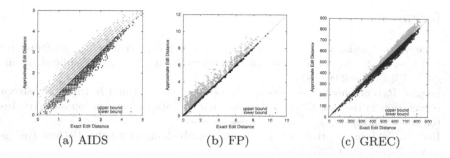

Fig. 1. Exact graph edit distance $d_{\lambda_{\min}}$ vs. upper bound d_ψ (gray points) and lower bound d'_ψ (black points) of the graph edit distance

the other two data sets by using the lower rather than the upper bound. That is, using d'_ψ rather than d_ψ the deviations can be reduced by 5.67% and 6.00% on the AIDS and FP data set, respectively.

Note the remarkable improvement of the approximation accuracy on the FP data set which can also be observed in the scatter plot in Fig. 1 (b). These scatter plots give us a visual representation of the accuracy of the suboptimal methods on all data sets. We plot for each pair of graphs their exact distance $d_{\lambda_{\min}}$ and approximate distance values d_ψ and d'_ψ (shown with gray and black points, respectively).

6 Conclusions

The main focus of the present paper is on theoretical issues. First, we give a formal prove that the existing approximation returns an upper bound of the true edit distance. Second, we show how the same approximation scheme can be used to derive a lower bound of the true edit distance. Both bounds are simultaneously computed in $O((n+m)^3)$, where n and m refer to the number of nodes of the graphs. In an experimental evaluation we empirically confirm our theoretical investigations and show that the lower bound leads to more accurate graph edit distance approximations on two out of three data sets.

In future work we aim at exploiting the additional lower bound in our approximation framework. For instance, a prediction of the true edit distance $d_{\lambda_{\min}}$ based on d_ψ and d'_ψ by means of regression analysis could be a rewarding avenue to be pursued. Moreover, we aim at using both bounds in a pattern recognition application (e.g. in database retrieval where both bounds can be beneficially employed).

Acknowledgements. This work has been supported by the *Hasler Foundation* Switzerland and the *Swiss National Science Foundation* project P300P2-151279.

References

1. Conte, D., Foggia, P., Sansone, C., Vento, M.: Thirty years of graph matching in pattern recognition. Int. Journal of Pattern Recognition and Artificial Intelligence 18(3), 265–298 (2004)
2. Foggia, P., Percannella, G., Vento, M.: Graph matching and learning in pattern recognition in the last 10 years. Int. Journal of Pattern Recognition and Art. Intelligence (2014)
3. Luo, B., Wilson, R., Hancock, E.: Spectral embedding of graphs. Pattern Recognition 36(10), 2213–2223 (2003)
4. Torsello, A., Hancock, E.: Computing approximate tree edit distance using relaxation labeling. Pattern Recognition Letters 24(8), 1089–1097 (2003)
5. Gärtner, T.: A survey of kernels for structured data. SIGKDD Explorations 5(1), 49–58 (2003)
6. Bunke, H., Allermann, G.: Inexact graph matching for structural pattern recognition. Pattern Recognition Letters 1, 245–253 (1983)
7. Neuhaus, M., Bunke, H.: An error-tolerant approximate matching algorithm for attributed planar graphs and its application to fingerprint classification. In: Fred, A., Caelli, T.M., Duin, R.P.W., Campilho, A.C., de Ridder, D. (eds.) SSPR&SPR 2004. LNCS, vol. 3138, pp. 180–189. Springer, Heidelberg (2004)
8. Justice, D., Hero, A.: A binary linear programming formulation of the graph edit distance. IEEE Trans. on Pattern Analysis ans Machine Intelligence 28(8), 1200–1214 (2006)
9. Emms, D., Wilson, R.C., Hancock, E.R.: Graph edit distance without correspondence from continuous-time quantum walks. In: da Vitoria Lobo, N., Kasparis, T., Roli, F., Kwok, J.T., Georgiopoulos, M., Anagnostopoulos, G.C., Loog, M. (eds.) S+SSPR 2008. LNCS, vol. 5342, pp. 5–14. Springer, Heidelberg (2008)
10. Riesen, K., Bunke, H.: Approximate graph edit distance computation by means of bipartite graph matching. Image and Vision Computing 27(4), 950–959 (2009)
11. Caetano, T.S., McAuley, J.J., Cheng, L., Le, Q.V., Smola, A.J.: Learning graph matching. IEEE Trans. on Pattern Analysis and Machine Intelligence 31(6), 1048–1058 (2009)
12. Hart, P., Nilsson, N., Raphael, B.: A formal basis for the heuristic determination of minimum cost paths. IEEE Transactions of Systems, Science, and Cybernetics 4(2), 100–107 (1968)
13. Burkard, R., Dell'Amico, M., Martello, S.: Assignment Problems. Society for Industrial and Applied Mathematics, Philadelphia (2009)

Linear Contrast Classifiers in High-Dimensional Spaces

Florian Schmid*, Ludwig Lausser*, and Hans A. Kestler**

Medical Systems Biology and
Institute of Neural Information Processing
Ulm University, 89069 Ulm, Germany
hans.kestler@uni-ulm.de

Abstract. Linear classifiers are mainly discussed in terms of training algorithms that try to find an optimal hyperplane according to a data dependent objective. Such objectives might be the induction of a large margin or the reduction of the number of involved features. The underlying concept class of linear classifiers is analyzed less frequently. It is implicitly assumed that all classifiers of this function class share the same common properties.

In this work we analyze the concept class of linear classifiers. We show that it includes different subclasses that show beneficial properties during the prediction phase. These properties can directly be derived from the structural form of the classifiers and must not be learned in a data dependent training phase.

We describe the concept class of contrasts, a class of linear functions that is for example utilized in variance analyses. Models from this concept class share the common property of being invariant against global additive effects. We give a theoretical characterization of contrast classifiers and analyze the effects of replacing general linear classifiers by these new models in standard training algorithms.

1 Introduction

The limited availability of samples is one of the major challenges of the development of diagnostic models for biomolecular data. Especially for high-dimensional gene expression profiles as obtained from microarray experiments the low cardinality of a dataset can lead to non generalizable models. Possible reasons for small sample sizes may be the rareness of a disease, high costs or technical burdens. A possible strategy for enlarging a dataset can be to merge the samples of different research groups. These collections might be affected by an additional source of variation. Being geographically separated the experimental conditions may not be identical in different labs. Environmental factors such as humidity and temperature are likely to vary for different locations. They can affect all simultaneous measurements of a sample (e.g. all measurements of a gene expression profile). These effects should be minimized during the development of

---

* F. Schmid and L. Lausser contributed equally.
** Corresponding author.

N. El Gayar et al. (Eds.): ANNPR 2014, LNAI 8774, pp. 141–152, 2014.
© Springer International Publishing Switzerland 2014

a classification model. In this work we propose a linear classification model that may be suitable for this scenario.

The research on linear classifiers mainly focuses on the development of training algorithms and optimization criteria [7]. Famous example are linear support vector machines that maximize the margin between training samples [14] or the nearest shrunken centroid classifier that utilizes embedded feature selection [13]. The structural properties of linear classifiers are often neglected. It is implicitly assumed that all classifiers of this function class share the same common properties.

We recently showed that subclasses of linear classifiers exist which share common structural properties that induce an invariance against a certain type of noise [8]. That is a classifier is guaranteed to be unaffected by this type of data transformation, if chosen from this subclass. In this work we present an additional subclass, which we call linear contrast classifiers. It consists of linear classifiers that are invariant against global transition. In the following we will prove the invariance property of this subclass. We will also characterize the benefit of contrast classifiers in different noise experiments.

2   Methods

In this work a classifier c will be seen as a function mapping that predicts the class label $y \in \mathcal{Y}$ of an object according to a vector of measurements $\mathbf{x} \in \mathcal{X}$. The components of a vector will be denoted as $\mathbf{x} = (x^{(1)}, \dots, x^{(n)})^T$. In the following we restrict ourselves to binary classification tasks (e.g. $\mathcal{Y} = \{0, 1\}$) and $\mathcal{X} \subseteq \mathbb{R}^n$. A classifier will be selected out of a concept class of functions sharing some structural property (e.g. linear classifiers) $c \in \mathcal{C}$. The selection or training of a classifier is normally done according to a performance measure evaluated on a set of labeled training examples $\mathcal{S}_{tr} = \{(\mathbf{x}_i, y_i)\}_{i=1}^m$. The generalization ability of the selected classifier is estimated on an independent set of test samples $\mathcal{S}_{te} = \{(\mathbf{x}'_i, y'_i)\}_{i=1}^{m'}$.

In the following the concept class of linear classifiers $\mathcal{C}_{lin}$ is analyzed.

$$\mathcal{C}_{lin} = \left\{ \mathbb{I}_{[\langle \mathbf{w}, \mathbf{x} \rangle \geq t]} \mid \mathbf{w} \in \mathbb{R}^n, t \in \mathbb{R} \right\} \tag{1}$$

A classifier $c \in \mathcal{C}_{lin}$ can be interpreted as a linear hyperplane separating the samples into two classes. It is typically parameterized by a weight vector $\mathbf{w} \in \mathbb{R}^n$ determining its orientation and a threshold or offset $t \in \mathbb{R}$ determining its distance to the origin (Figure 1).

2.1   Invariant Classification Models

Reducing the complexity of a concept class is not the only reason for searching for structural subgroups. Another reason might be the gain of invariances that allow to completely neglect a certain kind of influence on the label prediction.

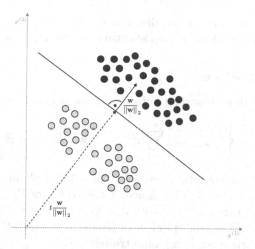

Fig. 1. Scheme of a linear classifier: The decision boundary of a linear classifier is shown. The shade of the samples indicates the corresponding classification. Samples in light gray are categorized as class 1 and samples in dark gray as class 0.

Definition 1. *A classifier $c : \mathcal{X} \to \mathcal{Y}$ is called* invariant *against a parameterized class of data transformations $f_\theta : \mathcal{X} \to \mathcal{X}$ if*

$$\forall\, \theta, \forall\, \mathbf{x} \in \mathcal{X} : \quad c(f_\theta(\mathbf{x})) = c(\mathbf{x}). \tag{2}$$

A concept class $\mathcal{C}$ is called invariant against f_θ if each $c \in \mathcal{C}$ is invariant against f_θ.

Choosing a classifier out of an invariant concept class ensures that a classification result is completely unaffected by a misleading transformation $f_\theta : \mathcal{X} \to \mathcal{X}$. Invariant classifiers are especially beneficial if the measurements of an object are known to be distorted by a technical process.

An example of an invariant concept class is the concept class of offset-free linear classifiers $\mathcal{C}_0$.

$$\mathcal{C}_0 = \left\{ \mathbb{I}_{[\langle \mathbf{w}, \mathbf{x} \rangle \geq 0]} \;\middle|\; \mathbf{w} \in \mathbb{R}^n \right\} \tag{3}$$

These classifiers omit the adaption of an offset ($t = 0$) [8]. Offset-free linear classifiers are invariant against the global scaling of the data $f_a(\mathbf{x}) \mapsto a\mathbf{x}, a \in \mathbb{R}_+$.

The invariance of a concept class is also of interest for classifier ensembles $\mathcal{E}$

$$\forall\, c \in \mathcal{E} : \mathcal{X} \to \mathcal{Y}, \quad h_{\mathcal{E}} : \mathcal{Y}^{|\mathcal{E}|} \to \mathcal{Y} \tag{4}$$

More precisely Equation 4 describes a fusion strategy $h_{\mathcal{E}}$ that operates on the predicted class labels of the base classifiers $c \in \mathcal{E}$ (late aggregation) [12]. Choosing all base classifiers from an invariant concept class $\mathcal{E} \subset \mathcal{C}$ the fusion strategy gets itself invariant against f_θ.

$$h_{\mathcal{E}}\left(c_1(f_{\theta_1}(\mathbf{x})), \ldots, c_{|\mathcal{E}|}(f_{\theta_{|\mathcal{E}|}}(\mathbf{x}))\right) = h_{\mathcal{E}}\left(c_1(\mathbf{x}), \ldots, c_k(\mathbf{x})\right) \tag{5}$$

If the base classifiers operate on distinct feature spaces, the invariance property is able to counteract local effects affecting only subgroups of measurements.

2.2 Contrast Classifiers

In this work we study an invariant subclass of linear classifiers $\mathcal{C}_{con} \subset \mathcal{C}_{lin}$

$$\mathcal{C}_{con} = \left\{ \mathbb{I}_{[\langle \mathbf{w}, \mathbf{x} \rangle \geq t]} \,\middle|\, \sum_{i=1}^{n} w^{(i)} = 0, \mathbf{w} \in \mathbb{R}^n, t \in \mathbb{R} \right\}. \tag{6}$$

For these linear classifiers, the orientation of a hyperplane $\mathbf{w}$ is additionally constraint by $\sum_{i=1}^{n} w^{(i)} = 0$. The choice of t is equal for $\mathcal{C}_{lin}$ and $\mathcal{C}_{con}$. Linear functions that fulfill this property are called *contrasts* in variation analysis [5]. We therefore call $\mathcal{C}_{con}$ the concept class of (linear) contrast classifiers. They can be shown to be invariant against global translation.

Theorem 1. *The concept class of contrast classifiers $\mathcal{C}_{con}$ as defined in Equation 6 is invariant against global translation. That is*

$$\forall c \in \mathcal{C}_{con}, \forall b \in \mathbb{R}, \forall \mathbf{x} \in \mathcal{X}: \quad c(f_b(\mathbf{x})) = c(\mathbf{x}), \tag{7}$$

where $f_b \mapsto \mathbf{x} + \mathbf{b}$ and $\mathbf{b} = (b, \ldots, b)^T$, $\mathbf{b} \in \mathbb{R}^n$.

Proof (Theorem 1). For proofing Theorem 1, it is sufficient to show that global translation does not affect the projection of a contrast.

$$\langle \mathbf{w}, \mathbf{x} + \mathbf{b} \rangle = \langle \mathbf{w}, \mathbf{x} \rangle + \langle \mathbf{w}, \mathbf{b} \rangle \tag{8}$$

$$= \langle \mathbf{w}, \mathbf{x} \rangle + \sum_{i=1}^{n} w^{(i)} b \tag{9}$$

$$= \langle \mathbf{w}, \mathbf{x} \rangle + b \underbrace{\sum_{i=1}^{n} w^{(i)}}_{=0} \tag{10}$$

$$= \langle \mathbf{w}, \mathbf{x} \rangle \tag{11}$$

$$\tag{12}$$

$\square$

Figure 2 gives a two-dimensional example of a contrast classifier. In this low dimensional space contrast classifiers are restricted to two possible orientations $w \in \{(-1, +1)^T, (+1, -1)^T\}$. Only the threshold t can be modified. The orientation of a contrast classifiers becomes more flexible in higher dimensions. Figure 2a shows the invariance property of a contrast classifier. If a sample is moved by a global transition, the sample is guaranteed to receive the same class label as before. Interestingly there exist some linear separable cases that can not be separated by a linear contrast classifier (Figure 2b).

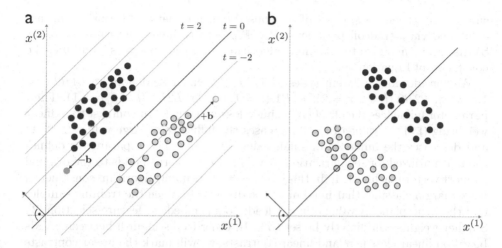

Fig. 2. Scheme of a linear contrast classifier: In a two-dimensional space a contrast can only be adapted by modifying its threshold t (Panel a). The contrast classifier is invariant against global transitions of type $(x^{(1)}, x^{(2)})^T \pm (b, b)^T$. Panel b indicates the limitations of a linear contrast classifier. It shows an example of a linear separable dataset that can not be separated by a linear contrast classifier in two dimensions.

3 Experiments

In order to investigate the properties of contrast classifiers we have conducted experiments on artificial datasets (Section 3.2) and real datasets (Section 3.3). In general all these experiments are performed with help of the TunePareto R package [9].

3.1 Support Vector Machines

The influence of the contrast constraint is investigated in the context of training linear support vector machines (SVM) [14]. Linear SVMs are linear classifiers that optimize following objective:

$$\min_{\mathbf{w},t,\xi} \quad \frac{1}{r}\|\mathbf{w}\|_r^r + C \sum_{i=1}^n \xi_i^l \tag{13}$$

$$\text{s.t.} \quad \forall i: y_i(\mathbf{w}^T \mathbf{x}_i - t) \geq 1 - \xi_i \tag{14}$$

$$\forall i: \xi_i \geq 0 \tag{15}$$

$$\sum_{i=1}^n w^{(i)} = 0 \text{ (only contrasts)} \tag{16}$$

The objective mainly consists of two terms (Equation 13). The first one is a regularization term based on a norm of $\mathbf{w}$. The second one is a loss term that summarizes the losses $\xi_i \in \mathbb{R}_+$ of the constraints in Equation 14. Both terms are combined via a tradeoff parameter $C \in \mathbb{R}_+$. The training algorithm for linear SVMs can be modified to a training algorithm for linear contrasts by adding the constraint of Equation 16.

We consider four different types of SVMs [1], which we denote by $L1R2$ ($l = 1, r = 2$), $L2R2$ ($l = 2, r = 2$), $L1R1$ ($l = 1, r = 1$), $L2R1$ ($l = 2, r = 1$). The l parameter influences the effects of a single loss variable ξ_i. In contrast to a linear weighting (l=1), a value of $l = 2$ increases the influence of large values ($\xi_i > 1$) and decreases the influence of small values ($\xi_i < 1$). The r parameter defines the norm utilized for regularization. A value of $r = 2$ corresponds to the original support vector machine [14]. In this case the SVM training algorithm constructs a large margin classifier that maximizes the distance between the training samples and the hyperplane. A value of $r = 1$ leads to an embedded feature selection [2]. Smaller weights will directly be set to 0. In order to distinguish between SVMs based on linear classifiers and linear contrasts, we will mark the linear contrasts with the subscript con (e.g. $L2R1_{con}$).

3.2 Experiments on Artificial Data

We compare the new contrast classifiers to the standard support vector machines in sequences of experiments on artificial data. The datasets are generated by drawing samples from two normal distributions $\mathcal{N}(\mathbf{c}_y, \mathbf{I})$, $y \in \mathcal{Y}$. Here, $\mathbf{c}_y \in \mathbb{R}^n$ denotes the class wise centroids and $\mathbf{I} \in \mathbb{R}^{n \times n}$ denotes the identity matrix. The first centroid is chosen according to $\mathbf{c}_1 \sim \mathcal{N}(\mathbf{0}, \mathbf{1})$. The second one is calculated by $\mathbf{c}_0 = \mathbf{c}_1 + d\mathbf{w}/\|\mathbf{w}\|_2$, where $\mathbf{w} \sim \mathcal{N}(\mathbf{0}, \mathbf{1})$. In this way, it is guaranteed that $\|\mathbf{c}_1 - \mathbf{c}_0\|_2 = d$.

We generated different training/test scenarios by varying the dimensionality of the datasets $n \in \{2, 5, 10, 250, 500, 1000\}$ and the distance between class centriods $d \in \{1, 1.1, \ldots, 5\}$. Each experiment is based on a training set of 50 samples per class. An independent test set of 2×50 samples is generated for evaluating the classifier. Each combination of n and d is tested for 5 different pairs of centroids.

3.3 Experiments on Real Data

The classifiers are also compared in experiments on real microarray datasets. The utilized datasets are given in Table 1. In these experiments, the classifiers are evaluated in an $r \times f$ cross validation [3]. That is, the data is split into f folds of approximately equal size. One of the folds is removed from the dataset before a classifier is trained. It is used for testing the classifier afterwards. The procedure is repeated for each combination of folds. The cross validation is repeated on r permutations of the original dataset. The error of the classifier is estimated by

$$R_{cv} = \frac{1}{rm} \sum_{i=1}^{r} \sum_{j=1}^{f} \sum_{(\mathbf{x}, y) \in \mathcal{S}_{te}^{i,j}} \mathbb{I}\left[c_{\mathcal{S}_{tr}^{i,j}}(\mathbf{x}) \neq y\right]. \tag{17}$$

Here $\mathcal{S}_{tr}^{i,j}$ and $\mathcal{S}_{te}^{i,j}$ denote the training set and test set of the ith run and the jth split. We have chosen to set $r = 10$ and $f = 10$.

Besides the noise free experiment, we have also conducted experiments with two different types of artificial noise.

Noisy Test Samples: In this setting the samples of a test set $\mathcal{S}_{te} = \{(\mathbf{x}_i, y_i)\}_{i=1}^{m'}$ are affected by a random global transition $f_b \mapsto \mathbf{x} + \mathbf{b}$.

$$\mathcal{S}_{te}' = \{(f_{b_i}(\mathbf{x}_i), y_i)\}_{i=1}^{m'}, \quad \forall i : b_i \sim \mathcal{U}(-p, p) \tag{18}$$

Here $\mathcal{U}$ denotes a uniform distribution. The upper and lower limit of b_i is chosen according to $p \in \{1, \ldots, 5\}$.

Noisy Training Samples: Here we assume that the training set $\mathcal{S}_{tr} = \{(\mathbf{x}_i, y_i)\}_{i=1}^{m}$ is affected by a global transition. We model an offset of the samples of class 1.

$$\mathcal{S}_{tr}' = \{(f_{b_i}(\mathbf{x}_i), y_i)\}_{i=1}^{m}, \quad b_i = \begin{cases} p & \text{if } y_i = 1 \\ 0 & \text{else} \end{cases} \tag{19}$$

Table 1. The table summarizes the four analyzed microarray datasets. The number of features n, the class-wise numbers of samples m_y, $y \in \{0, 1\}$ and the analyzed phenotypes are shown.

name	n	m_1	m_0	phenotypes
Buchholz [4]	169	37	25	pancreatic adenocarcinoma vs panctreatitis
Dyrskjøt [6]	7071	20	20	subclasses of bladder cancer
Pomeroy [10]	7071	9	25	classic vs desmoplastic medulloblastoma
Shipp [11]	7071	19	58	diffuse large B-cell lymphoma vs follicular lymphoma

4   Results

The results of the experiments on artificial data are shown in Figure 3. It can be seen that the standard linear classifiers perform better than their contrast equivalents in low dimensional spaces ($n \in \{2, 5\}$). If the dimensionality is increased it can be seen that the results of the contrasts get closer to those of the standard linear classifiers ($n \in \{10, 250\}$). For the datasets with highest dimensionality ($n \in \{500, 1000\}$) no difference between the standard linear classifiers and their corresponding contrasts can be observed.

For each combination of dimensionality and pair of standard linear classifier and corresponding contrast a two-sided Wilcoxon test has been performed. The resulting p-values are all greater 0.1308 for $n \in \{250, 500, 1000\}$.

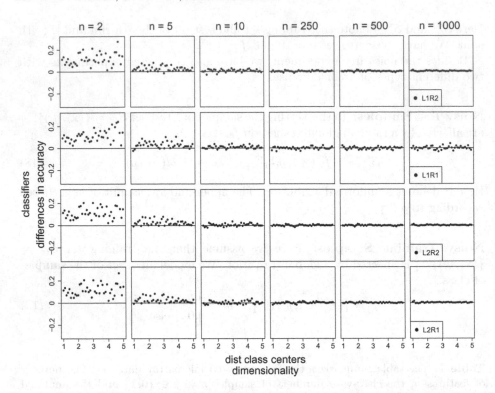

Fig. 3. Results of the artificial experiments. Each row shows the mean differences in accuracy between a support vector machine ($L1R1$, $L1R2$, $L2R1$, $L2R2$) and the corresponding contrast classifier ($L1R1_{con}$, $L1R2_{con}$, $L2R1_{con}$, $L2R2_{con}$). The single columns show experiments with different dimensionality $n \in \{2, 5, 10, 250, 500, 1000\}$. In each cell, the distance between the class centroids is varied $d \in \{1.0, 1.1, \ldots, 5.0\}$. A value above 0 indicates a better performance of the standard support vector machine. A value below 0 indicates a better performance of the contrast classifier.

Table 2 shows the 10×10 cross-validation errors of the noise free experiments on the real datasets. For each of the datasets, one of the contrast classifiers outperforms all other models. Over all datasets the $L2R1$ support vector machines perform better or equal than all other tested classifiers. Interestingly in most of the cases the contrast classifiers outperform the standard linear classifiers. Exceptions for this are the $L1R1_{con}$ on the Buchholz and the Pomeroy dataset.

The results for the experiments with noisy test samples are given in Figure 4. While the contrast classifiers are not affected by the added noise the performance of the standard linear classifiers varies. In most of the cases the error rates are increased or equal compared to the zero noise cases. Only in a few combinations the error rate is slightly decreased for the standard linear classifiers.

The results for noisy training samples are shown in Figure 5. It can be seen that the results of the standard linear classifiers (except the $L1R1$ on the Pomeroy dataset) become worse by an increasing noise level. The contrast classifiers perform equally across all experiments and are not affected by noise.

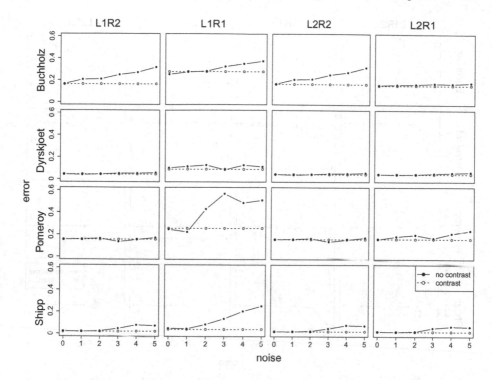

Fig. 4. Noisy test sample experiment on real data: The 10×10 cross-validation errors for different levels of noise are shown. Each cell includes the results for a combination of dataset and pair of classifiers (standard vs corresponding contrast classifier).

Table 2. Results of the noise free 10×10 cross validation experiments on the real datasets. The mean error rates are shown. Lowest errors per dataset are given in bold.

	Buchholz [4]	Dyrskjøt [6]	Pomeroy [10]	Shipp [11]
$L1R1$	25.16	10.25	24.70	4.94
$L1R1_{con}$	27.58	9.00	25.59	4.03
$L1R2$	16.45	**4.50**	**15.59**	2.08
$L1R2_{con}$	16.29	**4.50**	**15.59**	2.08
$L2R1$	15.65	**4.50**	**15.59**	2.08
$L2R1_{con}$	**15.32**	**4.50**	**15.59**	**1.95**
$L2R2$	16.45	**4.50**	**15.59**	2.08
$L2R2_{con}$	16.29	**4.50**	**15.59**	2.08

5   Conclusion

Contrast classifiers describe a functional subclass of linear classifiers that are invariant against global transition. This property can directly be related to the

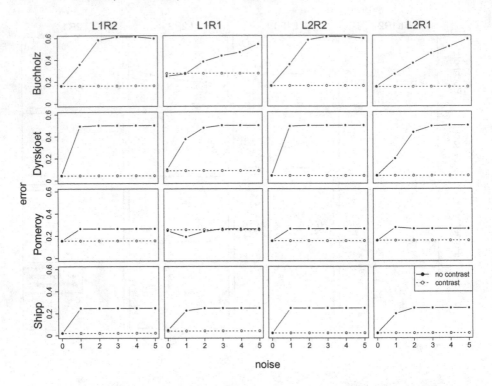

Fig. 5. Noisy training sample experiment on real data: The 10×10 cross-validation errors for different levels of noise are shown. Each cell includes the results for a combination of dataset and pair of classifiers (standard vs corresponding contrast classifier).

structure of a classifier and is independent from a particular training process or a particular dataset.

Our experiments show that invariances can be beneficial, if the samples can not be guaranteed to be affected by a unique source of noise. This is at least the case, if a classifier is trained under well defined conditions and afterwards applied in a real life scenario on noisy test samples. Invariances might also be of interest for clinical multi center studies, which are based on data collected in different laboratories. Here, global effects can also affect the training process of a classifier. Our experiments suggest that biased training samples can be even more corruptive for the prediction accuracy on further samples than noisy test samples.

Nevertheless, an invariance property also reduces the flexibility of a classifier and therefore decreases its complexity. In the case of contrast classifiers, our experiments show that this is only an limitation in low-dimensional settings. In high-dimensional spaces the accuracies of linear classifiers with and without invariance properties are comparable. For lower dimensional settings it may be an option to develop classifier ensembles of contrast classifiers as they inherit the invariance property of the base learners.

Acknowledgements. The research leading to these results has received funding from the European Communitys Seventh Framework Programme (FP7/2007–2013) under grant agreement n°602783 to HAK, the German Research Foundation (DFG, SFB 1074 project Z1 to HAK), and the Federal Ministry of Education and Research (BMBF, Gerontosys II, Forschungskern SyStaR, project ID 0315894A to HAK).

References

1. Abe, S.: Support Vector Machines for Pattern Classification. Springer (2010)
2. Bhattacharyya, C., Grate, L., Rizki, A., Radisky, D., Molina, F., Jordan, M., Bissell, M., Mian, I.: Simultaneous classification and relevant feature identification in high-dimensional spaces: application to molecular profiling data. Signal Processing 83(4), 729–743 (2003)
3. Bishop, C.: Pattern Recognition and Machine Learning (Information Science and Statistics). Springer (2006)
4. Buchholz, M., Kestler, H.A., Bauer, A., Böck, W., Rau, B., Leder, G., Kratzer, W., Bommer, M., Scarpa, A., Schilling, M.K., Adler, G., Hoheisel, J.D., Gress, T.M.: Specialized DNA arrays for the differentiation of pancreatic tumors. Clinical Cancer Research 11(22), 8048–8054 (2005)
5. Casella, G., Berger, R.: Statistical Inference. Thomson Learning (2002)
6. Dyrskjøt, L., Thykjaer, T., Kruhøffer, M., Jensen, J., Marcussen, N., Hamilton-Dutoit, S., Wolf, H., Orntoft, T.: Identifying distinct classes of bladder carcinoma using microarrays. Nature Genetics 33(1), 90–96 (2003)
7. Lausser, L., Kestler, H.A.: Robustness analysis of eleven linear classifiers in extremely high–dimensional feature spaces. In: Schwenker, F., El Gayar, N. (eds.) ANNPR 2010. LNCS, vol. 5998, pp. 72–83. Springer, Heidelberg (2010)
8. Lausser, L., Kestler, H.: Fold change classifiers for the analysis for the analysis of gene expression profiles. In: Gaul, W., Geyer-Schulz, A., Baba, Y., Okada, A. (eds.) Proceedings volume of the German/Japanese Workshops in 2010 (Karlsruhe) and 2012 (Kyoto), Studies in Classification, Data Analysis, and Knowledge Organization, pp. 193–202 (2014)
9. Müssel, C., Lausser, L., Maucher, M., Kestler, H.: Multi-objective parameter selection for classifiers. Journal of Statistical Software 46(5), 1–27 (2012)
10. Pomeroy, S., Tamayo, P., Gaasenbeek, M., Sturla, L., Angelo, M., McLaughlin, M., Kim, J., Goumnerova, L., Black, P., Lau, C., Allen, J., Zagzag, D., Olson, J., Curran, T., Wetmore, C., Biegel, J., Poggio, T., Mukherjee, S., Rifkin, R., Califano, A., Stolovitzky, G., Louis, D., Mesirov, J., Lander, E., Golub, T.: Prediction of central nervous system embryonal tumour outcome based on gene expression. Nature 415(6870), 436–442 (2002)
11. Shipp, M., Ross, K., Tamayo, P., Weng, A., Kutok, J., Aguiar, R., Gaasenbeek, M., Angelo, M., Reich, M., Pinkus, G., Ray, T., Koval, M., Last, K., Norton, A., Lister, T., Mesirov, J., Neuberg, D., Lander, E., Aster, J., Golub, T.: Diffuse large b-cell lymphoma outcome prediction by gene-expression profiling and supervised machine learning. Nature Medicine 8(1), 68–74 (2002)

12. Tulyakov, S., Jaeger, S., Govindaraju, V., Doermann, D.: Review of classifier combination methods. In: Marinai, S., Fujisawa, H. (eds.) Machine Learning in Document Analysis and Recognition, pp. 361–386. Springer (2008)
13. Tibshirani, R., Hastie, T., Narasimhan, B., Chu, G.: Diagnosis of multiple cancer types by shrunken centroids of gene expression. PNAS 99(10), 6567–6572 (2002)
14. Vapnik, V.: Statistical Learning Theory. Wiley (1998)

A New Multi-class Fuzzy Support Vector Machine Algorithm

Friedhelm Schwenker, Markus Frey, Michael Glodek, Markus Kächele,
Sascha Meudt, Martin Schels, and Miriam Schmidt

Ulm University, Institute of Neural Information Processing
89069 Ulm, Germany
friedhelm.schwenker@uni-ulm.de

Abstract. In this paper a novel approach to fuzzy support vector machines (SVM) in multi-class classification problems is presented. The proposed algorithm has the property to benefit from fuzzy labeled data in the training phase and can determine fuzzy memberships for input data. The algorithm can be considered as an extension of the traditional multi-class SVM for crisp labeled data, and it also extents the fuzzy SVM approach for fuzzy labeled training data in the two-class classification setting. Its behavior is demonstrated on three benchmark data sets, the achieved results motivate the inclusion of fuzzy labeled data into the training set for various tasks in pattern recognition and machine learning, such as the design of aggregation rules in multiple classifier systems, or in partially supervised learning.

1 Introduction

In real-world applications such as medical diagnosis or affective computing in an human-computer interaction scenario, the ground truth of the collected data is not always clearly defined, and even human experts have their difficulties to find a correct and unique class label, thus, labeling the collected data in such scenarios is not only expensive and time consuming [4], actually, in some cases it might be impossible to assign a unique label [10]. For instance, when asking a group of medical doctors one by one to categorize the status of a patient, these experts may disagree on the correct class label. Leaving out all such data when designing a training set may lead to small training sets and to classifiers of limited performance. One possible approach to avoid this, is to include all data into the training set, and to express the uncertainty of the class information in terms of fuzzy labels, so that a training set may be given by

$$\mathcal{S} = \left\{ (\mathbf{x}_i, \mathbf{y}_i) \mid \mathbf{x}_i \in \mathbb{R}^d, \ \mathbf{y}_i \in \Delta^L, \ i = 1, \dots, m \right\}$$

where L is the number of classes and $\Delta^L = \{ \mathbf{y} \in [0,1]^L \mid \sum_{j=1}^{L} y_j = 1 \}$ is the set of possible fuzzy memberships. Components of $\mathbf{y}_i \in \Delta^L$ are interpreted as class memberships to the L classes. In this paper the aim is to demonstrate how fuzzy

N. El Gayar et al. (Eds.): ANNPR 2014, LNAI 8774, pp. 153–164, 2014.

memberships can be incorporated into the overall learning process of support
vector machines in multi-class classification.

The paper is organized in the following way: In Section 2 we review the standard SVM approach for binary classification (in Section 2.1) and the multi-class
classification SVM (in Section 2.3). In Section 2.2) we report our previous work
on two-class fuzzy-input fuzzy-output SVM (F^2SVM) (see [12]), and in Section
2.4 the F^2SVM approach is extended to the multi-class classification setting.
In Section 3 we present a statistical evaluation of fuzzy SVM on three data sets
(two artificial and one from optical character recognition), finally we conclude
in Section 4.

2   SVM Learning with Fuzzy Labels

2.1   Review on Binary SVM Classification

Basic principles of SVM classification will be introduced before we consider fuzzy
SVM. Binary classification with crisp labels is the starting point for further
investigations on learning from fuzzy labeled data sets. In the crisp classification
framework, any given observation $\mathbf{x} \in X$ is associated with a corresponding
target label $y \in Y$. It is assumed that X is a compact subset of a real-valued
vector space (i.e., $X \subseteq \mathbb{R}^d$), and that $Y = \{y_1, \ldots, y_L\}$ is the set of L class
labels. The training set is given by

$$S = \{(\mathbf{x}_i, \mathbf{y}_i) \mid \mathbf{x}_i \in X, \mathbf{y}_i \in Y, \ i = 1, \ldots, m\}$$

In case of binary SVM classification we have $y_i \in \{-1, 1\}$. An introduction to
SVM may be found in [13] or [1]. A generalized linear discriminant function with
a fixed nonlinear transformation $\Phi : X \mapsto X'$

$$f(x) = \text{sgn}\left(\mathbf{w}^T \Phi(\mathbf{x}) + w_0\right) \tag{1}$$

classifies all data points correctly if the following conditions are satisfied

$$y_i(\mathbf{w}^T \Phi(\mathbf{x}_i) + w_0) \geq 1 \quad i = 1, \ldots, m. \tag{2}$$

Here $\mathbf{w}$ is a weight vector in X' and $w_0 \in \mathbb{R}$ is a bias parameter. The distance of
the transformed data points $\Phi(\mathbf{x}_i)$ to the separating hyperplane $H_{\mathbf{w},w_0} := \{\mathbf{x} \in X \mid \mathbf{w}^T\Phi(\mathbf{x}) + w_0 = 0\}$ is given by $1/\|\mathbf{w}\|_2$. In order to maximize this distance that
is the *margin* between the data points and the separating hyperplane, we seek for
a solution that is minimizing the cost function $\varphi(\mathbf{w}) := \|\mathbf{w}\|_2^2/2 = \mathbf{w}^T\mathbf{w}/2$ under
the constraints given in Eq. (2). The original SVM the optimization problem is
then formulated as *primal form*:

$$L_P(\mathbf{w}, w_0, \alpha) = \frac{\mathbf{w}^T\mathbf{w}}{2} - \sum_{i=1}^{m} \alpha_i(y_i(\mathbf{w}^T \Phi(\mathbf{x}_i) + w_0) - 1) \tag{3}$$

with Lagrange multipliers $\alpha_i \geq 0$, $i = 1, \ldots, m$. Differentiating L_P with respect to $\mathbf{w}$ and w_0 leads to the conditions $\mathbf{w} = \sum_{i=1}^m \alpha_i y_i \Phi(\mathbf{x}_i)$ and $\sum_{i=1}^m \alpha_i y_i = 0$, respectively. Substituting these conditions in Equation (3) leads to the dual form

$$L_D(\alpha) = \sum_{i=1}^m \alpha_i - \frac{1}{2} \sum_{i=1}^m \sum_{j=1}^m \alpha_i \alpha_j y_i y_j \Phi(\mathbf{x}_i)^T \Phi(\mathbf{x}_j) \tag{4}$$

which must be maximized with respect to the constraints $\alpha_i \geq 0$, $i = 1, \ldots, m$ and $\sum_{i=1}^m \alpha_i y_i = 0$.

Once the multipliers $\alpha_i \geq 0$ have been computed, the weight vector is given by

$$\mathbf{w} = \sum_{i \in \mathcal{SV}} \alpha_i y_i \Phi(\mathbf{x}_i), \tag{5}$$

were $\mathcal{SV}$ is the set of indices of data points with $\alpha_j \neq 0$, the support vectors. From the Karush-Kuhn-Tucker conditions $\alpha_j(y_j(\mathbf{w}^T \Phi(\mathbf{x}_j) + w_0) - 1) = 0$, $i = 1, \ldots, m$, the value w_0 can be determined by averaging over all support vector equations, with $\alpha_j > 0$:

$$\sum_{j \in \mathcal{SV}} y_j(\mathbf{w}^T \Phi(\mathbf{x}_j) + w_0) = |\mathcal{SV}| =: n_{\mathcal{SV}} \tag{6}$$

and therefore

$$w_0 = \frac{1}{n_{\mathcal{SV}}} \left(\sum_{j \in \mathcal{SV}} y_j - \sum_{j \in \mathcal{SV}} \sum_{i \in \mathcal{SV}} \alpha_i y_i \Phi(\mathbf{x}_i)^T \Phi(\mathbf{x}_j) \right). \tag{7}$$

The discriminant function is then determined by substituting Eqs. (5) and (7) into the discriminant function (1).

Since the separations constraints in Eq. (2) can not be fulfilled in realistic data sets they can be relaxed by introducing slack-variables ξ_i, $i = 1, \ldots, m$:

$$\mathbf{w}^T \Phi(\mathbf{x}_i) + w_0 \geq 1 - \xi_i \quad \text{for} \quad y_i = 1$$
$$\mathbf{w}^T \Phi(\mathbf{x}_i) + w_0 \leq -1 + \xi_i \quad \text{for} \quad y_i = -1 \tag{8}$$
$$\xi_i \geq 0 \quad i = 1, \ldots, m.$$

These soft-constraints are incorporated into the cost function $\varphi(\mathbf{w})$ by adding $C \sum_{i=1}^m \xi_i$, with a positive regularization parameter $C > 0$,

$$\varphi(\mathbf{w}, \xi) := \frac{\mathbf{w}^T \mathbf{w}}{2} + C \sum_{i=1}^m (\xi_i^+ + \xi_i^-). \tag{9}$$

The primal form is defined through

$$L_P(\mathbf{w}, w_0, \xi, \alpha, \mathbf{r}) = \frac{\mathbf{w}^T \mathbf{w}}{2} + C \sum_{i=1}^m \xi_i - \sum_{i=1}^m \alpha_i(y_i(\mathbf{w}^T \Phi(\mathbf{x}_i) + w_0) - 1 + \xi_i) - \sum_{i=1}^m r_i \xi_i \tag{10}$$

here $r_i \geq 0$ and $\alpha_i \geq 0$ are the Lagrange multipliers. Differentiating $L_P(\mathbf{w}, w_0, \xi, \alpha, \mathbf{r})$ with respect to $\mathbf{w}$ and w_0 leads again to $\mathbf{w} = \sum_{i=1}^{m} \alpha_i y_i \Phi(\mathbf{x}_i)$ and $\sum_{i=1}^{m} \alpha_i y_i = 0$, differentiating with respect to ξ_i gives the equations $C - \alpha_i - r_i = 0$, $i = 1, \ldots, m$. Substituting them into Eq. (10) yields the dual form:

$$L_D(\alpha, \mathbf{r}) = \sum_{i=1}^{m} \alpha_i - \frac{1}{2} \sum_{i=1}^{m} \sum_{j=1}^{m} \alpha_i \alpha_j y_i y_j \Phi(\mathbf{x}_i)^T \Phi(\mathbf{x}_j) \tag{11}$$

with constraints $C \geq \alpha_i \geq 0$, $i = 1, \ldots, m$ and $\sum_{i=1}^{m} \alpha_i y_i = 0$. Here the upper bound $C \geq \alpha_i$ derived from the equations $C - \alpha_i - r_i = 0$.

The bias term w_0 can be computed as in Eq. (7) by averaging over all support vector equations satisfying $0 < \alpha_j < C$.

At this point it should be mentioned that the optimization of (11) as well as the discriminating function relies only on dot products $\Phi(\mathbf{x}_i)^T \Phi(\mathbf{x}_j)$ which can be replaced in many cases by a kernel function $K(\mathbf{x}_i, \mathbf{x}_j) = \Phi(\mathbf{x}_i)^T \Phi(\mathbf{x}_j)$. This so-called kernel-trick makes the use of SVM very appealing.

2.2   Fuzzy SVM for the Two Class Classification Problem

In the fuzzy-input fuzzy-output Support Vector Machine fuzzy class memberships of the training data are used during training, and a fuzzy output is generated by using a logistic function [9,8,12]. For instance, in a two class classification problem, the class memberships y_i^+ and $y_i^- := (1 - y_i^+)$ for a data point $\mathbf{x}_i$ are incorporated in the SVM training in the following way.

$$\varphi(\mathbf{w}, \xi) := \frac{\mathbf{w}^T \mathbf{w}}{2} + C \sum_{i=1}^{m} (\xi_i^+ y_i^+ + \xi_i^- y_i^-) \tag{12}$$

using slack variables ξ_i^-, ξ_i^+ and constraints

$$\begin{aligned}
\mathbf{w}^T \Phi(\mathbf{x}_i) + w_0 &\geq 1 - \xi_i^+ \quad i = 1, \ldots, m \\
\mathbf{w}^T \Phi(\mathbf{x}_i) + w_0 &\leq -1 + \xi_i^- \quad i = 1, \ldots, m \\
\xi_i^- &\geq 0, \quad \xi_i^+ \geq 0 \quad i = 1, \ldots, m
\end{aligned} \tag{13}$$

as in Eq.(8). This yields the primal form

$$\begin{aligned}
L_P(\mathbf{w}, w_0, \xi, \alpha, \mathbf{r}) = \frac{\mathbf{w}^T \mathbf{w}}{2} + C \sum_{i=1}^{m} (\xi_i^+ y_i^+ + \xi_i^- y_i^-) \\
- \sum_{i=1}^{m} \alpha_i^+ (\mathbf{w}^T \Phi(\mathbf{x}_i) + w_0 - 1 + \xi_i^+) - \sum_{i=1}^{m} r_i^+ \xi_i^+ \\
+ \sum_{i=1}^{m} \alpha_i^- (\mathbf{w}^T \Phi(\mathbf{x}_i) + w_0 + 1 - \xi_i^-) - \sum_{i=1}^{m} r_i^- \xi_i^- \tag{14}
\end{aligned}$$

Differentiation of $L_P(\mathbf{w}, w_0, \xi, \alpha, \mathbf{r})$ with respect to $\mathbf{w}$ and w_0 yields

$$\mathbf{w} = \sum_{i=1}^{m} (\alpha_i^+ - \alpha_i^-)\Phi(\mathbf{x}_i) \quad \text{and} \quad \sum_{i=1}^{m} (\alpha_i^+ - \alpha_i^-) = 0,$$

and differentiation with respect to ξ^+ and ξ^- gives $Cy_i^+ - r_i^+ - \alpha_i^+ = 0$ and $Cy_i^- - r_i^- - \alpha_i^- = 0$ for $i = 1, \ldots, m$. Thus the dual form is given by

$$L_D(\alpha) = \sum_{i=1}^{m} \alpha_i^+ + \sum_{i=1}^{m} \alpha_i^- - \frac{1}{2} \sum_{i=1}^{m} \sum_{j=1}^{m} (\alpha_i^+ - \alpha_i^-)(\alpha_j^+ - \alpha_j^-)\Phi(\mathbf{x}_i)^T\Phi(\mathbf{x}_j) \quad (15)$$

subject to

$$\sum_{i=1}^{m} (\alpha_i^+ - \alpha_i^-) = 0, \quad \text{and} \quad 0 \le \alpha_i^+ \le Cy_i^+, \quad 0 \le \alpha_i^- \le Cy_i^-, \quad i = 1, \ldots, m.$$

The fuzzy SVM approach given in Eq. (12), (14), and (15) reduces to the crisp SVM Eq. (9), (10), and (11), in case of crisp labeled data.

2.3 Multi-class SVM for Crisp Labeled Data

The support vector optimization approach has been applied to the multi-class classification scenario, see for example [13,5,6,2]. In the case of L classes one is considering discriminant functions

$$f_l(x) = \text{sgn}\left(\mathbf{w}_l^T \Phi(\mathbf{x}) + w_{0l}\right) \quad l = 1, \ldots, L \tag{16}$$

with the aim to compute $\mathbf{w}_l^T$ and w_{0l} for $l = 1, \ldots, L$ such that by using the **argmax**-decision rule the training data is separated without error. The **argmax**-decision rule says that a data point $\mathbf{x}$ is assigned to class ω if $\omega = \text{argmax}_l f_l(\mathbf{x})$. Such a solution satisfies the crisp separation conditions

$$\mathbf{w}_k^T \Phi(\mathbf{x}_i) + w_{0k} - (\mathbf{w}_l^T \Phi(\mathbf{x}_i) + w_{0l}) \ge 1 \tag{17}$$

for all data points $\mathbf{x}_i$ where data point $\mathbf{x}_i$ is from class k (denoted by $\mathbf{x}_i \in C_k$), and for all classes $l \in \{1, \ldots, L\}$ with $l \ne k$. The maximal margin solution is then computed by minimizing the cost function

$$\varphi(\mathbf{w}_1, \ldots, \mathbf{w}_L) - \frac{1}{2} \sum_{k=1}^{L} \mathbf{w}_k^T \mathbf{w}_k \tag{18}$$

For non-separable classification problems slack-variables $\xi_i^{k,l}$ for all data points $i = 1, \ldots, m$, and or all classes $l = 1, \ldots, L$ with $l \ne k$ are introduced into the separation constraints. This leads to pairwise soft-constraints:

$$(\mathbf{w}_k^T \Phi(\mathbf{x}_i) + w_{0k}) - (\mathbf{w}_l^T \Phi(\mathbf{x}_i) + w_{0l}) \ge 1 - \xi_i^{k,l} \tag{19}$$

for all data points $\mathbf{x}_i$ from class k_i, and for all classes $j \neq k_i$. These slack-variables $\xi_i^{k_i,l}$ are then introduced into the cost function:

$$\varphi(\mathbf{w}_1, \ldots, \mathbf{w}_L) = \frac{1}{2} \sum_{k=1}^{L} \mathbf{w}_k^T \mathbf{w}_k + C \sum_{k=1}^{L} \sum_{l=1, l \neq k}^{L} \sum_{\mathbf{x}_i \in C_k} \xi_i^{k,l} \tag{20}$$

this leads to a primal from, which is then the starting point for further developments, e.g. derivation of the dural form. We stop at this point and will provide more details for the multi-class SVM in in the fuzzy multi-class setting.

2.4 Fuzzy Multi-class SVM

In the next step we consider the multi-class classification problem with fuzzified class labels, here it is assumed that a training data set is given

$$\mathcal{S} = \left\{ (\mathbf{x}_i, \mathbf{y}_i) \mid \mathbf{x}_i \in R^d, \ \mathbf{y}_i \in \Delta^L, \ i = 1, \ldots, m \right\}$$

where $\Delta^L = \{ \mathbf{y} \in [0,1]^L \mid \sum_{j=1}^{L} y^j = 1 \}$ and L is the number of classes. Following the idea of the two-class fuzzy SVM [8,12] we incorporate the fuzzy class memberships into the cost function in the following form.

For any data point $\mathbf{x}_i$ the values of membership vector $\mathbf{y}_i$ are considered, we assume that they are given in descending order $y_i^{k_1} \geq y_i^{k_2} \cdots \geq y_i^{k_L}$. Fuzzy memberships can be incorporated into the multi-class optimization procedure by pairwise constraints in the following way: For a given data points $\mathbf{x}_i$ and all classes such that $j \neq k(= k_1)$ ($k = k_1$ denotes the class with the largest class membership for $\mathbf{x}_i$) the following constraints are introduced:

$$(\mathbf{w}_k^T \Phi(\mathbf{x}_i) + w_{0k}) - (\mathbf{w}_j^T \Phi(\mathbf{x}_i) + w_{0j}) \geq 1 - \xi_i^{k,j} \tag{21}$$

Overall, for each data point $L-1$ constraints are defined, so $m(L-1)$ constraints in total. The fuzzy memberships can be introduced directly into the cost function:

$$\varphi(\mathbf{w}, \xi) = \frac{1}{2} \sum_{k=1}^{L} \mathbf{w}_k^T \mathbf{w}_k + C \sum_{k=1}^{L} \sum_{\mathbf{x}_i \in C_k} \sum_{l=1, l \neq k}^{L} \xi_i^{k,l} (y_i^k - y_i^l) \tag{22}$$

note that $y_i^k - y_i^l \geq 0$ for all possible combinations, because k denotes the class with the highest membership for data point $\mathbf{x}_i$. The primal form of the fuzzy multi-class SVM problem is then given by

$$L_P(\mathbf{w}, w_0, \xi, \alpha, \mathbf{r}) = \frac{1}{2} \sum_{k=1}^{L} \mathbf{w}_k^T \mathbf{w}_k \tag{23}$$

$$+ C \sum_{k=1}^{L} \sum_{\mathbf{x}_i \in C_k} \sum_{l=1, l \neq k}^{L} \xi_i^{k,l} (y_i^k - y_i^l) - \sum_{k=1}^{L} \sum_{\mathbf{x}_i \in C_k} \sum_{l=1, l \neq k}^{L} \xi_i^{k,l} r_i^{k,l}$$

$$+ \sum_{k=1}^{L} \sum_{\mathbf{x}_i \in C_k} \sum_{l=1, l \neq k}^{L} \alpha_i^{k,l} (1 - \xi_i^{k,l} - ((\mathbf{w}_k^T \Phi(\mathbf{x}_i) + w_{0k}) - (\mathbf{w}_l^T \Phi(\mathbf{x}_i) + w_{0l})))$$

Considering the largest class membership is just one possible approach for the fuzzy multi-class classification scenario. Another way to to take advantage from the class member ships is to define a constraint for each pair of classes k_p, k_q with $y_i^{k_p} \geq y_i^{k_q}$. But this yields $L(L-1)/2$ constraints per data point, so overall $mL(L-1)/2$ constraints. Differentiating with respect to $\mathbf{w}_k^T$ and w_{0k} gives the same constraints for the crisp multi-class classification case.

Differentiating with respect to $\mathbf{w}_k^T$ gives

$$
\mathbf{w}_k^T = \sum_{l=1,l\neq k}^{L} \overbrace{\left(\underbrace{\sum_{\mathbf{x}_i\in C_k} \alpha_i^{k,l}\Phi(\mathbf{x}_i)}_{=:u_k}\right)}^{=:u_k^l} - \sum_{l=1,l\neq k}^{L} \overbrace{\left(\underbrace{\sum_{\mathbf{x}_i\in C_l} \alpha_i^{l,k}\Phi(\mathbf{x}_i)}_{=:v^k}\right)}^{=:v_l^k} \qquad k=1,\ldots,L.
$$

(24)

Differentiating with respect to w_{0k} leads to

$$
0 = \sum_{l=1,l\neq k}^{L}\sum_{\mathbf{x}_i\in C_k} \alpha_i^{k,l} - \sum_{l=1,l\neq k}^{L}\sum_{\mathbf{x}_i\in C_l} \alpha_i^{l,k} \qquad k=1,\ldots,L. \tag{25}
$$

Differentiation with respect to $\xi_i^{k,l}$ gives the conditions

$$
C(y_i^k - y_i^l) - \alpha_i^{k,l} - r_i^l = 0 \quad \text{for} i=1,\ldots,m \text{with} l=1,\ldots,L \text{and} l\neq k. \tag{26}
$$

or as re-formulated as conditions to the $\alpha_i^{k,l}$

$$
C(y_i^k - y_i^l) \leq \alpha_i^{k,l} \leq 0 \quad \text{for} i=1,\ldots,m \text{with} l=1,\ldots,L \text{and} l\neq k. \tag{27}
$$

Now, substitution all these conditions and using shortcuts u_k, v^k for $k=1,\ldots,L$ and u_k^l, v_l^k for $k=1,\ldots,L$ and $l=1,\ldots,L$ $l\neq k$ and $u_l^k = v_l^k$ yields to the corresponding dual from.

$$
L_D(\alpha) = \sum_{k=1}^{L}\sum_{l=1,l\neq k}^{L}\sum_{\mathbf{x}_i\in C_k} \alpha_i^{k,l}
$$

$$
-\frac{1}{2}\sum_{k=1}^{L}\left((u_k)^T u_k + (v^k)^T v^k\right) - \sum_{k=1}^{L}(v^k)u_k \tag{28}
$$

here dot products given through the following equations.

$$
(u_k)^T u_k = \sum_{l=1,l\neq k}^{L}\sum_{\bar{l}=1,\bar{l}\neq k}^{L}\sum_{\mathbf{x}_i\in C_k}\sum_{\mathbf{x}_j\in C_k} \alpha_i^{k,l}\alpha_j^{k,\bar{l}}\left(\Phi(\mathbf{x}_i)\right)^T \Phi(\mathbf{x}_j) \tag{29}
$$

$$
(v^k)^T v^k = \sum_{l=1,l\neq k}^{L}\sum_{\bar{l}=1,\bar{l}\neq k}^{L}\sum_{\mathbf{x}_i\in C_l}\sum_{\mathbf{x}_j\in C_{\bar{l}}} \alpha_i^{l,k}\alpha_j^{\bar{l},k}\left(\Phi(\mathbf{x}_i)\right)^T \Phi(\mathbf{x}_j) \tag{30}
$$

$$(v^k)^T u_k = \sum_{l=1, l \neq k}^{L} \sum_{\tilde{l}=1, \tilde{l} \neq k}^{L} \sum_{\mathbf{x}_i \in C_l} \sum_{\mathbf{x}_j \in C_k} \alpha_i^{l,k} \alpha_j^{k,\tilde{l}} \left(\Phi(\mathbf{x}_i) \right)^T \Phi(\mathbf{x}_j) \qquad (31)$$

The dual form (28) has to be maximized with respect to the constraints (25) and (27).

3   Numerical Evaluation on Benchmark Data Sets

3.1   Data Sets

In this section the numerical evaluation of the proposed fuzzy SVM approach is presented on a realistic benchmark data set consisting of 20,000 hand-written digits (2,000 instances for each class). These digits, normalized in height and width, are represented through a 16×16 matrix G where the entries $G_{ij} \in \{0, \ldots, 255\}$ are values taken from an 8 bit gray scale, see Figure 1. Previously, this data set has been used for the evaluation of machine learning techniques in the STATLOG project and many other studies (see for instance [11]).

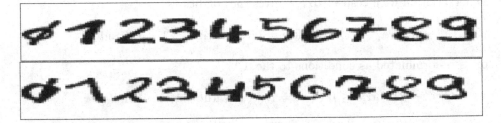

Fig. 1. Data set of hand-written digits. Each instance given through a 16×16 gray scale image (8-bit resolution).

In order to control the degree of fuzziness in the numerical experiments two different types of data sets have been prepared. For this, we define the ball of radius r in $\mathbb{R}^d$ in l_1-norm by $B_d^1(r) := \{\mathbf{x} \in \mathbb{R}^d \mid \|\mathbf{x}\|_1 := \sum_{i=1}^d |x_i| \leq r\}$.

Data set **A** has been sampled according to the uniform distribution of set $B_2^1(2)$ and fuzzy labels for the data points are assigned in the following way: Given an instance $\mathbf{x} = (x_1, x_2) \in B_2^1(2)$ then its corresponding fuzzy class label l is set to the following two-dimensional vector, representing the memberships of the two classes:

$$l := \left(\frac{e^{d(x_1 + x_2)}}{1 + e^{d(x_1 + x_2)}}, \frac{1}{1 + e^{d(x_1 + x_2)}} \right).$$

The parameter $d \geq 0$ is used to control the degree of overlap between the data of the two classes: For small values of parameter d the classes are overlapping, and for increasing d-values the data of the two classes becomes more more and more separated, thus d is reflecting the distance between the data of the class distributions. This data set is used to demonstrate how the fuzzy SVM works in case of weak class memberships.

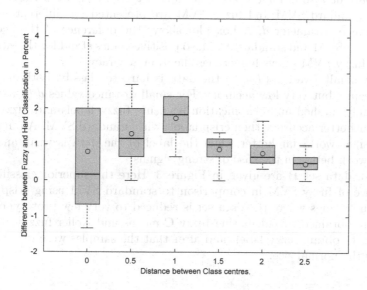

Fig. 2. Results for the artificially generated data set **A**, shown are differences between classification accuracy of crisp and fuzzy SVM for different values of distances d (see text). A box plot shows the difference of classification accuracy between fuzzy SVM and standard SVM; positive difference means that fuzzy SVM performs better than standard SVM. For medium class overlap fuzzy labels are beneficial; for well separated classes ($d = 2.5$) and for highly overlapping classes ($d = 0$) the SVM can not benefit from the fuzzy labels.

Data set **B** is a four-class data set, and has been generated by four bi-variate Gaussian distributions of spherical shape ($\sigma^2 = 1$ in both directions), where each distribution is located in one of the four corners of $B_2^1(2)$. The fuzzy labels are generated by data clustering and fuzzification of the prototypes. The data set is used to show how data set reduction by vector quantization and prototype fuzzification can be applied in classification tasks of big data sets by utilizing fuzzy SVM.

Learning classifiers in a *big data* application is a time consuming task, and thus, instance selection or vector quantization might help to reduce the overall complexity. Clustering or vector quantization are common approaches to compute a small set of representative prototypes out of a larger data set. We applied fuzzy c-means clustering algorithm to compute representative prototypes,

followed by Keller fuzzification [7] of the prototypes, then, the result of the procedure is a small set of prototypes, where each prototype has a fuzzy label derived
from the crisp labels weighted by the cluster membership of the data points.

3.2   Numerical Results

First we present results for the artificial data set **A**. In Figure 2 classification
results for standard SVM and fuzzy SVM are presented for different settings
of the distance parameter d. A box plot shows the difference of the accuracy
between fuzzy SVM and standard SVM, so positive values stand for the situation
where the fuzzy SVM shows higher classification accuracy.

For very small d-values ($d = 0$) the data is hard to classify, both classifiers
show the same, but very low accuracy. For small distance values d (x-axis) the
data is highly meshed and classification by using fuzzy labels and fuzzy SVM
provides far better accuracy then crisp labels with standard SVM. All in all the
fuzzy classifier works far better when the label of the data is weak and hard
classifiers work better in the case of strong signals.

Result of data set **B** are given in Figure 3. Here the superior classification
performance of fuzzy SVM in comparison to standard SVM using crisp labels
is shown in settings where the data set is reduced to very few prototypes. The
results were obtained by calculating fuzzy C-means and Keller fuzzification on
the dataset to obtain fuzzy labels and after that the samples were reduced to a
fraction of the normal size.

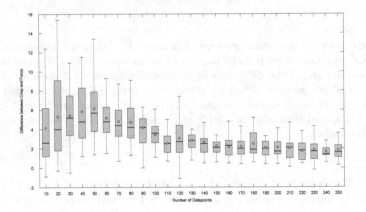

Fig. 3. Results for the artificially generated data set **B**, shown are differences between
classification accuracy of crisp and fuzzy SVM for different numbers of prototypes
$p = 10, \ldots, 250$. A box plot shows the difference of classification accuracy between
fuzzy SVM and standard SVM; positive difference means that fuzzy SVM performs
better than standard SVM. Fuzzy SVM using fuzzy labels is beneficial for a wide range
of degree of data reduction.

Similar behavior of the classification performance can be observed in the digit
dataset (see Figure 4 for the results). It shows the same behavior as dataset **B**

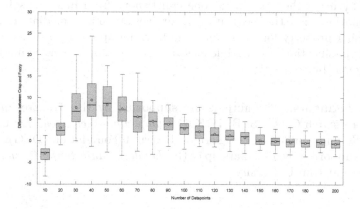

Fig. 4. Results for the digit data set, shown are differences between classification accuracy of crisp and fuzzy SVM for different numbers of prototypes $p = 10, \ldots, 250$. A box plot shows the difference of classification accuracy between fuzzy SVM and standard SVM; positive difference means that fuzzy SVM performs better than standard SVM. Fuzzy SVM using fuzzy labels is beneficial for a wide range of degree of data reduction.

in which for very few data samples the fuzzy approach has better generalization compared to the crisp one. As described above the dataset contains 256 features, corresponding to a grayscale image of a digit. The results were obtained by fuzzification of the labels with the fuzzy-c-means method and for labels which switched class we calculated the Keller algorithm. The digit data set is a real milt-class classification benchmark where a sub set of data points are difficult to classify, e.g. for instance data from the classes $0, 3, 8$ or 9.

4   Conclusion

We proposed a new SVM approach dealing with fuzzy or soft labels in multi-class classification applications. In contrast to other multi-class approaches we introduced a new technique where the fuzzy memberships of all classes are incorporated in an overall cost function. To gain results between the crisp and the fuzzy SVM we considered three datasets, in which two are artificial datasets. As shown above in dataset one the fuzzy approach has a better accuracy than the crisp one for some places were the signal level is weak. This could be helpful in cases where the crisp SVM has problems figuring out the seperation between classes. Furthermore the fuzzy SVM classifier has advantages over the crisp SVM in applications with very few samples as shown in the results for dataset 2 and 3. In these cases a good fuzzification approach can lead to better accuracy because each data point is optimized for each class present. This could also be useful for high dimension low sample size data, if the labels are fuzzified in a suitable way. This could happen either by applying the fuzzy-c-means algorithm or by obtaining the fuzzy labels by hand. In our benchmark data sets we could show that by using the fuzzy SVM one can benefit from fuzzy or soft labeled data in scenarios where the recognition accuracies are in intermediate range, this is a promising property for many machine learning applications, such as semi-supervised classification [3], multiple classifier systems, or in general information fusion systems [14].

Acknowledgements. The authors are supported by the Transregional Collaborative Research Centre SFB/TRR 62 *Companion-Technology for Cognitive Technical Systems* funded by the German Research Foundation (DFG); Markus Kächele is supported by a scholarship of the *Landesgraduiertenförderung Baden-Württemberg* at Ulm University.

References

1. Abe, S.: Support Vector Machines for Pattern Classification (Advances in Pattern Recognition). Springer-Verlag New York, Inc., Secaucus (2005)
2. Bordes, A., Bottou, L., Gallinari, P., Weston, J.: Solving multiclass support vector machines with larank. In: Proceedings of the 24th International Conference on Machine Learning, ICML 2007, pp. 89–96. ACM, New York (2007)
3. Chapelle, O., Schölkopf, B., Zien, A.: Semi-Supervised Learning, 1st edn. The MIT Press (2010)
4. Hady, M.F.A., Schwenker, F.: Semi-supervised learning. In: Bianchini, M., Maggini, M., Jain, L.C. (eds.) Handbook on Neural Information Processing. ISRL, vol. 49, pp. 215–239. Springer, Heidelberg (2013)
5. Hsu, C.W., Lin, C.J.: A comparison of methods for multiclass support vector machines. IEEE Transactions Neural Networks 13(2), 415–425 (2002)
6. Kahsay, L., Schwenker, F., Palm, G.: Comparison of multiclass SVM decomposition schemes for visual object recognition. In: Kropatsch, W.G., Sablatnig, R., Hanbury, A. (eds.) DAGM 2005. LNCS, vol. 3663, pp. 334–341. Springer, Heidelberg (2005)
7. Keller, J.M., Gray, M.R., Givens, J.A.: A fuzzy k-nearest neighbor algorithm. IEEE Transactions on Systems, Man and Cybernetics 4, 580–585 (1985)
8. Lin, C.F., Wang, S.D.: Fuzzy support vector machines. IEEE Transactions on Neural Networks 13(2), 464–471 (2002)
9. Platt, J.: Probabilistic outputs for support vector machines and comparisons to regularized likelihood methods. Advances in Large Margin Classifiers, pp. 61–74 (1999)
10. Scherer, S., Kane, J., Gobl, C., Schwenker, F.: Investigating fuzzy-input fuzzy-output support vector machines for robust voice quality classification. Computer Speech & Language 27(1), 263–287 (2013)
11. Schwenker, F., Kestler, H.A., Palm, G.: Three learning phases for radial-basis-function networks. Neural Networks 14(4-5), 439–458 (2001)
12. Thiel, C., Scherer, S., Schwenker, F.: Fuzzy-input fuzzy-output one-against-all support vector machines. In: Apolloni, B., Howlett, R.J., Jain, L. (eds.) KES 2007, Part III. LNCS (LNAI), vol. 4694, pp. 156–165. Springer, Heidelberg (2007)
13. Vapnik, V.: Statistical Learning Theory. John Wiley and Sons (1998)
14. Zhou, Z.H.: Ensemble Methods: Foundations and Algorithms. Chapman Hall/CRC (2012)

A Reinforcement Learning Algorithm to Train a Tetris Playing Agent

Patrick Thiam, Viktor Kessler, and Friedhelm Schwenker

Ulm University, Institute of Neural Information Processing
89069 Ulm, Germany
friedhelm.schwenker@uni-ulm.de

Abstract. In this paper we investigate reinforcement learning approaches for the popular computer game *Tetris*. User-defined reward functions have been applied to $TD(0)$ learning based on ε-greedy strategies in the standard Tetris scenario. The numerical experiments show that reinforcement learning can significantly outperform agents utilizing fixed policies.

1 Introduction

Tetris is a popular computer game originally invented by the Russian mathematician Alexey Pajitnov in the mid 1980's, and nowadays it is implemented on almost all operating systems and hardware platforms. The standard Tetris board has a size of 200 cells (arranged in 10 columns and 20 rows), where each cell has two states: free or occupied, and thus 2^{200} gaming board configurations are possible. During the game, gaming pieces (shapes of four connected cells, also called *tetrominos*) are dropped from the top of the gaming board into the board and stacked upon occupied cells or the bottom line of the gaming board. In the standard Tetris seven different tetrominos exist, and pieces to be dropped are selected with equal probability. The player can select one out of the ten columns and can rotate the current tetromino before dropping it. When a line of cells is occupied the line is removed and all cells above it are moved down by one line. Each removed line adds to the player's score, and multiple lines can be removed at the same time. The game is over when a cell in the top row is occupied by the current tetromino. The goal is to maximize the score. Because of its complex nature Tetris has been proven to be NP-complete [3]. The consequence of this result is that it is not possible to find an optimal policy effectively, and thus artificial intelligence methods could be of interest to find approximating solutions. Because of its popularity, standard Tetris [8] as well as variations, such as SZ-Tetris [7] have become popular benchmark tests for various machine learning algorithms during the last years.

Neural networks have been successfully applied to numerous real world applications, for instance in pattern recognition, data mining, time series prediction. In recent years several attempts have been made to train artificial neural networks for game playing tasks. For instance, Tesauro [10] has successfully applied feedforward neural networks to play Backgammon. In this scenario artificial neural networks are applied in conjunction with reinforcement learning (RL) algorithms. Combinations of reinforcement learning with artificial neural networks

N. El Gayar et al. (Eds.): ANNPR 2014, LNAI 8774, pp. 165–170, 2014.
© Springer International Publishing Switzerland 2014

and ensemble learning have been successfully applied to board games such as *Connect Four* or *English Draughts* [6,4,5]. Here in this paper we apply temporal difference learning - a well know RL algorithm - to train a Tetris playing agent.

The major goal of this work is to explore and evaluate the effectiveness of reinforcement learning techniques to train a Tetris playing agent. The paper is organized as follows: In Section 2 a brief introduction to RL is presented, then in Section 3 the $TD(0)$ implementation for the standard Tetris application is described. The numerical experiments are shown in Section 4, and finally we discuss results and draw conclusions in Section 5.

2   Introduction to Reinforcement Learning

A reinforcement learning scenario contains two interacting parties: an agent and its environment. We assume that the environment can be completely observed, so for any time step t the environment is in a particular state s_t. Given this sate, the agent can select an action a_t out of a set of possible actions $\mathcal{A}(s_t)$. After the agent has performed an action a_t the environment gives a particular reward $r_t(a_t, s_t)$ to the agent and performs a state transition $s_t \mapsto s_{t+1}$.

The agent's goal is to maximize the sum of rewards over time, for this, a state value function (in the following denoted by V) has to be estimated. Using this information allows the agent to choose appropriate actions with respect to the given task. A comprehensive guide on reinforcement learning can be found in [9].

The *greedy* action a_t^* is determined by taking the one with a maximum sum of reward and value of the following state.

$$a_t^* := \operatorname*{argmax}_{a_t \in \mathcal{A}(s_t)} r_t + \gamma V(s_{t+1})$$

here $\gamma \in (0,1)$ is some discounting factor.

The *policy* defines the strategy used by the agent to choose its next action. Obviously, only these greedy actions are used for testing. In training, it is useful to explore other states and actions. To allow other actions and states to be reached, a random action is taken with a rate of ε. In this work an ε-greedy policy will be use, with $\varepsilon \in [0,1)$. The agent plays a *greedy* action with a rate of $1 - \varepsilon$ (*exploitation*) and a *random* action with a rate of ε (*exploration*).

In case a *greedy* action is chosen, the value of the current state $V(s_t)$ has to be adjusted according to the *temporal difference learning rule* (see Eq. 2). The *greedy* action a_t^* is determined by taking the one with a maximum sum of reward and the weighted value of the following state $r_{t+1} + \gamma V(s_{t+1})$. The reward is a function that assesses the configuration of a state at each given time t giving it a numerical valuation r_t. This function is used to evaluate the next state s_{t+1} and the value obtained r_{t+1} is used in combination with the weighted value of the next state $\gamma V(s_{t+1})$ as a comparison parameter to select the *greedy* action.

By modelling the reinforcement learning scenario as an Markov decision process through $\mathcal{P}_{ss'}^a$, namely the propability of changing from state s to s' under action a, and $\mathcal{R}^a_{ss'}$, the respective reward, one could formulate the relationship between values of an optimal V-function:

$$V^*(s) = \max_{a \in \mathcal{A}(s)} \sum_{s' \in \mathcal{S}} \mathcal{P}^a_{ss'} (\mathcal{R}^a_{ss'} + \gamma V^*(s')) \tag{1}$$

These conditions are called *Bellman equations*, please see [1] for a detailed mathematical analysis.

There are many approaches for estimating such optimal solutions. In this work, we will use the simple *temporal difference learning rule*

$$V(s_t) := V(s_t) + \alpha [r_t + \gamma V(s_{t+1}) - V(s_t)] \tag{2}$$

where $\alpha > 0$ is a small positive learning rate.

3 $TD(0)$-Learning for Tetris

It has been shown in [2] that Tetris cannot be won. Therefore it is less promising to give some rewards at only at the end of the game. To avoid such weak rewards to the agent, a heuristic evaluation function for all the possible states are defined to get some more valuable rewards at any time step t.

The reward functions used in this work have been designed through linear combinations of weighted features. The first two features consist of the value of the highest used column (max^t_{height}) and the average of the heights of all used columns (avg^t_{height}) at each given time t. The next feature consists of the total number of holes between pieces at each given time t (cnt^t_{holes}). The last feature consists of the quadratic unevenness of the profile (U^t_{Pro}). This feature results from summing the squared values of the differences of neighboring columns.

$$r_{t+1} = 5 \times (avg^{t+1}_{height} - avg^t_{height}) + 16 \times (cnt^{t+1}_{holes} - cnt^t_{holes}) \tag{3}$$

$$r_{t+1} = 5 \times (avg^t_{height} - avg^{t+1}_{height}) + 16 \times (cnt^t_{holes} - cnt^{t+1}_{holes}) + (U^t_{Pro} - U^{t+1}_{Pro}) \tag{4}$$

Both reward functions take both next state s_{t+1} and current state s_t into consideration. They describe how good is the transition from the current state to the next state, whereby the higher the returned value the better the state. Furthermore the second reward function (cf. Eq. 4) uses the quadratic unevenness as an additional feature. Later we will see the impact of this particular feature in the performance of the agent.

A tabular representation of the V-Function is too large to be stored in any available memory. Just take into account every one of the 200 cells is allowed to be in 2 different states gives 2^{200} configurations. In order to tackle this problem and reduce the state space to a usable size, the height difference between adjacent columns was used to encode each state. For a given state the height difference between successive columns is computed. Prior to that, a threshold is set to limit the maximum and minimum height difference. In this work the threshold was set to ±3. The possible height differences form a set of 7 values: $\{+3, +2, +1, 0, -1, -2, -3\}$. All height differences outside this range are truncated to ±3. Subsequently, each state is represented as a 9-tuple of values taken from the previous set. Using this method results in reducing the state space to $7^9 \approx 40 \times 10^6$ possible states.

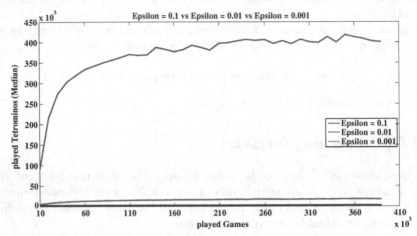

Fig. 1. Learning curves of the RL agent trained through $TD(0)$ with ε-greedy policy and reward function as defined in Eq. (4) $\varepsilon = 0.1$ (green) vs $\varepsilon = 0.01$ (blue) vs $\varepsilon = 0.001$ (red)

4   Numerical Evaluation

Several experiments were undertaken to assess the performance of the implemented agents. They consist primarily of alternating learning and test phases. Prior to that, values for the learning rate $\alpha = 0.1$ and the discount factor $\gamma = 0.9$ were set for the entire experiments. The total number of played gaming pieces per game is used as the performance indicator. At first, the agent is trained through a fixed number of games which depends on the experiment being undertaken. Subsequently, a test phase follows in which the agent is tested by 1000 games. These episodes of alternating learning and test phases are repeated several times in order to achieve a robust estimate of the agent's performance.

Figure 1 shows the performance of the reward function defined in Eq. (4). The agent is trained through 37 episodes of 10000 games each. The total number of played gaming pieces is collected for each played game. The median of these values is plotted at the end of each learning phase. This experiment is repeated for three different ε-values (0.1, 0.01, 0.001).

In order to perform a fair comparison between different ε-values a second experiment was conducted. Here the agent's performances are compared on the basis of *greedy* played gaming pieces instead of the number of played games. A test phase of 1000 games follows each training phase, and again the median value of the total number of played gaming pieces has been taken as performance measure. Figure 2 shows the performance of the second (cf. Eq. 4) reward function. For this experiment a threshold of 10^8 *greedy* played gaming pieces is set. During each training phase, the agent is trained with so many games until this threshold is reached. The experiment is repeated for two different values of ε (0.1, 0.01). The abscissas depict the total number of *greedy* played gaming pieces during the training phase, and the ordinates depict the median value of played gaming pieces during the test phase, whereby the value labeled zero shows the result of the untrained agent. Untrained agents take actions according to the evaluation function given in Eq. (4).

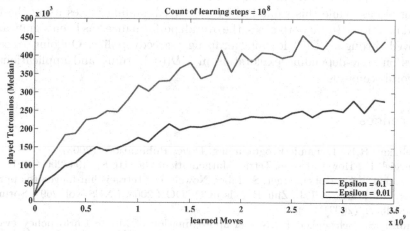

Fig. 2. Test curves of the RL agent trained through $TD(0)$ with ε-greedy policy ($\varepsilon = 0.1$ (blue) vs $\varepsilon = 0.01$ (green))

5 Discussion and Conclusion

After taking a closer look at the results plotted from the first experiment, it is clear that exploration increases with increasing ε values. Thus the number of played gaming pieces decreases. As the number of games per learning phase is constant, the number of *greedy* played game pieces decreases with increasing Epsilon. It follows that the agent needs to be trained with more games to achieve results comparable with those obtained with lower ε values. However, it is not possible through this experiment to determine the effect of ε on the learning performance of the agent. The next experiment serves this purpose.

Through the second experiment a comparison is done between the number of *greedy* played gaming pieces during the training phase and the number of played gaming pieces during the test phase. This comparison depicts at which extend the agent is able to learn when he follows solely its *greedy* policy. Thereby it is also possible to determine the effect of the *greedy* parameter ε on the learning performance of the agent. The results of the experiment for the second reward function (cf. Eq. 4), shows clearly that the agent yields better performances for lower values of ε. Both plotted curves seem to reach a saturation point after playing approximately 2.5×10^9 *greedy* gaming pieces. Consequently, Figure 3 seems to show that larger exploration doesn't seem to help the agent play better. At this point it has to be pointed out that the agent had not been trained further due to the time consumption of both training and test phases. Thus this observation could not be proven throughout further investigation.

For the first reward function given in Eq. 3 learning results are very poor (only approximately 2000 pieces after training with 3.5×10^9 *greedy* gaming pieces) at least in comparison with the reward function given in Eq. 4 (approximately $400,000$ pieces after training with 3.5×10^9 *greedy* gaming pieces). The untrained agent using the second reward function reached a median value of 14000 played gaming pieces, while this value lied by 60 played gaming pieces using the first reward function. But in both cases, the overall performance has been significantly improved through $TD(0)$ learning utilizing ε-greedy policy. Ongoing research focuses on time-depending exploration in $TD(0)$ learning and applications to real world scenarios.

References

1. Bellman, R.E.: Dynamic Programming. Dover Publications (2003)
2. Burgiel, II.: How to lose at Tetris. Mathematical Gazette 81, 194–200 (1997)
3. Demaine, E.D., Hohenberger, S., Liben-Nowell, D.: Tetris is hard, even to approximate. In: Warnow, T.J., Zhu, B. (eds.) COCOON 2003. LNCS, vol. 2697, Springer, Heidelberg (2003)
4. Faußer, S., Schwenker, F.: Neural approximation of Monte Carlo policy evaluation deployed in Connect Four. In: Prevost, L., Marinai, S., Schwenker, F. (eds.) ANNPR 2008. LNCS (LNAI), vol. 5064, pp. 90–100. Springer, Heidelberg (2008)
5. Fußer, S., Schwenker, F.: Learning a strategy with neural approximated temporal-difference methods in English Draughts. In: 2010 20th International Conference on Pattern Recognition (ICPR), pp. 2925–2928. IEEE (2010)
6. Fußer, S., Schwenker, F.: Ensemble methods for reinforcement learning with function approximation. In: Sansone, C., Kittler, J., Roli, F. (eds.) MCS 2011. LNCS, vol. 6713, pp. 56–65. Springer, Heidelberg (2011)
7. Fußer, S., Schwenker, F.: Neural network ensembles in reinforcement learning. Neural Processing Letters, 1–15 (2013)
8. Groß, A., Friedland, J., Schwenker, F.: Learning to play Tetris applying reinforcement learning methods. In: ESANN, pp. 131–136 (2008)
9. Sutton, R.S., Barto, A.G.: Introduction to Reinforcement Learning. MIT Press, Cambridge (1998)
10. Tesauro, G.: Temporal difference learning and TD-Gammon. Commun. ACM 38(3), 58–68 (1995)

Bio-Inspired Optic Flow from Event-Based Neuromorphic Sensor Input

Stephan Tschechne, Roman Sailer, and Heiko Neumann

Inst. for Neural Information Processing, Ulm University,
Albert-Einstein-Allee 1, 89069 Ulm, Germany
{stephan.tschechne,roman.sailer,heiko.neumann}@uni-ulm.de
http://www.uni-ulm.de/in/neuroinformatik.html

Abstract. Computational models of visual processing often use frame-based image acquisition techniques to process a temporally changing stimulus. This approach is unlike biological mechanisms that are spike-based and independent of individual frames. The neuromorphic Dynamic Vision Sensor (DVS) [Lichtsteiner et al., 2008] provides a stream of independent visual events that indicate local illumination changes, resembling spiking neurons at a retinal level. We introduce a new approach for the modelling of cortical mechanisms of motion detection along the dorsal pathway using this type of representation. Our model combines filters with spatio-temporal tunings also found in visual cortex to yield spatio-temporal and direction specificity. We probe our model with recordings of test stimuli, articulated motion and ego-motion. We show how our approach robustly estimates optic flow and also demonstrate how this output can be used for classification purposes. [1]

Keywords: Event-Vision, Optic Flow, Neural Model, Classification.

1 Introduction

Event-based representation of visual information differs from a frame-based approach by providing a continuous stream of changes in the recorded image instead of full frames at fixed intervals. This approach provides numerous benefits: For one, local changes are indicated with very low latency and temporal resolution is immensely increased. The response to local relative luminance changes allows the construction of sensors with an extremely high dynamics range. Simultaneously, data output is reduced to relevant information only, ignoring image regions where no changes occur. Neural modellers profit from the biologically inspired transmission of information based on events or spikes, which allows new and more realistic models of visual processing, because the visual system also operates with a series of spikes generated by light-sensitive retinal neurons. This neural activity is integrated by neurons at subsequent processing stages

---

[1] This work has been supported by grants from the Transregional Collaborative Research Center SFB/TRR62 'A Companion Technology for Cognitive Technical Systems' funded by the German Research Foundation (DFG).

N. El Gayar et al. (Eds.): ANNPR 2014, LNAI 8774, pp. 171–182, 2014.

utilizing increasingly larger receptive fields to build feature representations of higher complexity. In the dorsal pathway of visual processing in cortex, neurons are tuned to patterns of motion in the visual field while they are tuned to static form patterns along the ventral pathway [Ungerleider and Haxby, 1994]. [Lichtsteiner et al., 2008, Liu and Delbrück, 2010] present a neuromorphic approach for a vision sensor that models the mechanisms of spike-generation of a mammalian retina. Their Dynamic Vision Sensor (DVS) generates events whenever a local change of illumination occurs without being bound to temporal frames. Events are marked with the spatial position, phase of illumination change (*on* or *off*) and time-stamp of appearance. This approach provides an energy-efficient and temporally very accurate method to represent temporal changes.

In this work, we propose a novel approach to estimate the apparent motion of visual features (optic flow) using the events generated by the DVS. We model a simplified version of the initial stages of cortical processing along the visual dorsal pathway in primates. In particular, initial responses are generated to represent movements in the spatio-temporal domain, corresponding to V1 direction sensitive cells. This introduces a new approach how biologically inspired models of motion estimation process input.

2   Previous Work

Several research investigations took advantage of the high-speed response properties of event-based sensor design for applications in various domains [Lichtsteiner et al., 2008]. For example, in an assisted living scenario a bank of such sensors can successfully be utilized to detect and classify fast vertical motions as unintended occasional falls, indicative of an preemptive alarm situation [Fu et al., 2008]. Tracking scenarios have been investigated to steer real-world interaction [Delbrück, 2012] as well as in microbiology set-ups [Drazen et al., 2011]. Basic research investigations have been reported as well. For example, stereo matching approaches have been suggested that consider partial depth estimation on the basis of calibrated cameras and the associated epipolar geometry constraints [Rogister et al., 2011]. Using monocular event-based sensing different approaches to optical flow computation have been investigated. In [Benosman et al., 2012] a standard optical flow integration mechanism has been applied [Lucas and Kanade, 1981]. Their approach approximates spatial and temporal derivatives from the event stream to estimate a least-squares solution of flow vectors from intersection of constraint lines. The results show precise tracking of sample inputs, like a bouncing ball or waving hands. More recently, the authors propose to estimate flow directly by evaluating the event-cloud in the spatio-temporal space [Benosman et al., 2014]. In a nutshell, the surface geometry of a small patch of a surface fitted to the cloud of sensor events is evaluated. The local gradient information directly yields an estimate of the ratio of space-time changes and, thus, a local speed measure. In sum, these approaches to motion detection rely on mechanisms that eliminate outliers in order to get reliable speed estimates, either indirectly or directly. [Benosman et al., 2014] rely on local fitting of the data by planar patches such that, movement components orthogonal to the grey value structure can be measured (aperture problem).

We, here, suggest an alternative approach to initial motion detection that is motivated by the first stages of spatio-temporal processing in biological vision. We make use of data about non-directional and directional cells in primary visual cortex and how these generate a first representation of local movements [De Valois et al., 2000]. Below, we outline the general mechanism that makes full use of the event-based input. We demonstrate that the approach is capable to generate spatio-temporal flow estimates for extended boundaries and for intrinsic features. Such estimates can already be used to classify different motion patterns in space and time without the need of sophisticated processing.

3 Methods

Estimation of Optical Flow Using Spatio-Temporal Filters. Spatio-temporal changes of different speed generate oriented structure in x-y-t-space with varying off-axis angles as measured against the temporal axis. These spatio-temporal structures need to be analysed in order to estimate optic flow. The method that we propose herein is inspired by mechanisms found in visual cortex. Our contribution is the adaption of these principles to be compatible with event-based represenation of visual events.

Movement detection in visual cortex is based on cells that are either directionally selective or non-selective [De Valois et al., 2000]. Directional selectivity in cortical cells is generated by linear combination of spatial even-symmetric bi-phasic and odd-symmetric mono-phasic cells as measured orthogonal to the locally oriented contrast. Based on theoretical considerations, [Adelson and Bergen, 1985] have argued in favor of combining cells with even and odd spatial symmetry each having different temporal band-pass characteristics. The final calculation of the response energies lead to demonstrate that such filtering is formally equivalent to spatio-temporal correlation detectors as proposed by [Hassenstein and Reichard, 1956] as a model of motion detection in the fly visual system. We focus here on the findings of [De Valois et al., 2000] and apply a set of spatio-temporal filters on the event stream to yield a selectivitiy to different motion speeds and directions while maintaining the sparse representation provided by the address-event coding. Figure 2 shows the principle of this estimation process. Two spatial filter functions of different class (even and odd-symmetric) and two temporal filter functions (monophasic and biphasic) are combined to build two different spatio-temporal filters. Those can be added to generate a spatio-temporally tuned filter that is able to respond to spatio-temporal structures. Parameters of the filters contribute to the speed- and direction selectivity of the resulting filter. In the following we describe how such filters are modelled and applied to the event-based representation.

Event Representation and Data Structure. A conventional camera projects light onto an image sensor (CCD or CMOS) and reads out the measurement of all light- and colour-sensitive pixels at a fixed frame rate. Several limitations concerning the representation can be identified: The temporal sampling rate is limited, thus

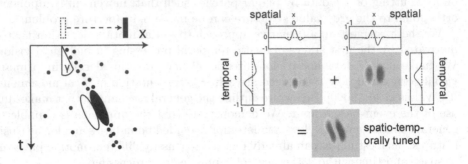

Fig. 1. *Left:* In an event based representation, an object leaves a trail of events when moving across the visual array, along with some spurious events. Depending on the speed, the resulting structure is slanted with angle γ towards the t axis. *Right:* Filters tuned to structures can be generated using two spatially and two temporally tuned functions, motivated by recordings of individual cells in visual cortex.

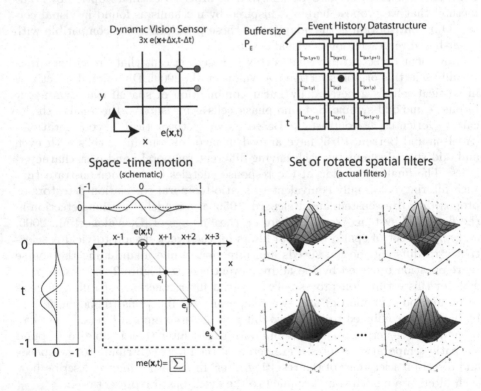

Fig. 2. *Top:* The DVS produces set of events which are stored in an Event History Datastructure for efficient processing. *Bottom:* Spatial and temporal filters are applied to yield sensitivity for motion caused by events.

when all pixel values are sampled in a fixed time interval, fast local changes are integrated and thus lost for further processing. The photo sensitivity of the sensor is limited, thus different illumination conditions force the camera system to globally adapt the operating range using different exposure times or apertures. Finally, from the perspective of sampling the plenoptic function $P(x, y, \lambda, t, V_x, V_y, V_z)$ [Adelson and Bergen, 1991] the corresponding continuous volume is regularly sampled along the t-axis, irrespective of the structure along this dimension. In case that no changes happen to occur redundant information is generated that is carried to the subsequent processing steps.

In an *event-based* system individual pixels generate events in the moment they measure a change in contrast. A single DVS event encodes the characteristic phase of luminance change, namely *on* or *off*, the pixel coordinates of its origin and a timestamp (in microseconds). In this way only luminance changes in the observed scene are recorded and bandwidth can be saved while the dynamic range can be increased [Lichtsteiner et al., 2008]. This happens with temporal sampling rate that is orders of magnitude higher than those of a conventional camera. The way it represents the captured information is referred to as address-event representation [Lichtsteiner et al., 2008].

For the discrete implementation, we suggest the following simple buffering structure to achieve an event-based sampling: On a basis of the x, y-resolution of the sensor (128×128 in case of the DVS) lists with space for P_E timestamps are managed in a first in, first out principle. Whenever a DVS event $e\,(\mathbf{x}, t)$ (with $\mathbf{x} = (x, y)$) occurs, its timestamp t is pushed to the list $L_{(x,y)}$ at the reference grid-position. In case that P_E elements are already held in $L_{(x,y)}$, the last (and hence, oldest) event is pruned. The buffering structure subsequently holds the last P_E timestamps per DVS pixel in a chronological order at each point of time.

Generation of Spatio-Temporal Filters. To model spatial filters, we use a population of rotated two-dimensional Gabor functions, that are calculated using the common definition:

$$G(x, y) = \frac{1}{2\pi\sigma^2} exp(-\pi \left[\frac{(\hat{x} - x_0)^2}{\sigma^2} + \frac{(\hat{y} - y_0)^2}{\sigma^2} \right]) \cdot exp(i[\xi_0 x + \xi_0 y]) \quad (1)$$

$$with \quad \begin{pmatrix} \hat{x} \\ \hat{y} \end{pmatrix} = \begin{pmatrix} cos\theta & -sin\theta \\ sin\theta & cos\theta \end{pmatrix} \begin{pmatrix} x \\ y \end{pmatrix} \quad (2)$$

Even and odd components of this function are used and parameterized to generate 11×11 filter kernels ($x, y \in [-2\pi..2\pi]$) using standard deviation $\sigma = 2.5$, frequency tuning $\xi_0 = 2\pi$ and rotation $\theta = [0, \frac{1}{4}\pi, .., \pi]$. Figure 2 depicts some of the filters used. Temporal filter functions are generated using the following equation:

$$f_t = w_{m1} \cdot exp(-\frac{x - \mu_{m1}^2}{2\sigma_{m1}^2}) - w_{m2} \cdot exp(-\frac{x - \mu_{m1}^2}{2\sigma_{m2}^2}) \quad for \quad x \in [0..1] \quad (3)$$

For monophasic temporal kernels T_1 these values are set to [$w_{m1} = 1.95$ $\mu_{m1} = 0.55$, $\sigma_{m1} = 0.10$, $w_{m2} = 0.23$, $\mu_{m1} = 0.55$, $\sigma_{m2} = 0.16$. For biphasic temporal kernels T_2 the values are set to $w_{m1} = 0.83$ $\mu_{m1} = 0.44$, $\sigma_{m1} = 0.12$, $w_{m2} = -0.34$, $\mu_{m1} = 0.63$, $\sigma_{m2} = 0.21$. The resulting spatio-temporal filter is then generated by $F = G \cdot T_1 + G \cdot T_2$ and applied to incoming events. Due to the sparse representation, a full convolution with the filter is not necessary. Weights are only calculated for the events in the spatial and temporal neighborhood of the newest event, thus saving computational power.

For each application of a spatio-temporal filter the model generates a confidence for the occurrence of this motion direction and speed $C_{s,\theta}$. To maintain a single motion hypothesis, we generate a weighted sum of the fundamental N directions using the estimated confidence value:

$$\begin{pmatrix} u \\ v \end{pmatrix} = \sum_{s=1}^{N} \sum_{\theta=0}^{2\pi} C_{s,\theta} \cdot \begin{pmatrix} cos\,\theta \\ -sin\,\theta \end{pmatrix} \tag{4}$$

Flow Features for Action Recognition. To demonstrate the usability of event-based optic flow features for classification purposes, we perform a small classification scenario. We train classifiers on features extracted from optic flow for the classification of sequences of articulated motions. Our feature extraction roughly follows concepts proposed by [Escobar and Kornprobst, 2012]. Optic flow is estimated using the methods developed in this paper, directly resulting in hypotheses for 16 motion directions, which are binned into 8 directions for classification. From this input, 12 spatial and 4 temporal regions of interest (ROIs) are selected and flow is summed in these regions to calculate motion available motion energy. We perform a principal component analysis (PCA) on the feature vectors to identify the dominant dimensions of the distribution. Energy responses are projected onto the first 10 eigenvectors for a more compact representation. We trained four support vector machines (SVM), one for each class using 32 training samples.

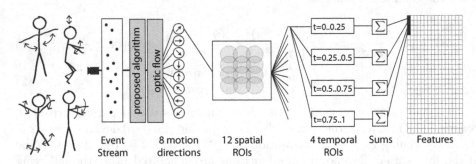

Fig. 3. *Left:* Four actions used for classification demo: *Butterfly, Jumpfrog, JumpJack* and *Harvester. Right:* Schematic of feature extraction. For the classification we extract 384 feature values from the sequences. Eight directional parts are extracted from the motion estimation stage. These are subsequently split into 12 spatial and 4 temporal ROIs, before a summation is performed.

4   Results

We recorded a set of different stimuli to test the performance of our model. Probes were mounted on simple mechanical devices that produced linear or rotational motion. The DVS sensor was mounted on a tripod and the sensor surface aligned parallel and centered to the motion plane. The recorded event streams were caused by translational and rotational motion as well as articulated motion. The sensor as well as our algorithm does not process complete frames but individual sensor events. Individual events cannot sensibly be visualized in printed form. For this reason we integrate estimated motion events, however, over a short period of time. All results presented here used an integration window of $T = 50 \cdot 10^{-3} s$. T therefore lies within a range of commonly used integration times (per frame) used in video technologies which is long enough to ensure that also tiny structures produce enough DVS events for a rich flow visualization. Where adequate we calculated an error measure for our results as follows: With the type of motion known, we synthesize a ground truth vector field of a linear or rotational motion for each sequence. The estimated error is the angular error between 0° and 180° between the synthetic vector field and the estimated motion direction at this position.

Motion events are estimated whenever a new event is processed in our model, in contrast to frame-based approaches where all pixels are processed for one synchronised point in time. For illustration purposes, our model offers the freedom to decide how often a visual output should be generated, ranging from outputs visualising every motion event to an output integrating the events of a complete sequence.

Translatory Motion. We selected three stimuli for tests with a translatory motion, which we label *in italic letters* for future identification: (i) a black bar that moved *orthogonal* to its longer side, (ii) a *tilted* black bar that was rotated 45° relative to its movement direction and (iii) a natural *photography*. The movement speed was approximately $170 mm \cdot sec^{-1}$. Ground truth of this sequence is a constant motion to the right. Results of our tests for translatory motion are displayed in Fig. 4, *Top*. In the figure, we displayed a sketch of the stimulus, the events used for the current motion estimation, ground truth and errors, with the large image depicting the estimated motion vectors.

In *orthogonal*, the results correctly show motion vectors to the right. In the *tilted* case the results shows that our mechanism suffers from the aperture problem because it processes motion events only in a local surround. Motion is only estimated in a direction orthogonal to the contrast orientation in the absence of intrinsically two-dimensional features. However, our algorithm produces correct motion vectors at the ends of the bar. The movement of *photography* shows correct motion estimations but also spurious ones, which are caused by locally tilted contrasts and aliasing problems due to the low resolution of the sensor. The error measure reveals that many estimations are still correct.

Rotational Motion. The second test set was used to acquire events caused by rotational motion. The stimuli were rotated with approx. $85 deg \cdot sec^{-1}$ using a

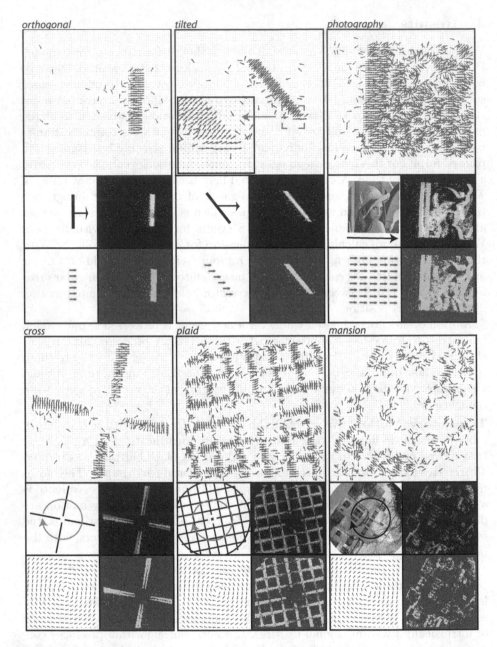

Fig. 4. Processing results for six example stimuli. Each image contains the results of flow detection (*large image*), a stimulus sketch, events used for the estimation, ground truth and illustration of error.(*small images, from top left to bottom right*). *Top:* Three test cases for translational motion: A bar moving *orthogonal* to its orientation, a *tilted* bar and a *photographic* image. *Bottom:* Rotational motion for three test cases: *cross*, *plaid* and *mansion*. Angular errors were estimated using an artificially generated ground truth in all cases.

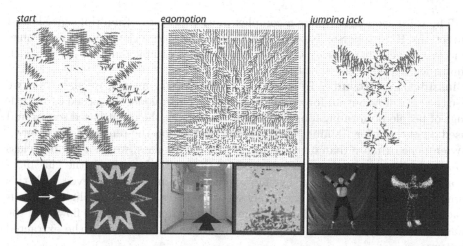

Fig. 5. Test cases for complex motion. *Left:* Correct motion direction is estimated at line endings and normal flow at lines in a complex translational pattern. *Middle:* Results for *ego-motion*. With increased distance to the center, flow quality increases. *Right:* Snippets from a sequence containing articulated motion.

DC motor with adjustable speed. The image plane of the sensor was parallel to the stimulus in a distance of $150mm$. We selected two artificial stimuli, a rotating *plaid* and a rotating *cross* as well as a natural photography of a *mansion*. We also generated a rotational vector field as ground truth. The results for these patterns are more complex and desire more explanations. The arms of the rotating *cross* move orthogonally to their orientation, which is depicted in the figure. The structures of the moving *plaid* are excentric and thus motion estimation locally again suffers from the aperture problem at some positions. The rotating photography of the *mansion* is challenging for our algorithm due to the high rotational speed and the large number of produced visual events. Our model nevertheless achieves to estimate a number of correct estimates.

Complex Motion Sequences. The first sequence highlights the ability of our algorithm to estimate to real flow for local image features. We generate sequences of a moving *star*. At the line endings, the correct direction of optic flow is estimated while normal flow is estimated along the boundaries.

For the second sequence, the event sensor was mounted on a mobile platform and pushed across a hallway to imitate *egomotion*. Forward motion with parallel optical axis causes an expansion field of optic flow, with the focus of expansion centered in the image, see Fig. 5. Our results show that our model can estimate such an expansion field reliably. However, the low spatial resolution of the sensor seems inadequate to differentiate small expanding features close to the focus of expansion. Here, local transitions of width cause spurious local expansion patterns. To overcome these challenges we integrated motion events over the complete sequence length. We applied a 3×3 median filter on the results of this sequence.

The third sequence depicts a complex motion pattern of multiple speeds and directions. Here, a person is performing a *jumping jack* action, consisting of huge and fast movements of arms and legs. This causes a complex motion pattern with multiple directions and speeds.

Classification Results. To round up our line of arguments, we show how we use the optic flow output for classification of simple actions. We picked 64 recordings of people performing four different actions using the event based sensor and used 32 of them for training. Tests were performed on the remaining sequences. When none or more than one classifier voted for its class, we assigned one class randomly. The evaluation was repeated 50 times. Table 1 shows the classification results we achieved with our approach. We emphasize that even with this simple approach, the classes are nicely separated. We want to highlight that our main contribution is the biologically motivated estimation of optic flow, and the classification is just a case study to demonstrate the richness of the extracted features.

Table 1. Confusion matrix for a simple action recognition experiment using the estimated optic flow algorithm. We trained four classifiers on four different actions to demonstrate the feasibility of the approach.

↓Signal	Detection			
	butterfly	*jumpfrog*	*jumpjack*	*harvester*
butterfly Action	100.0%	0.0%	0.0%	0.0%
jumpfrog Action	5.5%	73.0%	6.75%	14.75%
jumpjack Action	7.0%	0.0%	93.0%	12.5%
harvester Action	0.0%	0.0%	0.0%	100.0%

5   Discussion

In this work we presented a biologically inspired mechanism to estimate optic flow using an event-based neuromorphic vision sensor. Our model implements mechanisms of motion detection reminiscent of early stages of processing in visual cortex. Initial motion hypotheses are estimated by detectors tuned to a set of directions. Their responses are integrated at a subsequent level to yield a tuning in the spatio-temporal domain that allows estimation of motion in different directions with increased robustness. Many approaches exist for detecting and integrating motion, among those are biologically motivated [Hassenstein and Reichard, 1956, Adelson and Bergen, 1985] and other algorithms of error minimisation [Lucas and Kanade, 1981, Brox et al., 2004]. Our model differs from other approaches in a number of ways. First, we make use of the event-based representation of visual input instead of processing input in a frame-based manner. This allows our model to generate responses with short latency to the input, while other frame-based models process input at fixed intervals given by the frame rate. In time-critical applications like robotics or mechatronics, immediate availability of flow signals clearly is beneficial. Second, our model profits from the sparse representation that comes with an event-based approach and only processes flow where visual changes occur in the image, leaving those regions unconsidered in the first place where no optic flow is expected. Third, our model incorporates detectors that allow a temporal tuning function with an increased temporal resolution compared to a frame-based approach and thus allows to model the temporal characteristics with increased precision.

Our model was evaluated with a series of stimuli and exhibited the desired behaviour. The model robustly estimated visual motion in a series of test cases with an identical parameter set. Along elongated visual features, the model estimates normal flow due to the aperture problem, but at line endings it generates the correct motion direction under certain conditions. These correct local motion estimates have earlier been used to solve the aperture problem in a model of motion integration [Bayerl and Neumann, 2004] and the mechanisms used therein can be used for our model as well. However, the limited spatial resolution of the sensor handicaps the estimation of accurate optic flow in conditions where shapes are sampled having small and many visual features and complex motions. In many cases, sampling artefacts generate spurious estimations. This problem is however not a conceptual one of our approach and will most likely be solved with future versions of the vision sensor. Optic flow can be used to aquire an enriched feature set because it provides the trace of temporal correspondences of image features. It thus serves well for the extraction of more meaningful features from a scene containing motion and has been used in many domains like action recognition or navigational problems. With the event based processing approach, our models builds the basis for more sophisticated processing models where even the early processing stages have a high level of biologically plausibility. We also briefly demonstrated the power of motion feature extraction for classifying actions from articulated motions. The classification test case should be seen as demonstration and feasibility study, a serious evaluation would undoubtly contain more than 64 samples in four classes.

References

Adelson and Bergen, 1991. Adelson, E.H., Bergen, J.H.: The plenoptic function and the elements of early vision. In: Landy, M., Movshon, J.A. (eds.) Computational Models of Visual Processing, pp. 3–20. MIT Press (1991)

Adelson and Bergen, 1985. Adelson, E.H., Bergen, J.R.: Spatiotemporal energy models for the perception of motion. J. Opt. Soc. Am. 2(2), 284–299 (1985)

Bayerl and Neumann, 2004. Bayerl, P., Neumann, H.: Disambiguating visual motion through contextual feedback modulation. Neural Computation 16(10), 2041–2066 (2004)

Benosman et al., 2014. Benosman, R., Clercq, C., Lagorce, X., Ieng, S.H., Bartolozzi, C.: Event-Based visual flow. IEEE Trans. on Neural Networks and Learning Systems 25(2), 407–417 (2014)

Benosman et al., 2012. Benosman, R., Ieng, S.-H., Clercq, C., Bartolozzi, C., Srinivasan, M.: Asynchronous frameless event–based optical flow. Neural Networks 27, 32–37 (2012)

Brox et al., 2004. Brox, T., Bruhn, A., Papenberg, N., Weickert, J.: High Accuracy Optical Flow Estimation Based on a Theory for Warping. In: Pajdla, T., Matas, J(G.) (eds.) ECCV 2004. LNCS, vol. 3024, pp. 25–36. Springer, Heidelberg (2004)

De Valois et al., 2000. De Valois, R., Cottaris, N.P., Mahon, L.E., Elfar, S.D., Wilson, J.A.: Spatial and temporal receptive fields of geniculate and cortical cells and directional selectivity. Vision Research 40, 3685–3702 (2000)

Delbrück, 2012. Delbruck, T.: Fun with asynchronous vision sensors and processing. In: Fusiello, A., Murino, V., Cucchiara, R. (eds.) ECCV 2012 Ws/Demos, Part I. LNCS, vol. 7583, pp. 506–515. Springer, Heidelberg (2012)

Drazen et al., 2011. Drazen, D., Lichtsteiner, P., Häflinger, P., Delbrück, T., Jensen, A.: Toward real-time particle tracking using an event-based dynamic vision sensor. Exp. Fluids 51, 1465–1469 (2011)

Escobar and Kornprobst, 2012. Escobar, M.J., Kornprobst, P.: Action recognition via bio–inspired features: The richness of center-surround interaction. Computer Vision and Image Understanding 116(5), 593–605 (2012)

Fu et al., 2008. Fu, Z., Delbrück, T., Lichtsteiner, P., Culurciello, E.: An address-event fall detector for assisted living applications. IEEE Trans. on Biomedical Circuits and Systems 2(2), 88–96 (2008)

Hassenstein and Reichard, 1956. Hassenstein, B., Reichard, W.: Functional structure of a mechanism of perception of optical movement. In: Proc. 1st Intl. Congress on Cybernetics, pp. 797–801 (1956)

Lichtsteiner et al., 2008. Lichtsteiner, P., Posch, C., Delbrück, T.: A 128×128 120dB 15μs Latency Asychronous Temporal Contrast Vision Sensor. IEEE Journal of Solid–State Circuits 43(2), 566–576 (2008)

Liu and Delbrück, 2010. Liu, S.C., Delbrück, T.: Neuromorphic Sensory Systems. Current Opinion in Neurobiology 20, 288–295 (2010)

Lucas and Kanade, 1981. Lucas, B., Kanade, T.: An iterative image registration technique with an application to stereo vision. In: Proceedings of the 7th International Joint Conference on Artificial Intelligence, IJCAI, pp. 674–679 (1981)

Rogister et al., 2011. Rogister, P., Benosman, R., Ieng, S.H., Posch, C.: Asynchronous event–based binocular stereo matching. IEEE Transactions on Neural Networks 22(11), 1723–1734 (2011)

Ungerleider and Haxby, 1994. Ungerleider, L.G., Haxby, J.V.: 'what' and 'where' in the human brain. Current Opinion in Neurobiology 4(2), 157–165 (1994)

Prediction of Insertion-Site Preferences
of Transposons Using Support Vector Machines
and Artificial Neural Networks

Maryam Ayat and Michael Domaratzki

Bioinformatics Lab, Department of Computer Science, University of Manitoba,
Winnipeg, MB R3T 2N2, Canada
{ayatmary,mdomarat}@cs.umanitoba.ca

Abstract. Transposons are segments of DNA that are capable of moving from one location to another within the genome of a cell. Understanding transposon insertion-site preferences is critically important in functional genomics and gene therapy studies. It has been found that the deformability property of the local DNA structure of the integration sites, called V_{step}, is of significant importance in the target-site selection process. We considered the V_{step} profiles of insertion sites and developed predictors based on Artificial Neural Networks (ANN) and Support Vector Machines (SVM), and trained them with a Sleeping Beauty transposon dataset. We found that both ANN and SVM predictors are excellent in finding the most preferred regions. However, the SVM predictor outperforms the ANN predictor in recognizing preferred sites, in general.

1 Introduction

Transposons, or jumping genes [1], are short mobile DNA sequences that can insert themselves into the genome of the cell (i.e., host genome) and replicate. They are used in transferring genes of interest into the genome of the target cell and have applications in discovering function of genes (especially those that cause cancer) as well as in gene therapy (e.g., therapy of genetic disorders in humans). However, the applicability of a transposon for these uses depend highly on the target-site selection properties, which are not well understood. Predicting hotspots, or most preferred insertion sites of transposons helps in determining the risks of adverse effects that a transposon insertion may have.

There may be many factors that affect preferences in transposon integration [2], but among the studied factors the local DNA structure has a more effective role in the target-site selection process, as Liu et al. [3] showed for the Sleeping Beauty transposon (SB). Liu et al. found that there is a relationship between the natural deformability property of target sites, which is described by a parameter called V_{step} [4], and the mechanism of the target-site selection of the SB transposon. The composite parameter V_{step} represents the physical relationships of any two planar base pairs in term of their relative displacements and angular orientation in the 3D-structure of DNA [2]. It is a measure of dimer deformability: the higher the V_{step} value, the more deformable the dimer step is, where steps

N. El Gayar et al. (Eds.): ANNPR 2014, LNAI 8774, pp. 183–192, 2014.

Table 1. V_{step} values for dimer steps

dimer	AA	AC	AG	AT	CA	CC	CG	CT	GA	GC	GG	GT	TA	TC	TG	TT
V_{step}	2.9	2.3	2.1	1.6	9.8	6.1	12.1	2.1	4.5	4.0	6.1	2.3	6.3	4.5	9.8	2.9

refer to dinucleotides along a DNA sequence. Table 1 shows the V_{step} values for all possible dimer steps. For the details on V_{step} calculations, see [3,4].

Based on the work done by Liu et al., Geurts et al. [5] analyzed integration-site preferences of SB, piggyBac and Drosophila P-element transposons to detect V_{step} patterns for preferred sites. However, they did not succeed in finding consistent V_{step} pattern for all of the studied transposons. The main drawback of Geurts et al.'s method, in our opinion, is in the way of developing the preference rules, which is more ad-hoc than structured: first, they find the preferred sites based on observations; then, they try to infer the general form of V_{step} patterns of a transposon preferred sites by visually comparing the V_{step} diagrams of observed integration sites. To resolve this weakness, we used machine learning methods for predicting transposon insertion sites.

We considered the insertion site prediction problem as a classification problem, and constructed two types of predictors: one based on Support Vector Machines (SVMs) and the other based on Artificial Neural Networks (ANNs). Both SVMs and ANNs have applications in classification and regression problems, and have been widely used in bioinformatics [6,7]. We employed these predictors for identifying preferred regions (100 bp sequences) in a host genome based on the V_{step} profile of the individual insertion sites (12 bp sequences). To evaluate each predictor, we used a five-fold cross-validation on a SB transposon dataset. Finally, we compared the results of SVM and ANN predictors to each other as well as to Geurts et al.'s results.

2   Materials and Methods

2.1   Dataset

We used an SB transposon integration dataset for training and testing our predictors from the Hackett lab [8]. The main preference of the SB transposon is TA dinucleotides sites in a host genome. We possessed a 7758 bp plasmid pFV/Luc sequence, the actual SB transposon TA integration sites, and the number of hits per integration site in the host sequence. Therefore, our dataset consisted of all TA sites of the 7758 bp plasmid pFV/Luc sequence along with the insertion frequencies. In the 7758 bp plasmid pFV/Luc sequence, there were 489 TA sites with 193 total number of insertions in 97 sites. Similar to Geurts et al. [5], we used the V_{step} profile of a window of 12 bp, including 5 bp flanking each side of a target TA dinucleotide. A V_{step} vector has 11 elements, as there are 11 consecutive dinucleotides in a 12 bp subsequence.

We normalized the V_{step} values and scaled them to the range $[0, 1]$ using the min-max normalization technique. Also, we normalized the integration frequencies to the range $[0, 1]$, since their actual values were very small (less than 0.05).

2.2   Performance Measures

For each predictor, we measured sensitivity (SN), specificity (SP), and the overall accuracy (ACC). They are defined as:

$$SN = \frac{TP}{TP + FN},$$

$$SP = \frac{TN}{TN + FP},$$

$$ACC = \frac{TP + TN}{TP + FP + TN + FN},$$

where TP, FN, TN, and FP refer to the number of true positives, false negatives, true negatives, and false positives, respectively.

To evaluate the strength of a classification, we also generated a Receiver Operating Characteristic (ROC) curve and computed the area under the ROC curve (AUC). An AUC close to 1 indicates a strong test, and an AUC close to 0.5 represents a weak test.

2.3   Support Vector Machine

We used an SVM [9] with a Gaussian Radial Basis Function (RBF): $K(\mathbf{x}_i, \mathbf{x}_j) = e^{-\gamma(\mathbf{x}_i - \mathbf{x}_j)^2}$. SVM is used in its basic form for binary classification, and a version of SVM for function estimation from a set of training data is Support Vector Regression (SVR) [10]. In an epsilon-SVR model the margin of error tolerance, i.e., ϵ, should be set as a parameter. Our designing model for predicting preferences of transposon insertion sites has two phases. In the first phase, we construct the best binary SVM for predicting preferred individual insertion sites, and in the second phase, we construct an SVR with the same architecture as the best SVM resulting from the first phase. This SVR predicts the insertion distribution along the sequence bins (i.e., regions).

To implement our SVM predictior, we used the SVM package LIBSVM [11] in MATLAB environment. In following, we illustrate the SVM architecture for finding preferred individual sites and regions, respectively.

Preferred Individual Sites. In this case, the SVM-based tool predicts if a given 12 bp insertion site is a preferred SB transposon integration site or not. The input data for the SVM is the V_{step} vector for an insertion site of 12 bp, and the output is the label of the class, i.e., +1 or -1 corresponding to a preferred or not-preferred insertion site. We used a binary SVM with a Gaussian kernel for classification. In this architecture, there are two parameters: the soft margin parameter C and the kernel parameter γ. We applied a grid-search [12] on C and γ for selecting parameters. Using this method, we tried various pairs of (C, γ), within the range $1 \leq \log_2 C \leq 5$ and $-3 \leq \log_2 \gamma \leq 1$, and chose the one with the best 5-fold cross-validation accuracy which we measured by the area under the

ROC curve. We also set the parameter C for both positive and negative classes by different weights due to having unbalanced data. The final Gaussian-kernel SVM has the following configuration: the best parameters $(C, \gamma) = (2.0, 0.25)$, and weights for C in positive and negative classes $(w_{+1}, w_{-1}) = (20, 1)$.

To find the best results, we also tried different definitions of preferred insertion sites (i.e., positive class) in terms of the number of hits (i.e., the integration frequency). The best results of SVM were obtained when we assumed a preferred insertion site as a site with more than two integrations.

Preferred Regions. Assuming that we are given the V_{step} scores of a sequence, which is divided into bins of size 100 bp, the SVM predicts the most preferred SB transposon insertion bins (i.e., it predicts the insertion distribution along the sequence bins). In this case, we took advantage of support vector regression. We constructed an epsilon-SVR, with a Gaussian kernel and the same parameters we had found in the binary SVM, to model the relationship between the V_{step} vectors and integration frequencies of insertion sites. We also set the tolerance criterion, ϵ, to 0.001. Then, we ran a 5-fold cross validation over all insertion sites, and obtained the predicted frequency for each insertion site. Afterward, we computed the summation of frequencies for each bin, scaled them to the range [0,1], and obtained the distribution of predicted integration frequencies.

2.4   Radial Basis Function Neural Network

We took advantage of a three-layer RBF neural network [13] for prediction. RBF networks are suited for pattern recognition problems such as this research wherein the pattern dimension is sufficiently small. Similar to the SVM solution, our designing model for predicting preferences of transposon insertion sites has two phases. In the first phase, we construct the best RBF network for predicting preferred individual insertion sites, and in the second phase, we obtain the insertion distribution along the sequence bins based on the best RBF architecture resulting from the first phase.

We constructed our ANN predictor using Open Desire package [14]. In following, we illustrate the ANN architecture for finding preferred individual sites and regions, respectively.

Preferred Individual Sites. In this case, our RBF neural network predicts if a given 12 bp insertion site is a preferred SB transposon insertion site or not. The input data for the ANN is the V_{step} vector for an insertion site of 12 bp, and the output is the insertion frequency which is converted to a binary value 1/0 corresponding to a preferred or not-preferred site. For this purpose, we constructed a set of RBF networks with different configurations and applied a 5-fold cross validation over each to find the best neural network. Finding the best RBF neural network requires searching for the optimal number of hidden units, as well as the parameter σ (,or $\gamma^{-0.5}$) in the radial basis function and the threshold values. We used a destructive method in design. We started with a network with a maximal number of hidden units and connections, and gradually deleted hidden units to reach the optimal network. Meanwhile, we tried to find the best σ by a random selection for each network. We also benefited from ROC curve analysis to find a threshold or cut-off for generating binary outputs. The best RBF network has 262 hidden units and parameter $\sigma = 2.75$.

Similar to the SVM solution, we tried different definitions of preferred insertion sites in terms of the number of hits. The best ANN was obtained from the situation in which we defined a preferred insertion site as a site with more than two integrations.

Preferred Regions. Having the V_{step} scores of a sequence, which is divided into 100 bp bins, the ANN predicts the insertion distribution along the sequence bins and recognizes the most preferred SB transposon insertion regions. Here, we used the same constructed RBF neural network for individual sites, and ran a 5-fold cross validation over all insertion sites, but we did not apply any threshold on output frequencies. Instead, we computed the summation of frequencies for each bin, scaled them to the range [0,1], and obtained the distribution of predicted integration frequencies.

3 Results

3.1 SVM Results

In Individual Sites. Figure 1 shows the ROC curve for the final SVM predictor. The cut-off point recognizes the best binary SVM which has 83% sensitivity and 72% specificity. The area under the curve is 0.85, which indicates that the SVM predictor has a good performance in finding preferred individual insertion sites.

In Regions. Figure 3(a) shows a plot for comparing distribution of observed and predicted insertion sites in the 7758 bp pFV/Luc sequence. The sequence is divided into 77 bins of 100 bp. The plot illustrates an apparent overlap between the two distributions. For example, it shows that the SVM could predict the four most preferred bins (bins #16, #17, #47, and #69) successfully. Therefore, if our concern is to predict the preferred regions in the sequence, then the epsilon-SVR will produce better results compared to the binary SVM for individual sites. Using ROC curve analysis, we found that the epsilon-SVR predictor has 100% SN, 94% SP, and AUC=0.97 in recognizing the most preferred insertion regions. We considered the most preferred insertion regions in our observed data as the bins in which the number of insertions is more than three (i.e., bins in which the scaled insertion frequency is more than 0.4). Also, we found that the epsilon-SVR predictor has 85% SN, 90% SP, and AUC=0.89 in recognizing the preferred insertion regions, which we considered them in our observed data as the bins in which the number of insertions is more than two (i.e., bins in which the scaled insertion frequency is more than 0.3). The AUC values indicate an excellent discriminatory power of the SVM in finding preferred insertion regions.

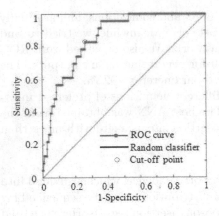
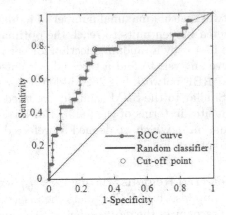

Fig. 1. ROC curve for the best Gaussian kernel SVM in individual sites, AUC=0.85

Fig. 2. ROC curve for the best RBF network in individual sites, AUC=0.71

3.2 ANN Results

In Individual Sites. Figure 2 shows the generated ROC curve for the best network. It revealed 78% SN and 71% SP for the cut-off point. The AUC is 0.71, which shows the classifier has a fairly good discriminatory power.

In Regions. Figure 3(b) shows a plot for comparing distribution of observed and predicted insertion sites in the 7758 bp pFV/Luc sequence. The sequence is divided into 77 bins of size 100 bp. The plot illustrates an apparent overlap between the two distributions. For example, it shows that the neural network could predict the four most preferred bins (bins #16, #17, #47, and #69) successfully. Therefore, if our concern is to predict the preferred regions in the sequence, then our RBF neural network will produce better results compared to predicting individual sites. Using ROC curve analysis, we found that the RBF network predictor has 100% SN, 97% SP, and AUC=0.98 in recognizing the most preferred insertion regions (i.e., bins in which the scaled insertion frequency is more than 0.4). Also, we found that the ANN predictor has 100% SN, 72% SP, and AUC=0.90 in recognizing the preferred insertion regions (i.e., bins in which the scaled insertion frequency is more than 0.3). Both AUCs indicate an excellent discriminatory power of the network in finding preferred insertion regions.

4 Discussion

4.1 SVM versus ANN

We have summarized the performance measures of the ANN- and SVM-based predictors for finding preferred individuals sites and regions in two different tables. Table 2 contains the 5-fold cross-validation result of the predictors in

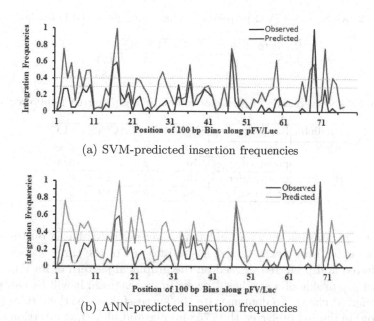

(a) SVM-predicted insertion frequencies

(b) ANN-predicted insertion frequencies

Fig. 3. Distribution of observed versus distribution of SVM and ANN-predicted insertion frequencies in the 7758 bp plasmid pFV/Luc. The sequence is divided into 77 bins of size 100 bp. Dashed lines show threshold values 0.3 and 0.4, used in defining preferred and most preferred insertion bins in observed data, respectively.

individual sits. According to these results, the SVM-based predictor outperforms the ANN-based predictor in identifying preferred individual insertion sites due to having higher AUC, SN and SP. Table 3 contains the 5-fold cross-validation result of the predictors in identifying preferred 100 bp insertion regions. Based on these results, both predictors are excellent in recognizing most preferred insertion regions (similar values for AUC, SN, and SP), but the SVM performs better in identifying preferred regions (similar AUCs, but higher ACC).

Also, it is interesting that comparing the results of either of ANN or SVM predictor in individual sites and regions shows that both ANN and SVM predictors performs better in regions than individual sites. An explanation for this fact might be that the amount of preferability of an insertion site not only depends on the local sequence itself, but also depends on the region that encompasses the insertion site. Therefore, it is worth considering larger sequences of insertion sites than 12 bp as inputs for the ANN and SVM predictors, or adding some region-related features (e.g., the number of TA sites in a region) to the current models in the future.

4.2 Comparison with Related Work

Geurts et al. [5] developed rules for describing the insertion-site preferences of the SB transposon. They did not report any SN, SP or ACC, neither for individual

Table 2. ANN versus SVM predictor in identifying preferred individual sites

Predictor	SN(%)	SP(%)	ACC(%)	AUC
ANN	78	71	72	0.71
SVM	83	72	95	0.85

Table 3. ANN versus SVM predictor in identifying preferred 100 bp regions

Predictor	Prediction	SN(%)	SP(%)	ACC(%)	AUC
ANN	most preferred	100	97	96	0.98
	preferred	100	72	75	0.90
SVM	most preferred	100	94	94	0.97
	preferred	85	90	89	0.89

sites nor for regions, as they had not made any predictor based on their rules. These rules categorizes each 12 bp TA site into one of the three classes - basal, semi-preferred and preferred - based on the graphical pattern of its V_{step} profile (e.g., if a V_{step} profile of a site has 4 peaks in its diagram, it will be categorized to the preferred class). To demonstrate the successfulness of their rules (for SB transposon) in finding preferred the 7758 bp plasmid pFV/Luc insertion regions, Geurts et al. classified all the TA sites based on their rules. Then, according to the ratio of the actual number of insertions in each class to the number of TA sites of that class in the sequence, they provided a formula for calculating the total V_{step} score of a bin of given length in the sequence. Next, they divided the sequence into 100 bp bins, produced the distribution of total V_{step} scores in bins, and compared the result with the distribution of observed insertion sites. In this way, Geurts et al. succeeded to identify the three most preferred bins (bins #17, #47, and #69) in the sequence.

To be able to compare our results with Geurts et al.'s, we used the distribution of total V_{step} scores in the pFV/Luc sequence, and measured the classification power of Geurts et al.'s rules. Consequently, similar to the SVM predictor in regions, we benefited from ROC curve analysis and found that the rules have 100% SN, 89% SP, and AUC=0.97 in finding most preferred bins, and 85% SN, 91% SP and AUC=0.91 in finding preferred bins.

Based on these results, we conclude that:

1. Both SVM and ANN predictors identify the four most preferred bins, while Geurts et al.'s rules recognized the three tops. Due to the higher SP, our predictors perform better in recognizing most preferred regions, compared to Geurts et al.'s rules; and
2. The SVM predictor performs as well as Geurts et al.'s rules in identifying preferred regions.

5   Conclusion

In this paper, we demonstrated how machine learning methods such as SVMs and ANNs can be used for predicting insertion sites of transposons based on the deformability property of the local DNA structure of the integration sites, or their V_{step} profiles. We constructed two predictors based on ANN and SVM methods for identifying insertion-site preferences of SB transposon in a genome, knowing that the main preference of SB trasposon is TA sites. Our model, either for SVM or ANN predictor, had two phases for predicting. In the first phase, we constructed a binary classifier for identifying preferred individual insertion sites (12 bp sites), and in the second phase, we constructed a predictor with the same architecture as the best classifier resulting from the first phase, but this time for regression purposes, or in other words, for predicting the insertion distribution along the sequence bins (100 bp regions). Using five-fold cross validation, we performed the parameter optimization process and evaluation of our SB predictors. However, measuring the performance of the final predictors by testing other host genomes remains as the next step.

We also compared our approach to Geurts et al.'s rule-based method. Our results show that both ANN and SVM predictors outperform Geurts et al.'s heuristic rules in finding the most SB preferred regions. Also, the SVM predictor outperforms the ANN predictor and is as good as Geurts et al.'s rules in recognizing preferred sites in general. However, the main preference of machine learning solutions such as ANNs and SVMs over Geurts et al.'s rule-based method is that these predictors are able to extract the rules, or the relations between inputs and outputs, themselves. It is for this reason Geurts et al.'s ad-hoc rules were not successful in identifying preferred insertion sites of the other transposons in general. Moreover, ANN and SVM predictors are scalable, so some other factors that may influence the insertion-site selection process can easily be modeled in them as new features. Therefore, it is worth constructing other transposon-specific predictors based on these methods as a future work. Such predictors can help direct experiments by helping researchers focus on potential regions of high likelihood of insertion before beginning experiments.

References

1. Pray, L.A.: Transposons: The jumping genes. Nature Education 1(1) (2008)
2. Hackett, C.S., Geurts, A.M., Hackett, P.B.: Predicting preferential DNA vector insertion sites: Implications for functional genomics and gene therapy. Genome Biology 8(S12 Suppl. 1) (2007)
3. Liu, G., Geurts, A.M., Yae, K., Srinivasan, A.R., Fahrenkrug, S.C., Largaespada, D.A., Takeda, J., Horie, K., Olson, W.K., Hackett, P.B.: Target-site preferences of sleeping beauty transposons. Journal of Molecular Biology 346(1), 161–173 (2005)
4. Olson, W.K., Gorin, A.A., Lu, X.J., Hock, L.M., Zhurkin, V.B.: DNA sequence - dependent deformability deduced from protein - DNA crystal complexes. Proceedings of the National Academy of Sciences of the United States of America: PNAS 95(19), 11163–11168 (1998)
5. Geurts, A.M., Hackett, C.S., Bell, J.B., Bergemann, T.L., Collier, L.S., Carlson, C.M., Largaespada, D.A., Hackett, P.B.: Structure-based prediction of insertion-site preferences of transposons into chromosomes. Nucleic Acid Research 34(9), 2803–2811 (2006)

6. Baldi, P., Brunak, S.: Bioinformatics: The machine learning approach, 2nd edn. MIT Press, Cambridge (2001)
7. Seiffert, U., Hammer, B., Kaski, S., Villmann, T.: Neural Networks and Machine Learning in Bioinformatics - Theory and Applications. In: European Symposium on Artificial Neural Networks, ESANN, pp. 521–532 (2006)
8. Hackett, P.B.: Sleeping Beauty transposon insertion data in the 7758 bp plasmid pFV/Luc, [Personal Communication] (2011)
9. Vapnik, V.: Statistical Learning Theory. John Wiley & Sons, Inc. (1998)
10. Smola, A.J., Schölkopf, B.: A tutorial on support vector regression. Statistics and Computing 14(3), 199–222 (2004)
11. Chang, C.C., Lin, C.J.: LIBSVM: A library for support vector machines. ACM Transactions on Intelligent Systems and Technology 2, 27:1–27:27 (2011), Software available at http://www.csie.ntu.edu.tw/~cjlin/libsvm
12. Hsu, C.W., Chang, C.C., Lin, C.J.: A practical guide to support vector classification (2010), http://www.csie.ntu.edu.tw/~cjlin/papers/guide/guide.pdf
13. Haykin, S.S.: Neural Networks: A Comprehensive Foundation. Prentice Hall (2009)
14. Korn, G.A.: Advanced Dynamic-System Simulation: Model-Replication Techniques and Monte Carlo Simulation. John Wiley & Sons, Inc. (2007)

Automatic Bridge Crack Detection –
A Texture Analysis-Based Approach

Sukalpa Chanda[1], Guoping Bu[2], Hong Guan[2], Jun Jo[1], Umapada Pal[3],
Yew-Chaye Loo[2], and Michael Blumenstein[1]

[1] School of Information and Communication Technology,
Griffith University, Queensland 4222, Australia
[2] School of Engineering,
Griffith University, Queensland 4222, Australia
[3] Computer Vision and Pattern Recognition Unit,
Indian Statistical Institute, Kolkata-700108, India

Abstract. To date, identifying cracks in bridges and determining bridge conditions primarily involve manual labour. Bridge inspection by human experts has some drawbacks such as the inability to physically examine all parts of the bridge, sole dependency on the expert knowledge of the bridge inspector. Moreover it requires proper training of the human resource and overall it is not cost effective. This article proposes an automatic bridge inspection approach exploiting wavelet-based image features along with Support Vector Machines for automatic detection of cracks in bridge images. A two-stage approach is followed, where in the first stage a decision is made as whether an image should undergo a pre-processing step (depending on image characteristics), and later in the second stage, wavelet features are extracted from the image using a sliding window-based technique. We obtained an overall accuracy of 92.11% while conducting experiments even on noisy and complex bridge images.

Keywords: Crack Detection, Wavelet-based Crack Detection, Automatic Bridge Inspection, SVMs.

1   Introduction

Physically investigating bridge conditions sometimes becomes unfeasible due to several factors such as the physical surroundings of the bridge, lack of expert knowledge and human resources. Bridges for the purpose of for maintenance and repair requires timely decision-making. Many bridge authorities consult Bridge Management Systems (BMSs) to manage their routine inspection information and to decide on consequent maintenance services. With the advent of sophisticated devices and powerful computers, an effort to automatically conduct bridge inspection has been noted in the recent past. Unfortunately, the proposed methods were not fully capable of addressing the challenges in automatic crack detection. The main difficulties encountered in automatic crack detection methods are variable lighting conditions, random camera/view angles, and random resolution of bridge images. Moreover, we found that

N. El Gayar et al. (Eds.): ANNPR 2014, LNAI 8774, pp. 193–203, 2014.

automatic crack detection gets even harder where the background texture randomly changes and hence segmentation of background and foreground elements becomes very challenging. This article proposes a non-trivial method which addresses the above mentioned challenges efficiently. It relies on a two-stage approach. At first, upon initially analyzing the characteristics of the pixel values in 'R' , 'G' and 'B' channels, the image is identified as either a 'complex image' or a 'simple image'. If the image is identified as a 'complex image' then we need to execute a pre-processing step, otherwise the image is directly processed for feature extraction. Using a non-overlapping sliding window, texture analysis–based features are extracted from the image region beneath the sliding window. Later those features are passed to a Support Vector Machine (SVM) classifier to decide whether the region beneath the sliding window contains a crack or not.

The rest of the article is organized as follows. In Section 2, we discuss related published work, and in Section 3 we describe methodology including the data acquisition process and experimental framework. In Sections 4, the pre-processing step is discussed. Feature extraction and classification approaches are discussed in Section 5. In Section 6 we discuss our experimental results, and finally Section 7 puts forward the conclusions of our paper.

2   Related Work

This section describes some of the existing works in automated crack detection using image processing and pattern recognition techniques. Lee et al. [1] proposed an algorithm for automatic detection of cracks. Their proposed method consisted of crack detection and crack tracing using the difference between the intensity of a crack and its background. Ehrig et al. [2] introduced three different crack detection algorithms namely template matching, sheet filtering based on Hessian eigenvalues, and percolation based on the phenomenon of liquid permeation. Their study focused on determining the suitability of each for crack detection. Mohajeri and Manning [5] proposed a method to identify cracks in concrete using directional filter. They stated that the crack is longitudinal if there is a high concentration of object pixels in a narrow interval of x (transverse) coordinates, and it is transverse if there is a high number of object pixels in a narrow interval of y (longitudinal) coordinates. Tong et al. [6] developed a new method of crack image processing using a pre-processing step which separates crack pixels from the background of the image. Abdel-Qader et al. [3] compared edge-detection algorithms in the context of bridge crack detection using a threshold based approach. Jahanshahi and Masri [4] proposed a morphological operations and Otsu's thresholding based method for segmentation. The purpose of the segmentation process was to reduce unnecessary data in the original image. The appropriate structural element size (in pixels) for the morphological operation was set based on camera focal length, the distance from the object to the camera, camera sensor resolution and size, as well as crack thickness. Oh et al. [8] developed an automatic system for bridge inspection that used median filter in order to remove noise for effective crack detection. Later, morphological operations were applied to determine the connections between crack segments. Lee et al. [9] developed a bridge inspection system that consisted of a robot transportation vehicle, a hydraulic transportation boom and a remote-controlled robot. The remote-controlled robot was used to acquire images of bridge slabs and girders. These images were then sent to an embedded computer for crack detection. Miyamoto et al. [10] developed an automatic crack recognition system to detect crack width on concrete surfaces, where the system could recognize the location and width of cracks. The crack width was computed using the information of difference in brightness in the cracked and non-cracked areas. Flash thermography was used for detecting cracks on concrete surfaces by Sham et al. [11]. In this article we perform a comparative analysis between two different forms of texture analysis features for the purpose of bridge crack detection.

3   Methodology

3.1   Image Acquisition and Dataset Details

We used 50 different images of bridges with a resolution of 5616 × 3744 (21 megapixels), all with random lighting conditions. Based on certain image characteristics those images can be categorized distinctly into two types- "Normal" and "Complex". In "Normal" images we noticed a near consistent background all along the image with a high contrast between the foreground and the background. Whereas in "Complex" images we noted a rapid change in intensities in both foreground and background, or the background was fused with the foreground. We considered 1369 "window" regions (image patches/sub-images) of type 'crack' and 'non-crack' from those images and our experiment was done on these 1369 sub-images.

3.2   Method Overview

At first using a heuristic we automatically try to determine whether an image is of type "normal" or "complex". For "normal" images no further pre-processing is required but for "complex" images we had to perform certain pre-processing steps before features were extracted. In order to analyse the bridge image locally, we deployed a sliding window strategy. For better computational efficiency, a non-overlapping 30x40 pixel window is glided over the entire image and from the region beneath each window (we call them 'window regions') wavelet (and also Gabor filter) features were extracted. Such feature from each window region is classified into a 'crack' or 'non-crack' region by an SVM classifier. The size of the sliding window is set after an empirical analysis of the images. It is noted that the cracks in the images were approximately 25 pixels in width, so a 'crack' region is supposed to contain the crack with the background part, whereas the 'non-crack' region should have the background element only.

3.3   Challenges with "Complex" Image and Their Characteristics

During initial experiments we noticed that our features were able to perform well when the images consisted of a near consistent background all along the image with a high contrast between the foreground and the background, which we term as 'normal' images. However, our features did not perform well with 'complex' images, which had a rapid change in intensities in both foreground and background, or when the images were dull in nature (where the background was fused with the foreground). We noticed that for the 'normal' images, the values of the 'R', 'G' and 'B' channels for a pixel were very close to each other (low standard deviation for all 3 values) and the range of those values was quite extensive. However for 'complex' images, the values of the 'R', 'G' and 'B' channels for a pixel were quite different to each other (high standard deviation for all 3 values) and the range of those values was quite narrow. Using this information (heuristic) we can easily cluster all input images broadly into two groups - 'complex' and 'normal'. Example of a 'complex' and a 'normal' bridge image are shown in Figure 1.

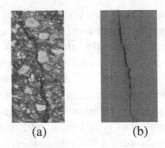

(a)   (b)

Fig. 1. (a) A 'complex' bridge image (b) A 'normal' bridge image

3.4   Motivation Behind Pre-processing Step

After an image is categorized as a 'complex' image, we needed to further process it so that the crack mark became more prominent with respect to its surroundings. After undergoing a series of colour space conversions and filtering of values in various colour space channels, we could obtain an equivalent grey scale image of the complex image. Further on we noticed that if we could process this grey scale equivalent image of the 'complex' images by a contrast stretching algorithm then the same features becomes effective. So we took a two stage approach to deal with this process. In the first stage using our heuristics we decided whether an image is of type 'complex' or type 'normal'. If 'complex' then we process the whole image using our pre-processing methods and then forward it to the feature extraction process. If an image is 'normal' then we do not process it with any pre-processing method and directly start extracting features from it. Since at the current time we could not acquire a large number of bridge images, instead of dividing our entire corpus into training and test subsets we implemented a five-fold cross-validation scheme. The features extracted from all 'crack' and 'non-crack' windows were put together and then we divided all feature vectors into five sets; we used 4 sets for training and the remaining one for testing. The process is repeated 4 more times so that each of the remaining 4 sets in the last training set is used for testing. We also noticed that if we implemented a five-fold cross-validation scheme involving feature vectors from both 'complex' and 'normal' images simultaneously then the accuracy diminishes. We investigated further and found that mostly feature vectors from 'complex' images were being incorrectly classified. Even through tuning values of the SVM parameters the situation did not change. Only after removing all feature vectors that belonged to 'complex' images, the accuracy improved. However, upon implementing a five-fold cross-validation scheme for feature vectors obtained exclusively from 'complex' images, we obtained almost similar accuracy as we achieved on 'normal' images. It is worth mentioning here that the best optimized parameters for feature vectors from two different types of images were quite different.

4   Pre-processing

We executed our pre-processing step on only those images that in our first stage were identified as 'complex' images. The images were in RGB format, but we transformed that to a HSV colour space. The reason behind this is in HSV space the image intensity can be separated from the colour information. Also this transformation for 'complex' type images provided us robustness against lighting changes, and shadows. In the HSV colour space 'Hue' defines the colour component and ranges between 0-1.0 (another scale is 0-360 degree), 'Saturation' describes how white the colour is; whereas the 'value' defines the lightness component in a pixel (0 means white and 1 means completely black). During our initial experiments we noticed that highlighting the crack can be achieved by analysing the Hue and Saturation channel value in the image, and then manipulating them to our desired values. If in a pixel the Hue value is >=0.9 and the corresponding Saturation value is <=0.2 then we change them to Hue=0.6, Saturation=1.0 and Value (intensity/brightness) =0.1; otherwise we set saturation to 0.2 and keep the rest of the two channel values intact. With the first option we are ensuring that the crack pixels are labelled as a proper blue colour with a dark shade (see figure 2b; in the Hue axis 0.6 resembles blue and a Saturation of 1 ensures that the pixel can be visually perceived as the true blue, the low intensity value ensures darkness with respect to the surroundings). With the second option we try to ensure that rest of the pixels become a more grey-like shade by selecting a low saturation value. From the final output image in figure 2(f) it is clearly evident that using our pre-processing step we can easily convert a 'complex' image type as shown in figure 2(a) to appear like a 'normal' image. If we compare figure 1(b) with figure 2(f), it is evident that they look visually similar.

Our pre-processing steps can be outlined as follows:

(i)   RGB to HSV colour space transformation;
(ii)   Check the range of Hue and Saturation values of a pixel and set the values of all H,S,V channels accordingly.
(iii) Conversion to RGB.
(iv)   Then convert the RGB to Grey scale.

(v)   Enhance contrast of the grey scale image by applying histogram equalization technique on the grey scale image.

(vi) Perform final filtering on grey scale values (fix all grey scale values above a certain threshold to one particular high grey scale value) to get the desired output image.

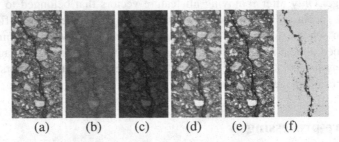

(a)   (b)   (c)   (d)   (e)   (f)

Fig. 2. (a) Extreme left - an original input image, (b)-(e) same image in various intermediate stages; (f) extreme right - corresponding final output image after pre-processing

5   Feature Extraction and Classification

We were keen to study and perform a comparative analysis between different texture analysis-based methods for the purpose of crack detection. All those different features were extracted from the sliding window that glided over the image. We experimented with two different texture-analysis based features, which included Gabor filter features and Daubchies Wavelet features. Wavelet features outperformed the Gabor filter features in our experiments. Since description on all those features are easily available we are not describing them further vividly but only providing a short description of the Gabor and wavelet feature.

5.1   Gabor Filter

The Gabor filters are band-pass filters which essentially do texture analysis. The response of these filters is the product of a Gaussian envelope function multiplied with a complex oscillation [17].

The Gabor filter response image with respect to a crack region and a non-crack region is shown in figure 3. Details about Gabor filter can be found in [18].

5.2   Wavelet Features

The wavelet transform is a useful technique used to analyze non-stationary signals in time-frequency domain. Daubechies wavelets are a family of orthogonal wavelets defining a discrete wavelet transform. This consists of 4 scaling function coefficients and 4 wavelet function coefficients. The four scaling function coefficients used in our experiments were as follows:

$$\frac{1+\sqrt{3}}{4\sqrt{2}}, \frac{3+\sqrt{3}}{4\sqrt{2}}, \frac{3-\sqrt{3}}{4\sqrt{2}}, \frac{1-\sqrt{3}}{4\sqrt{2}}$$

The sliding window is glided all over the grey scale image. The image region beneath each window is copied. We extracted features after size normalizing each such grey scale image regions obtained from the sliding windows to 32 x 32 dimensions. The wavelet response image with respect to a crack region and a non-crack region is shown in figure 4. Details of the feature can be found in [14].

5.3   Classifier Details

In our experiments, we have used Support Vector Machines (SVMs) for classification. In our experiments, we noted that the Gaussian kernel SVM outperformed other non-linear SVM Kernels; hence we are reporting our classification results based on the Gaussian kernel only. The best Kernel parameters were selected for each class type by means of a series of validation experiments. The best optimized results were obtained when ($1/2\sigma 2$) in the Gaussian kernel was set to values such as 80.00 (while dealing with 'normal' images) and 9.00 (while dealing with 'complex' images) with the penalty multiplier value set to 1.

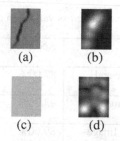

(a)   (b)

(c)   (d)

Fig. 3. (a) Top left - an original input image with crack, (b) Corresponding Gabor response of crack image.(c) Bottom left - an original input image without crack, (d) Corresponding Gabor response of non-crack image.

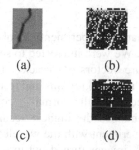

(a)   (b)

(c)   (d)

Fig. 4. (a) Top left - an original input image with crack, (b) Corresponding Wavelet response of crack image.(c) Bottom left - an original input image without crack, (d) Corresponding Wavelet response of non-crack image.

6   Results and Discussion

We did some analysis on our experimental results to provide more insight to our proposed method. As we have mentioned earlier, when five-fold cross validation was implemented separately on feature vectors from 'complex' and 'normal' images we obtained higher accuracy compared to when we implemented five-fold cross validation on feature vectors from both window image types together. Here we are reporting accuracy only on Wavelet features. This fact is depicted in Table 1. We can see that we obtained 87.06% accuracy when feature vectors from 'complex' and 'normal' images were considered together. Similarly while considering feature vectors from only 'normal' images we obtained 93.26% (873 correctly classified considering 936 samples from 'normal' images during five-fold cross validation) whereas while considering feature vectors from only 'complex' images (388 correctly classified considering 433 samples during cross validation) we obtained 89.60%. Thus the average accuracy of our systems becomes 92.11%. In Table 2 we try to inspect the performance of our system by training it using feature vectors exclusively obtained out of one particular image type ('complex'/ 'normal') and testing it on the other image type. In Table 3 we provide a comparative analysis of two different features and followed by an error analysis in Sub-section 6.3.

Table 1. Five-fold cross validation accuracy

Image Type	Accuracy	
Complex Image and Normal Image	87.00%	
Only Normal Image	93.26%	92.11%
Only Complex Image	89.60%	

6.1   Effect on Performance Due to Complex (Normal) Images in Training (Test) Set

We have mentioned earlier that our experiments involved two different types of images: 'complex' and 'normal'. We were interested to see what happens when we only train our classifier with feature vectors of "crack" and "non-crack" image regions obtained from all 'complex' ('normal') regions and test them with "crack" and "non - crack" image regions obtained from all 'normal' ('complex') images. Since during our earlier experiments we obtained the highest accuracy while using wavelet features, we are reporting this experiment with the wavelet feature only. From the results we can conclude that 'normal' images turned out to be much better as a training set and provides us a more generalized learning model.

Table 2. Effect of training image types on accuracy

Train set type	Test Set type	Accuracy
Complex Image	Normal Image	78.95% (739 out of 936)
Normal Image	Complex Image	87.06% (377 out of 433)

6.2   Comparison between Texture Analysis Based Features

Here in Table 3 we provide a comparison between the two different feature extraction methods. Note here we are reporting the accuracy while dealing with all feature vectors simultaneously irrespective of the image type (complex/normal). We obtained highest accuracy with Wavelet features.

Table 3. Comparison between two different features

Gabor Filter	74%
Wavelet	87%

6.3   Error Analysis

Upon analyzing the errors we noticed that most of the time 'window regions' with a blurred appearance were misclassified to the wrong class. This happened to 'window regions' obtained from both 'complex' and 'normal' image types where the foreground element was not prominent compared to the background element in the images and that they tend to fuse with each other. Nevertheless, it is worth mentioning here that in such images our contrast stretching algorithm did not perform well, which is one of the reasons behind not recognizing the cracks. An example of such an image is shown in figure 5. It should be noted that the region marked within the rectangular area highlights a crack mark, which is almost invisible there; however the crack mark is more visible in regions above the rectangular area.

Fig. 5. An invisible crack mark within the rectangular region

7   Conclusions and Future Works

In this paper, we have investigated the problem of automatic bridge crack detection in bridge images. Two different features (Gabor filter and Wavelet) were evaluated for this purpose. The novel issue with our proposed system is that we have obtained encouraging accuracy even while dealing with complex bridge image types with heterogeneous background and foreground characteristics. However, the present preprocessing is based on a threshold approach, which prevents it to be equally efficient under all kind of images. In future we shall look for a more robust technique to accomplish our objective. Further future work includes autonomous image data acquisition using devices such as robots or UAVs. Obtaining an image at a specific position in high precision is not a trivial task, when using an autonomous device. Various sensors, such as optical, acoustic and magnetic sensors, may aid in this task. Multiple sensors, based on individual specialties, are commonly used in order to complement limitations imposed by certain sensors and thus enrich the perception of single sensors. However, it is challenging to integrate the heterogeneous types of sensory information and produce useful results. A pilot study of the likelihood-based data fusion system has been implemented for robot positioning [15] [16]. This system integrates a Light Detection And Range (Lidar), a vision sensor (a webcam) and an Inertial Measurement Unit (IMU). The implementation outcomes showed promising results [15].

References

1. Lee, J.H., Lee, J.M., Park, J.W., Moon, Y.S.: Efficient algorithms for automatic detection of cracks on a concrete bridge. In: The 23rd International Technical Conference on Circuits, Systems, Computers and Communications, Shimonoseki, Japan, pp. 1213–1216 (2008)
2. Ehrig, K., Goebbels, J., Meinel, D., Paetsch, O., Prohaska, S., Zobel, V.: Comparison of Crack Detection Methods for Analysing Damage Processes in Concrete with Computed Tomography. In: International Symposium on Digital Industrial Radiology and Computed Tomography, Berlin, Germany, P2 (2011)
3. Abdel-Qader, I., Abudayyeh, O., Kelly, M.E.: Analysis of edge-detection techniques for crack identification in bridges. Journal of Computing in Civil Engineering, American Society of Civil Engineers 17(4), 255–263 (2003)
4. Jahanshahi, M.R., Masri, S.F.: A novel crack detection approach for condition assessment of structures. In: ASCE International Workshop on Computing in Civil Engineering, Miami, Florida, pp. 388–395 (2011)
5. Mohajeri, M.H., Manning, P.J.: ARIA: An operating system of pavement distress diagnosis by image processing. Transportation Research Record (1311), 120–130 (1991)
6. Tong, X., Guo, J., Ling, Y., Yin, Z.: A new image-based method for concrete bridge bottom crack detection. In: Image Analysis and Signal Processing, Hubei, China, pp. 568–571 (2011)
7. Yamaguchi, T., Hashimoto, S.: Practical image measurement of crack width for real concrete structure. Electronics and Communications, Japan 92(10), 605–614 (2009)
8. Oh, J.K., Jang, G., Lee, J.H., Yi, B.J., Moon, Y.S., Lee, J.S., Choi, Y.: Bridge inspection robot system with machine vision. Automation in Construction 18(7), 929–941 (2009)
9. Lee, J.B., Shin, D.H., Seo, W.J., Jung, J.D., Lee, Y.J.: Intelligent bridge inspection using remote controlled robot and image processing technique. In: International Symposium on Automation and Robotics in Construction (ISARC), Seoul, Korea, pp. 1426–1431 (2011)

10. Miyamoto, M., Konno, M.A., Bruhwiler, E.: Automatic Crack Recognition System for Concrete Structures Using Image Processing Approach. Asian Journal of Information Technology 6, 553–561 (2007)

11. Sham, F.C., Chen, N., Long, L.: Surface crack detection by flash thermography on concrete surface. Insight - Non-Destructive Testing and Condition Monitoring 50(5), 240–243 (2008)

12. Burges, C.: A Tutorial on support Vector machines for pattern recognition. Data Mining and Knowledge Discover 2, 2 (1998)

13. Vapnik, V.: The nature of statistical learning theory. Springer (1995)

14. Daubechies, I.: The wavelet transform, time-frequency localization and signal analysis. IEEE Transactions on Information Theory 36(5), 961–1005 (1990)

15. Jo, J., Tsunoda, Y.: A Data Fusion Model based on ROI and Likelihood for the Integration of Multiple Sensor Data. Accepted and will appear in the Proceedings of the 2nd International Conference on Robot Intelligence Technology and Applications 2013. Springer, Germany (2013)

16. Jo, J., Tsunoda, Y., Sullivan, T., Lennon, M., Jo, T., Chun, Y.: BINS: Blackboard-based Intelligent Navigation System for Multiple Sensory Data Integration. In: The 17th International Conference on Image Processing, Computer Vision, & Pattern Recognition, Nevada, USA (2013)

17. http://homepages.inf.ed.ac.uk/rbf/CVonline/LOCAL_COPIES/TRAPP1/filter.html

18. http://mplab.ucsd.edu/tutorials/gabor.pdf

Part-Based High Accuracy Recognition of Serial Numbers in Bank Notes

Bo-Yuan Feng[1], Mingwu Ren[1], Xu-Yao Zhang[2], and Ching Y. Suen[3]

[1] School of Computer Science, Nanjing University of Science and Technology,
Nanjing 210094, P.R. China
fengboyuannj@gmail.com
[2] National Laboratory of Pattern Recognition, Institute of Automation of Chinese
Academy of Sciences, Beijing 100190, P.R. China
xyz@nlpr.ia.ac.cn
[3] Centre for Pattern Recognition and Machine Intelligence, Concordia University,
Montreal H3G 1M8, Canada
suen@cenparmi.concordia.ca

Abstract. This paper proposes a novel part-based character recognition method for a new topic of RMB (renminbi bank note, the paper currency used in China) serial number recognition, which is important for reducing financial crime and improving financial market stability and social security. Given an input sample, we first generate a set of local image parts using the Difference-of-Gaussians (DoG) keypoint detector. Then, all of the local parts are classified by an SVM classifier to provide a confidence vector for each part. Finally, three methods are introduced to combine the recognition results of all parts. Since the serial number samples suffer from complex background, occlusion, and degradation, our part-based method takes advantage of both global and local character structure features, and offers an overall increase in robustness and reliability to the entire recognition system. Experiments conducted on a RMB serial number character database show that the test accuracy boosted from 98.90% to 99.33% by utilizing the proposed method with multiple voting based combination strategy. The part-based recognition method can also be extended to other types of banknotes, such as Euro, U.S. and Canadian dollars, or in character recognition applications with complex backgrounds.

Keywords: RMB seiral number, part-based, character recognition, classifier outputs combination.

1 Introduction

In the community of handwriting recognition, much attention has been paid to online and offline handwriting recognition and printed character recognition. As RMB circulation management becomes a serious problem to China's financial industry in recent years, a high reliability RMB serial number recognition system is needed. However, little research has been done on bank notes serial number

N. El Gayar et al. (Eds.): ANNPR 2014, LNAI 8774, pp. 204–215, 2014.

Fig. 1. RMB images scanned by contact image sensor

recognition (e.g. [1] [2]). In this paper, we investigate the recognition of RMB serial numbers which consist of 10 printed characters (including alphabetic and numeru characters). Fig. 1 shows two scanned RMB images with serial numbers (marked by red rectagles), which have been designed uniquely and used as the identification of RMB.

In recent years, a few RMB character classification techniques have been proposed to improve the recognition performance by artificial neural network (ANN) and support vector machine (SVM). However, no previous study has achieved high accuracy so far. In [1], the back propagation (BP) artificial neural network based on genetic algorithm training method achieved the accuracy of 95%. The serial number identification system based on SVM [2] yields high recognition result on brand new printed RMB, however, the recognition of used RMB serial number with complex background is much more complicated and not studied in that paper.

To obtain RMB serial number characters, we proposed a RMB serial number extraction method in [3]. First, the skew correction and orientation identification are used to detect the region containing RMB serial number. Then we binarize the text region and extract RMB characters by a local contrast average scheme. Overlap recall rate of 79.68% and precision rate of 98.10% are achieved. Recently, a RMB serial number database has been released [4]. In [4], we comprehensively compared different types of feature extraction methods, classifiers, multiple classifier combination strategies, distortion methods, and rejection schemes on the new database, and provided a large amount of experiment results, which could greatly profit further research of RMB banknote recognition.

The character samples in the RMB serial number database contain complex security texture and suffer from inaccurate extraction, various illumination and contamination. Specifically, there are some small circles located around the character strokes, which will strongly affect the classification process and increase the challenge on the recognition task. These color-marked circles shown in Fig. 2 called "EURion constellation" [5] are designed for the prevention of counterfeiting using color photocopiers. As we observed, there are always some parts in the character sample without circles or complex background textures which can provide much more discriminative features for classification than the other parts. It is possible to recognize the input sample by taking advantage of these local parts.

(a) Circles in serial number region.

(b) Circles in character samlpes.

Fig. 2. EURion constellation circles

In this paper, we present a novel part-based method for the recognition of RMB serial number characters. For each training sample, a set of local image parts are synthesized by Difference-of-Gaussians (DoG) keypoint detector [6]. We train an SVM classifier [7] with both the original and local-part images. During the test, each part of the test sample is recognized by SVM, and the final category is determined via combining the classification results of all parts. Suen *et al.* [8] [9] proposed an alphanumeric handprints recognition method by parts, in which they produce a more effective character recognizer based on the probability of occurrence of the patterns. However, they manually choose the local image parts instead of using an automatic interest point detector.

Compared with the previous part-based handwritten character recognition methods [10] [11], our method has some advantages. First, given an interest point, instead of describing the local part by complicated speeded-up robust features (SURF) [12], we extract eight-direction gradient features directly from the fixed-size local image, which is more effective. Second, to speed up the training process, we only generate eight local image parts for each training sample, while the method in [10] uses 60 local parts. Third, the structure of our recognition system is more efficient. In the training step, we do not need to extract the SURF feature vector and build a dictionary. Here, the generated local parts are simply trained together with the original data. In the test step, the SVM model provides a confidence vector for each image part, and the final category is produced by considering all of these local recognition results.

The rest of the paper is organized as follows. In Section 2, we briefly introduce the collected RMB character database. Section 3 details the strategy of part-based character recognition. Section 4 summarizes the experimental results. Finally, we conclude our paper in Section 5.

2   Database

To evaluate the performance of various algorithms for RMB serial number recognition, we collect a database from daily-used RMB images. Scanned RMB images

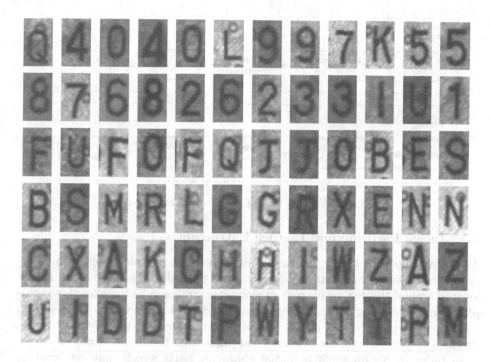

Fig. 3. NUST-RMB2013 dataset

(Fig. 1) are captured by the contact image sensor (CIS) which installed in the money counting machine with an output resolution of 200×180 dpi. RMB serial numbers are located on the bottom left corner of the scanned image and the characters are extracted straightforward by variance contrast and normalized by BMN [6] to the size of 36×60. As the extraction process is not so accurate, we manually choose the complete and human readable extraction results to compose the RMB database.

We name our RMB serial number database as NUST-RMB2013 [4] which represent all the different categories of RMB characters by separating into a training set of about 500 samples for each of the 35 classes (numeral 0-9 and alphabet A-Z except V) and a testing set of 200 samples for each class. Fig. 3 shows some RMB character samples containing circles in a complex background.

3 Part-Based Character Recognition

The flow of part-based character recognition approach is depicted in Fig. 4. The recognition procedure has three main steps. We first utilize the DoG keypoint detector to locate the interest points, and extract gradient features of the corresponding local image parts. Then, the trained SVM provides a confidence vector for each part. Finally, the recognition results of all parts are aggregated via three types of combination strategies. In the following, the procedures of image partition, feature extraction, and the integration of local parts are detailed.

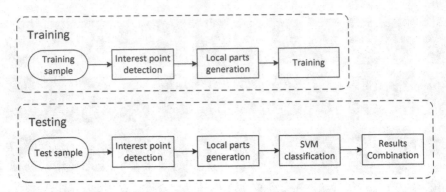

Fig. 4. Working flow of the proposed method

3.1   Image Partition and Feature Extraction

As we metioned, the character samples may contain a complex background, uneven illumination, smear, and contamination. Specially, there are small circles located around the distinctive part of character strokes, which makes these samples hard to be recognized. However, it is difficult to remove these circles from character samples. In this paper, we propose an alternative way to solve this problem by producing some "clean" local parts from the input samples.

In order to obtain some potential character parts which may not contain circles, we employ DoG keypoint detector [6] to locate the interest points, which are the scales-pace extremas in the convolutional result of Difference-of-Gaussian function and the image.

$$DoG(x,y,\sigma) = (\frac{1}{2\pi k\sigma^2}e^{-(x^2+y^2)/2k\sigma^2} - \frac{1}{2\pi\sigma^2}e^{-(x^2+y^2)/2\sigma^2}) \otimes I(x,y), \quad (1)$$

where $I(x,y)$ is an input image, σ represents the standard deviation of the Gaussian function, and k denotes a constant multiplicative factor. We estimate σ and k according to paper [6]. The keypoints are the local maxima and minima of $DoG(x,y,\sigma)$, specifically, the sample point will be selected as keypoint only if it is larger or smaller than its eight neighbors in the current image and nine neighbors in the images of scale above and below. Fig. 5 shows some keypoints detected in character samples.

Each training sample can generate W (depends on the internel structure of character) keypoints, which could enlarge the size of the training dataset W times. However, there are some keypoints located close to each other, and the relevant synthetic parts are very similar. To remove those redundant interest points and reduce the training time of SVM model, for each sample, we cluster the number of keypoints to K using the k-means [13] clustering algorithm (Fig. 5). Considering the trade-off between training cost and recognition rate, we empirically set K as eight. In the experiment, we also tested K with larger values such as 10 and 12, however, both of them barely provide any improvement.

The local image parts are extracted centered by the interest points. As we aim to remove the circles while keeping the integrity of character strokes, the

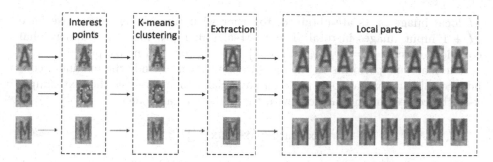

Fig. 5. Interest points detection and local parts generation

size of local parts can not be too small. We intentionally fixed it to 5/6 of the original image (30 × 50) by considering the size of characters and the circles in the RMB database. For the interest points which are located at the corners of the sample, we shift the extraction window according to the image boundries. As shown in Fig. 5, some local parts without circles can be generated, which will help to promote the recognition performance in the subsequent procedure. To make sure the features extracted from each local part has the same dimension, and also facilitate the classification process, we stretch all local parts to the same size of 36 × 60.

For each part of the image, the eight-direction gradient features are extracted [14]. First, eight gradient direction feature maps of the input sample are obtained by Sobel operator. Then, each gradient direction image is divided into 6 × 10 blocks. Gradient magnitude of each block is extracted by Gaussian blurring, hence the feature dimensionality is 480 (60 × 8). The feature vector is reduced to 34-dimensional (class number minus one) learned by Linear Discriminant Analysis (LDA) [15].

3.2 Local Parts Combination

System Training. We trained the SVM classifier with eight direction gradient features extracted from both the original and local images. The SVM models with Linear and RBF kernel are tested for fair comparison and analysis. We implemented the SVM via LibSVM software [16] which can provide the confidence weight of each category for the test sample. As our database has 35 classes, the output of SVM classifier is a 35-dimensional confidence vector.

Classifier Outputs Combination. The classifier combination applies on the outputs of SVM without knowledge of the internal structure of classifiers and their feature vectors. Given a test sample, we extract gradient features from the local parts. By feeding these features to SVM, a 35-dimensional confidence vector can be obtained from each part. We combine the classifier outputs of both the original and local images at abstract and measurement levels [17].

Three methods are used to aggregate the output measurements, namely, max rule, major voting, and multiple voting, respectively. For a test image **x**, assume

M local image parts are obtained, together with the original image, we have $M + 1$ input images in total. The SVM classifier provides an N-dimensional confidence vector $V_i = \{v_{i1}, v_{i2}, \ldots, v_{iN}\}, N = 35$ for each image part.

The max rule performs on the measurement level. It finds the maximum confidence value among all local parts for each class, then classifies the text sample **x** to the corresponding category with the highest measurement.

$$\mathbf{x} \in \arg \max_{j=1}^{N}\{\max_{i=1}^{M+1} v_{ij}\} \tag{2}$$

The major voting strategy combines the outputs of classifier on abstract level. Assuming each image part only belongs to one class, it counts the recognition results of all parts, and sorts the sample to the class which has the maximum votes. Let $R = \{r_1, r_2, \ldots, r_{M+1}\}$ denotes the classification results for all local parts, the value of r_i is set between 1 to N. The major voting method can be described as:

$$\mathbf{x} \in \arg \max_{j=1}^{N} \sum_{i=1}^{M+1} I(r_i), \tag{3}$$

where

$$I(\cdot) = \begin{cases} 1 & if \ \ r_i = j \\ 0 & else \end{cases}. \tag{4}$$

Both of these two methods introduced above provide outstanding recognition results in spite of their simplicity. As the local parts which are classified to the correct category always belong to the side of the majority, the major voting can restrain the influence of the circles by taking advantage of statistical distribution. Meanwhile, the lower complexity combination approach will compensate for the representational information lost during combinations.

The multiple voting method [11] can also be called as the sum rule strategy, which intergrates the recognition results on measurement level. Unlike the major voting method which assumes each part only belongs to one class, given a local image, multiple voting method considers its distribution to all categories. In the test, we obtain 35-dimensional confidence vectors $V_i = \{v_{i1}, v_{i2}, \ldots, v_{iN}\}, i = 1, 2, \ldots, M + 1$ for the local parts of sample **x**. Then, the confidence weights of different image parts are summed according to their categories. **x** belongs to the class with the maximum sum.

$$\mathbf{x} \in \arg \max_{j=1}^{N}\{\sum_{i=1}^{M+1} v_{ij}\} \tag{5}$$

Individual recognition results of all image parts contribute to the final decision by summing the confidence weights for each class. The benefit of using such strategy is that it effectively suppresses the influences of circles and other factors to the classification. For instance, the SVM classifier may have difficulties in distinguishing the original sample contaminated by circles while some synthetic

Table 1. Test accuracies (%) of local parts combination methods on NUST-RMB2013

Combination	SVM kernel	
	Linear	RBF
Conventional	98.90	99.31
Max rule	99.27	99.43
Major voting	99.07	99.30
Multiple voting	**99.33**	**99.51**

local parts may not. They can provide correct recognition results with high confidence weights helping to find the proper category after sum operation.

4 Experiments

4.1 Baseline Recognition Results

Both the Linear and RBF kernel based SVM classifiers have been used to construct the baseline classifier for the recognition of RMB serial number. Before classification, the features were reduced to 34 dimensions by LDA. The hyperparameters of SVM were selected via cross-validation on the training data. The NUST-RMB2013 database contains the serial number characters of 35 categories with 17,262 training samples and 7000 testing samples in total. The recognition rate on the original database using SVM is 98.90% with Linear kernel and 99.31% with RBF kernel, respectively [4].

4.2 Part-Based Recognition Results

In the training process, the part-based method enlarged the training data about eight times by generating eight local parts for each training sample. Along with the original data, the total size of training dataset became 155,358.

Since a larger number of local parts implies a higher probability that there exists some local images without circles, given a test sample, we created as many local images as we can to ensure the robustness of our method. Eight-direction gradient features of these local images were fed to SVM classifiers (Linear and RBF kernel) to get the 35-dimensional confidence vectors. After that, three combination methods aggregated the individual recognition results to produce higher accuracies. Table 1 compares the performances of different combination methods on NUST-RMB2013 database.

We find out that all of the part-based methods lead to approving results. Multiple voting is the most competitive strategy, which achieves the highest classification accuracy on SVM trained with both Linear and RBF kernels. The accuracy on test data boosts by 0.43 % and 0.2% using Linear and RBF kernel, respectively. The max rule also helps to improve the recognition rates to reach the accuracies of 99.27% and 99.43% by different SVM kernels. However, the advantages of major voting is relatively inferior to the other methods. It merely provides 0.17% improvement on the model trained by Linear SVM, and barely works on the model with RBF kernel because the accuracy is identical to the conventional recognition method.

4.3   Discussion

Since the max rule and multiple voting methods combine the outputs of the SVM classifier on the measurement level by utilizing confidence weights of each local part, it is reasonable that they outperform the major voting scheme which assumes each local part only belongs to one category and intergrates the recognition results on the abstract level. The experimental results prove our part-based method works well on the database which contains uneven illumination, contrast variation, smear, and complex background texture. Especially, the misclassification problem caused by the anti-counterfeiting circles appeared in the RMB character samples are properly solved. The best recognition rate of 99.51% is given by multiple voting combination method with the SVM model trained with RBF kernel.

According to Table 1, the part-based strategies is less sensitive to noise than the conventional recognition method. The reasons for the superior performance are twofold. First, we train the SVM classifier with both the full-size and local image part samples, which makes the SVM model considering not only the global but also local character features. Second, the part-based method makes capital of the statistical distribution of the recognition outputs obtained from individual local parts. The various local parts generated from the test sample can eliminate the influences caused by the circles and complex background, and help the classfier to make the right determination.

The distortion method has been proved very helpful in [4] and [18]. Compared with the distortion method which also expands the training dataset by generating additional synthetic training samlpes, our part-based method has its own merit. Since the elastic distortion randomly chooses its scaling and rotation parameters, and the translation distortion shifts the input pattern one or two pixels towards eight directions, neither of them considers the internal structure of training sample. On the contrary, we make use of the samples' characteristic structure information by utilizing DoG keypoint detector to locate the interest points and extracting the surrounding area as local parts. The distinctive principle of the DoG detector helps to extract more distinguishing local parts than distortion methods, which promises a higher recognition rate.

Fig. 6 demonstrates some samples misclassified by the part-based recognition method using multiple voting combination strategy. There are some circles touching the distinctive part of character strokes, which makes these samples extremely hard to be recognized. We are not able to remove these circles even by extracting the local parts. To deal with this problem, we plan to find a method to detect the circles in character samples using image processing technique in

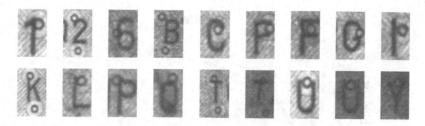

Fig. 6. Misrecognition samples

furture studies. After removing the circles from the data samples, a new recognition system with higher accuracy can be conducted on these "clean" character samples.

4.4   Rejection

High reliability is required in the problem of RMB serial number recognition which tolerates no classification error. Even a 0.1% error rate will cause significant economic loss in the banking system. The reliability [19] is defined by:

$$\text{Reliability} = \frac{\text{Number of correct recognitions in accepted samples}}{\text{Number of accepted samples}} \times 100\%, \quad (6)$$

We use the first two rank measurement (FTRM) rejection method to improve the reliability by rejecting confusing patterns. The FTRM is based on the gap between the top-2 outputs of classifiers. Fig. 7 shows the rejection tradeoff curves of the recognition methods using conventional SVM and the part-based scheme with multiple votiong combination strategy. It indicates that the part-based method produces a higher reliability than the traditional recognition method, which achieves 99.92% reliability with 2.91% rejection rate.

5   Conclusions

In this paper, a novel part-based character recognition approach for RMB serial number recognition has been proposed. According to the characteristics of the samples in RMB dataset, we automatically generate a set of local parts for each input sample using the Difference-of-Gaussians (DoG) keypoint detector. The feature vectors are extracted by eight-direction gradients and reduced to 34 dimensions by LDA. Both the original and local image parts are used to train an SVM classifier. In the test step, we first obtain the confidence vector for each part. Then, three different types of methods are investigated to combine the recognition results of all image parts.

Experiments conducted on a large serial number character database named NUST-RMB2013 show the superior performance of part-based character recognition method. It exploits both the global and local character structure features,

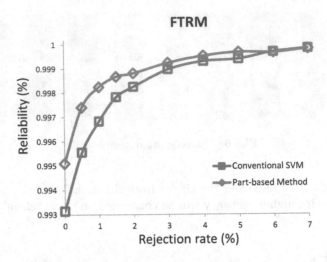

Fig. 7. Rejection-reliability tradeoff for FTRM

and offers an overall increase in robustness, performance and reliability to the entire recognition system. Apart from its obvious advantages of simplicity and completeness, it has been proved to be able to recognize the imprecisely extracted characters which have lost their global structure by occlusion, decoration, contamination, and other degradations (Fig. 3).

Three local parts combination methods have been compared and analysed. Since the measurement level combination approaches such as max rule and multiple voting consider the confidence vectors of test samples, they show better performances than the method based on abstract level (major voting). The highest recognition rate 99.51% is achieved by the RBF kernel based SVM cooperating with the multiple voting combination scheme. Compared to the conventional recognition method, the error reduction rate is 39.09% and 28.99% using Linear and RBF kernel based SVM, respectively. We find that the major voting is not a promising solution to our problem since it only leads to a slight improvement of the recognition accuracy.

What is more, the concepts of the proposed part-based character recognition method can also be used in other types of banknotes, such as Euro, U.S. and Canadian dollar, or in the recognition of document character samples with complex backgrounds.

Acknowledgments. This work was supported by the National Natural Science Foundation of China (NSFC) under Grant 61231014 and 60875010. We also wish to thank Prof. Cheng-Lin Liu, the colleagues of Centre for Pattern Recognition and Machine Intelligence (CENPARMI) and National Laboratory of Pattern Recognition (NLPR) for their great support and help in our work.

References

1. Zhao, T.T., Zhao, J.Y., Zheng, R.R., Zhang, L.L.: Study on RMB number recognition based on genetic algorithm artificial neural network. In: Proc. International Congress on Image and Signal Processing, pp. 1951–1955 (2010)
2. Li, W., Tian, W., Cao, X., Gao, Z.: Application of support vector machine (SVM) on serial number identification of RMB. In: Proc. World Congress on Intelligent Control and Automation, pp. 6262–6266 (2010)
3. Feng, B.Y., Ren, M., Zhang, X.Y., Suen, C.Y.: Extraction of serial numbers on bank notes. In: Proc. International Conference on Document Analysis and Recognition, pp. 698–702 (2013)
4. Feng, B.Y., Ren, M., Zhang, X.Y., Suen, C.Y.: Automatic recognition of serial numbers in bank notes. Pattern Recognition 47(8), 2621–2634 (2014)
5. Nieves, J., Ruiz-Agundez, I., Bringas, P.G.: Recognizing banknote patterns for protecting economic transactions. In: Proc. International Workshop on Database and Expert Systems Applications, pp. 247–249 (2010)
6. Lowe, D.: Distinctive image features from scale-invariant keypoints. International Journal of Computer Vision 60(2), 91–110 (2004)
7. Cortes, C., Vapnik, V.: Support-vector networks. Machine Learning 20(3), 273–297 (1995)
8. Suen, C.Y., Guo, J., Li, Z.: Analysis and recognition of alphanumeric handprints by parts. IEEE Trans. Syst., Man, Cybern. 24(4), 614–631 (1994)
9. Suen, C.Y., Kim, J., Kim, K., Xu, Q., Lam, L.: Handwriting recognition–the last frontiers. In: Proc. ICPR, pp. 1–10 (2000)
10. Uchida, S., Liwicki, M.: Part-based recognition of handwritten characters. In: Proc. International Conference on Frontiers in Handwriting Recognition, pp. 545–550 (2010)
11. Wang, S., Uchida, S., Liwicki, M.: Comparative study of part-based handwritten character recognition methods. In: Proc. International Conference on Document Analysis and Recognition, pp. 814–818 (2011)
12. Bay, H., Tuytelaars, T., Van Gool, L.: SURF: Speeded up robust features. In: Leonardis, A., Bischof, H., Pinz, A. (eds.) ECCV 2006, Part I. LNCS, vol. 3951, pp. 404–417. Springer, Heidelberg (2006)
13. Hartigan, J.A., Wong, M.A.: A k-means clustering algorithm. Applied Statistics 28(1), 100–108 (1979)
14. Bai, Z.L., Huo, Q.: A study on the use of 8-directional features for online handwritten chinese character recognition. In: Proc. International Conference on Document Analysis and Recognition, pp. 262–266 (2005)
15. Mika, S., Ratsch, G., Weston, J., Scholkopf, B., Muller, K.: Fisher discriminant analysis with kernels. In: Proc. Signal Processing Society Workshop, pp. 41–48 (1999)
16. Chang, C.C., Lin, C.J.: LIBSVM: A library for support vector machines. ACM Trans. Intell. Syst. and Technol. 2(3), 27:1–27:27 (2011)
17. Suen, C.Y., Lam, L.: Multiple classifier combination methodologies for different output levels. In: Proc. International Workshop on Multiple Classifier Systems, pp. 52–66 (2000)
18. Simard, P., Steinkraus, D., Platt, J.C.: Best practices for convolutional neural networks applied to visual document analysis. In: Proc. International Conference on Document Analysis and Recognition, pp. 958–963 (2003)
19. Wang, W.-N., Zhang, X.-Y., Suen, C.Y.: A novel pattern rejection criterion based on multiple classifiers. In: Zhou, Z.-H., Roli, F., Kittler, J. (eds.) MCS 2013. LNCS, vol. 7872, pp. 331–342. Springer, Heidelberg (2013)

Comparative Study of Feature Selection for White Blood Cell Differential Counts in Low Resolution Images

Mehdi Habibzadeh, Adam Krzyżak, and Thomas Fevens

Dept. of Computer Science & Software Engineering, Concordia University, Montréal, Québec
{me_habi,krzyzak,fevens}@encs.concordia.ca

Abstract. Features that are widely used in digital image analysis and pattern recognition tasks are from three main categories: shape, intensity, and texture invariant features. For computer-aided diagnosis in medical imaging for many specific types of medical problem, the most effective choice of a subset of these features through feature selection is still an open problem. In this work, we consider the problem of white blood cell (leukocyte) recognition into their five primary types: Neutrophils, Lymphocytes, Eosinophils, Monocytes and Basophils using a Support Vector Machine classifier. For features, we use four main intensity histogram calculations, set of 11 invariant moments, the relative area, co-occurrence and run-length matrices, dual tree complex wavelet transform, Haralick and Tamura features. Global sensitivity analysis using Sobol's RS-HDMR which can deal with independent and dependent input variables is used to assess dominate discriminatory power and the reliability of feature models in presence of high dimensional input feature data to build an efficient feature selection. Both the numerical and empirical results of experiments are compared with forward sequential feature selection. Finally, the results obtained from the preliminary analysis of white blood cell classification are presented in confusion matrices and interpreted using Cohen's kappa (κ) with the classification framework being validated with experiments conducted on poor quality white blood cell images.

1 Introduction and Complete Blood Count (CBC) Interpretation

The examination of peripheral blood smears represents the cornerstone of hematologic diagnosis. Plainly, the examination of the peripheral blood smear is an important indicator of haematological and other abnormal conditions that affect the body of an organism. Blood cells are classified as erythrocytes (Red Blood Cells), leukocytes (White Blood Cells) or platelets (not considered real cells). In all mammals species including human beings, leukocytes, which are less numerous than red blood cells, are divided in two main categories: *granulocytes* and *lymphoid* cells. Granulocyte white blood cell types are Neutrophil, Eosinophil (or acidophil) and Basophil. The lymphoid cells, are separated in Lymphocytes and Monocytes. Expressing the number of white blood cells (WBC) carries many quantitative and informative clues. For example, the increase or decrease of leukocytes is very critical and may prompt detailed medical attention.

The first attempts to build automated laboratory equipment to perform complete blood counts (CBC) about 60 years ago, in the 1950-1960s [44]. Automatic counting system have been available in the medical laboratories for the last 25 years.

N. El Gayar et al. (Eds.): ANNPR 2014, LNAI 8774, pp. 216–227, 2014.

The instruments used for performing cell counts are based on mix of mechanical, electronic and chemical approaches. Current hematology analyzers used routinely in medical laboratories are such as Sysmex XE-series [35] and also in the Abbott CELL-DYN range [11]. These known systems for white blood cell differential counts reveal good correlation with the manual ground truth reference analysis for neutrophils, lymphocytes, and eosinophils (accuracies of 0.925, 0.922, and 0.877, respectively) and lower accuracy for monocytes and basophils (accuracies of 0.756 and 0.763, respectively). The commonly used approach across biological disciplines and the ground truth is manual WBC counting and type sorting by a trained pathologist or skilled hematology expert, looking at the shape, e.g, nucleus and cytoplasm, occlusion, and degree of contact between cells.

Although the manual inspection method is adequate, it has *three* inevitable types of error: statistical, distributional and also human error [5] such as may happen in poor quality, low magnification view of the slides. Poor magnification and distribution of leukocytes adversely affect the accuracy of the differential count in manual counting. The computerized techniques are the best potential choices to carry out and moderate the load of these regular clinical activities for more efficiency and also to describe the frequency and spatial distribution, and portion of blood smear particles. Hematologists and hematopathologists study and analyze blood smears by looking at cells under an optical microscope. Accordingly, since haematology is a visual science, machine learning and digital image processing have great potential to develop ways to improve haematology research. Computer-aided diagnosis (CAD) also establish methods for accurate, robust and reproducible measurements of blood smear particles status while reducing human error and diminishing the cost of instruments and material used.

In this work, white blood cell analysis of an unfavourable low resolution data set via a feature extraction and selection framework to classify the five mature types of white blood cells is provided. There are no reliable and general comparative studies of feature selection strategies in white blood smear detection with high dimensional input feature data in particular and also in the presence of low quality and unfavourable conditions. This work unifies and extends primary feature vector sets introduced in our earlier work [12, 13], based on using the dual-tree complex wavelet transform (DT-CWT) and few textural features, to high dimensional comprehensive invariant feature sets that also include different invariant shape features such as 11 invariant moments, different histogram calculations, different efficient textual feature such as Tamura and so on. Furthermore, this paper critically examines and compares two feature selection strategies, random sampling-high dimensional model representation (RS-HDMR) and sequential forward selection (SFS), for the white blood cell classification problems in presence of small number of sample set.

2   Background and Literature Survey

The first published paper on blood processing is leukocyte pattern recognition by Bacusmber and Gose in 1972 [2]. In this primary work, classification of white blood cells using some shape features and a multivariate Gaussian classifier into their categories are presented. One decade after, the first fully automated processing of blood smear slides was introduced by Rowan [34] in 1986. The background on WBC classification by using computer vision concepts is substantial and involves feature extractors, classifiers, quantitative and qualitative process. Ramoser *et al.* [31] used hue, saturation and luminance values to locate WBCs and then leukocytes are classified using a 26-dimensional color feature vector and a classification polynomial support vector machine (SVM). Xiao-min *et al.* [46] introduced method based on threshold segmentation followed by mathematical morphology (TSMM). Sobrevilla *et al.* [40] used fuzzy logic to segment white blood cells from a digital blood smear image. However, in both TSMM [46] and fuzzy logic [40], parameter settings need to set by statistics and experience. Shitong *et al.* [37] proposed white cell detection based on fuzzy cellular neural networks (FCNN). Mukherjee *et al.* [26] proposed a leukocyte detection using image-level sets computed via threshold decomposition. Further, Theera-Umpon *et al.* [43] used four white blood cell nucleus features, and Bayes and artificial neural networks were the classifiers.

Ongun *et al.* [28] proposed an approach using active contours to track the boundaries of white blood cells although occluded cells were not precisely handled. Lezoray [24] introduced region-based white blood cells segmentation using extracted markers (or seeds). Kumar [22] applied a novel cell edge detector while trying to perfectly determine the boundary of the nucleus. Sinha and Ramakrishnan [38] suggested a two-step segmentation framework using k-means clustering of the data mapped to HSV color space and a neural network classifier using shape, color and texture features. Furthermore, in other work, WBC segmentation was achieved by means of mean-shift-based color segmentation in [7] by Comaniciu and Meer while in [19] Jiang *et al.* used watershed segmentation.

Ramesh *et al.* [8] proposed a two-step framework: segmentation and classification of normal white blood cells in peripheral blood smears. Color information and morphological processing were basis functions for segmentation part which was almost close to already published paper in [14]. Latter, WBC classification followed using 19 features such as area, perimeter, convex area, and so on. To lessen the computational burden, Fishers linear discriminant was also applied to trim a multi-dimensional set to six dimensions. In more recent work (2012) Dorini *et al.* [9] introduced automatic differential cell system in two levels to segment WBC nucleus and identify the cytoplasm region. In that work, five mature WBC types were classified using a K-Nearest Neighbor (K-NN) classifier with geometrical shape features and a reasonable accuracy (78% performance vs 85% classified manually by a specialist) was achieved. As a result, despite its long history in cell classification, questions have been raised about the reliability and feature selection in an appropriate white blood cell classification system. On the other hand, one major drawback of these aforementioned approaches is that no general attempt was made to quantify the association between low resolution cell appearance and their classification. Therefore, this latter work would have been more reliable if the framework considered these concerns.

3   Primary Feature Extraction

Continuing previous work [12, 13], the process of feature extraction and parameter estimation is carried out in this extended work. These candidate descriptors have appropriate potential for dealing effectively with challenges and problem in multi-distortion data set such as blurred, noisy and low magnification of a white blood cell image where internal white blood cell structure is not obvious to detect. All invariant features are scaled to the [0 1] range to simplify computational complexity and have consistent inputs for measurement. As a result with all three main feature types in this case, final features vector gives a total of 12140 coefficients for each white blood cell with 28×28 low image size. More details are addressed as below.

Intensity Features: This article examines the mean (μ), standard deviation (σ), skewness (γ_1), and kurtosis (K) in white blood cells classification. However, intensity features may prove inadequate for specially low quality white blood cell data set. A short mathematical background is addressed in our previous research [13]. Eventually, in this case, intensity features gives a total of 788 divided into 784 for raw gray intensity value and 4 measures for histogram calculation feature coefficients for each cell sample.

Shape Features: In terms of pattern recognition, shape descriptors can be classified into two descriptors; *contour-based* and *region-based* shape signifiers. The contour-based descriptors reviewed so far cannot represent ideally white blood cell shapes for which the complete and continuous boundary information is not ideally available with granular and non-uniform borders. Also, questions have been raised about the validity and reliability concern under the constraint of translation, rotation and uniform-scaling invariance properties. In reviewing the literature, the current study found that invariant moment as a region-based calculation which can provide invariant characteristics under different condition are likely occur in translation, changes in scale, also rotation and unique characteristics of a white blood cell that represent its heterogenous shape. Although moment algorithms and theory have been well established in mathematics, far too little attention has been paid to use invariant moment in computer-aided diagnosis (CAD) in medical imaging and for blood smear analysis in particular. This paper has given an account the reasons for the widespread use of *(11)* different invariant moments listed into: M_1 with 332 elements which are moment coefficients for all combined 11 following different moments, $M_2 = 36$ to Radial Tchebichef [27], $M_3 = 36$ to Fourier-Chebyshev magnitude [29], $M_4 = 36$ to Gegenbauer [16], $M_5 = 36$ to Fourier-Mellin magnitude [36], $M_6 = 36$ to Radial Harmonic Fourier magnitude [32], $M_7 = 36$ belong to Generalized Pseudo-Zernike [45], $M_8 = 36$ to Dual Hahn moments [21], $M_9 = 7$ belong to Hu set of invariant moments [17], $M_{10} = 36$ to Krawtchouk [47], $M_{11} = 36$ to Legendre [10, 48], $M_{12} = 1$ to Zernike [25]. In following shape feature category, the relative area (A_r) is also computed [13]. In conclusion, selective shape features provides a total of 333 feature coefficients for each white blood cell sample composed of (332) invariant moment coefficients and one measure for A_r.

Texture Features: This section extends the types of features considered in our earlier work [12, 13]. The vector includes features associated with the Laplace transform, gradient-based, flat texture features [33], and also co-occurrence matrix [15] which is defined over a white blood cell image to be the distribution of co-occurring values at a given offset. Various combinations of the matrix are taken to generate features called *Haralick* features [15] (namely, the angular second moment, contrast, correlation, sum of squares: variance, inverse difference moment, energy, and entropy). Afterwards, six parameters approximating visual perception is used based on the *Tamura* feature [41].

In addition, run-length is an another texture coarseness measurement at typical directions such as 0, 45, 90, and 135 degrees [42]. 11 features for a given gray-level for each individual white blood cell image are extracted. Dual-tree complex wavelet is also examined in this research. It calculates coefficients along rows and columns, and in *six* directions and angles at each individual pixel. The setting, details and proposed framework using DT-CWT is addressed to our previous work [12,13]. It follows that, textural features gives a total of 11019 feature coefficients for each white blood cell sample. This textural feature vector has been divided into seven aforementioned parts. The first part deals with gradient, Laplacian and flat texture features with 784 for each of them respectively. Then it will go on to Haralick vector and also Tamura textural features with 13 and 6 elements respectively. Finally gray-level run length matrix in four orientations provides 6296 coefficients where dual-tree complex wavelet in six directions also gives a total of 2352 features for each sample.

4   Global Feature Sensitivity and Feature Selection

This work address feature selection algorithm to trace effectiveness of aforementioned high dimensional invariant descriptors in white blood cell classification performance. Feature selection and discriminatory power is achieved using high dimensional model representation (RS-HDMR) and sequential forward feature selection (SFS) along with support vector machine classifier (see section. 5).

RS-HDMR/ Sensitivity Feature Analysis: Lastly, we look at the effect of each individual three multiple features (see section 3) contribution upon the corresponding supervised white blood cell classification. Several studies investigating high-dimensional model representation (HDMR) [1] have been carried out on input and output relationship analysis. High dimensional model representations (HDMR) is a statical approach that depicts the individual or cooperative contributions of the aforementioned features upon the corresponding white blood cell classes. To date, little evidence has been found associating HDMR with image processing and pattern recognition such as Kaya *et al.* research work [20]. Then, future studies on the current topic are therefore recommended. In this work, RS-HDMR approach with a random sample input over the entire domain is used where determination of expansion components is based on shifted Legendre polynomials approximation and Monte Carlo integration [1, 30, 49]. Following that, the influence of individual each input feature variables is computed using global sensitivity approach in which Sobol index is the basis function of calculation [39]. Therefore, global sensitivity indices are denoted by: $S_{i_1,...,i_s}$ where total of the summation $\sum_{s=1}^{n} s_{i_1} + \sum_{1<i<j\leqslant n}^{n} S_{ij},... + S_{1,2,...,n}$ is equal 1. The first order index S_i is fractional contribution of x_i (each individual feature coefficient) to the variance of $f(x)$ (five main white blood cell classes) where the second order shows the interaction power between x_i and x_j on the classification outcome and these sensitivity analysis terms will be continued. Rabitz *et al.* [1] demonstrated that often the low order interactions of input variables have the dominant impact upon the output assignment. It means that quite often the high ranked global sensitivity feature variable input in mathematical models are first order terms. In the current study, first order S_i for all each individual intensity, shape and texture coefficients are calculated to reach the most effective set.

Sequential Feature Selection: Sequential Feature Selection is an iterative method to select the most informative coefficient by choosing the next feature depending on the already selected features. The method removes redundant and irrelevant features while preserving the efficient features in order to optimize the subset combination of features by considering their predictive efficiency with a given classifier. The method has two distinctive variants: sequential forward selection (SFS), and in contrast, sequential backward selection (SBS) [18] where SFS is taken in this work. In SFS, new added feature x^+ should maximize $J(Y_k + x^+)$ where new component combined with the features Y_k that have already been selected in an iterative and incremental procedure $(x^+ = \arg_{x \notin Y_k} max\, J(Y_k + x^+))$. Despite its simplicity, questions have been raised about the update procedure used by sequential feature selection. For example, SFS is unable to revise optimal feature vector to remove feature variables after the addition of other features. It's also seen that its performance is related to an appropriate criterion to determine the iterative stop point. In this work the optimum criterion value means the minimum error rate in SVM supervised classification where each candidate feature is placed in the new revised subset vector. Several studies investigating SFS have been carried out on medical imaging [4, 6]

5   Discriminant Functions and Support Vector Machine

A linear SVM classifier [3] with 10-fold cross-validation is examined in this work. 10-fold cross-validation is commonly used in presence of a small size (140 samples) of the training and testing data set and with large number of parameters (12140 feature coefficients) to avoid over fitting and to cover all observations for both training and validation. The details of the proposed SVM settings and configuration are addressed in our previous work [12].

6   Experimental Results and Classification

In this section, a set of 140 8−bit gray scale poor images with low magnification $(28 * 28)_{px}$ in five balanced dataset (see fig. 1) are used. We have randomly chosen the data to construct the training set after removing almost 20% of the data to be used for testing the SVM classifier.

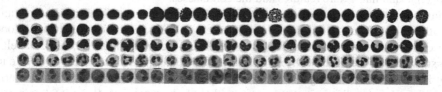

Fig. 1. WBC testing data, each row, top to bottom: Basophil(B), Lymphocyte(L), Monocyte(M), Neutrophil(N), and Eosinophil(E)

Sobol HDMR Analysis: In practice, in the initial configuration for this experiment all 140 samples are used for the RS-HDMR accuracy test. Also, the maximum order for approximation of the first order $\{f_i(x_i)\}$ terms is 5 where 3 is maximum assigned order for second order $\{f_{ij}(x_i, x_j)\}$. Also a ratio control variate (see section 2.1 in [49]) to supervise and regulate the Monte Carlo integration error with 10 iterations is set for the first and second order RS-HDMR component functions. It also should be noted that in the initial setting to ignore insignificant component functions from the HDMR expansion where the current white blood cell classification system has a high number of input features, a threshold mechanism set to 10% (see section 2.2 in [49]) is also used. During sensitivity analysis, first an intensity feature vector with 788 members composed of 1-784 raw gray scale intensity value, 785 mean, 786 standard deviation, 787 skewness, and 788 kurtosis features is considered. In this case, S_i analysis shows that 38 coefficient out of 788 are computed as non-zero. Sensitivity calculations indicate that indices: 711, 443, 284, 191 and 456 (in range of gray scale intensity value) and 785 (mean value) out of 38 have the first five most discriminative power of $S_i = 0.38$ in this current input - output relationship. Secondly, a shape feature vector with 333 members composed of 1-7 Hu set, 8 Zernike, 9-44 Hahn, 45-80 generalized pseudo-Zernike, 81-116 Chebyshev, 117-152 Legend, 153-188 Krawtchouk, 189-224 Fourier-Mellin, 225-260 Radial Harmonic Fourier, 261-296 Fourier-Chebyshev, 297-332 Gegenbauer, and 333 for relative area is considered. Global Sobol - HDMR Sensitivity calculations demonstrate that 18 of the above feature indices have the highest S_i of 0.82 where in that case, first six indices are: 44 (Hahn coefficient), 191,192 (in range of Fourier-Mellin), 225, 226 (in range of Radial Harmonic Fourier) and 290 (in range of Fourier-Chebyshev).

Then a texture feature vector with 11019 members composed of 1-784 gradient, 785-1568 Laplacian, 1569-2352 flat texture, 2352-2365 Haralick texture features, 2365-2371 Tamura, 2372-8667 Gray Level Run Length, and 8667-11019 for dual tree complex wavelet transform features is considered. To provide in-depth analysis of the Sobol index calculation, each of above individual ranges of features is used separately to estimate global sensitivity values. In the case of the gradient features, it can be seen that 43 out of 784 elements have the highest $S_i = 0.44$ where first five indices including 589, 185, 266, 658 and 659 have the most discriminatory power with total $S_i = 0.41$. Next, global sensitivity on the Laplacian features shows that just only 4 elements have non-zero values where these are indices including 421, 309, 337 and 365 with $S_i = 0.17$. Further in flat texture feature analysis result revealed that 13 elements with S_i equal to 0.17 have the dominant power. This suggests that a weak link may exist between Laplacian and flat texture features and the cell classes.

Further, a consequence of the analysis on Haralick features, Tamura shows 9 and 3 with $S_i = 0.7$ and $S_i = 0.6$ have most effective elements in feature - white blood cell class relationship. In terms of Gray Level Run Length feature set, result labeled the subset of 34 elements with $S_i = 0.62$ provides the good predictive power in current HDMR meta-modeling. Global sensitivity in dual tree complex wavelet transform identifies adequate discriminatory power with 111 elements with $S_i = 0.64$ as a major effective subset among all these feature coefficients. In this work based on above explanation 273 elements with exact addressed indices among all 12140 coefficients (almost 2.2%) which are the most convincing set on HDMR input - output relationship in current white blood cell classification system are selected (FV_{HDMR}).

Sequential Feature Selection: For comparison of the results of Sobol HDMR feature selection and to compare the performance on classification accuracy, sequential forward selection (SFS) is used. Sequential forward selection initialized using 10-fold cross-validation by repeatedly calling a criterion based SVM setting (see section 5) with different training and testing subsets of x_{in} and y_{out} where selected feature are saved into a logical matrix in which row (i) indicates the features selected at step (i) with minimum criterion value. In connection with sequential forward selection, many feature indices should be listed here but an exhaustive review is beyond the scope of this current work. Eventually, to do a comparative sensitivity analysis, a feature vector (FV_{SFS}) with the exact number of (FV_{HDMR}) is created. Therefore, this study may leads a difference between classification performance rate (see table 1) for these feature selection algorithms.

Confusion Matrices: A 5×5 confusion matrix is used to represent the different possibilities of the set of instances. The matrices are built on five rows and five columns: Neutrophil; Monocyte; Lymphocyte; Eosinophil; and Basophil representing the known WBC classes whereas for each matrix, each row the values are normalized to sum to 1. Several standard performance terms such as true positive, false positive, true negative, false negative rate, accuracy, precision have been extracted for the confusion matrix. This work addresses kappa (κ) measure as it provides accuracy (AC) versus precision (P) interpretation across class categories [23]. Common Cohen's unweighted κ interpretation is: $\leq 0 \Rightarrow Poor$, $[0, 0.20] \Rightarrow Slight$, $[0.21, 0.40] \Rightarrow Fair$, $[0.41, 0.60] \Rightarrow Moderate$, $[0.61, 0.80] \Rightarrow Substantial$, $[0.81, 1.00] \Rightarrow Almost Perfect$. The experiments are categorized into set of named selected features $(FV_{SFS}$ and $FV_{HDMR})$ also with a total high dimensional feature vector with 12140 members (FV_{Total}).

Statistical performance measure is analyzed using analysis of confusion matrices for each named feature & SVM summarized in tables 1a, 1b, and 1c. Further statistical tests revealed that given a small number of input samples (140) in high dimensional feature sets ($= 12140$) using non-linear SVM kernels leads to over-fitting. The result, as shown in table 1, indicates that for normal low resolution white blood cells using linear SVM & all feature vector FV_{Total} 85% of known white blood cells were classified as such, with this classification rate decreasing to 83% for (FV_{HDMR}) (see table 1c) where the efficiency of (FV_{SFS}) is also 81% which is less than proposed Sobol - HDMR with 83%. RS-HDMR classification performance with 273 elements is less and more similar where classification accuracy is also found with all 12140 coefficients are selected. As confusion matrix tables illustrate, in this poor imaginary database there is not a significant difference between for example the all high dimensional data set and feature selected group with RS-HDMR expansion. The results, as shown in confusion matrix tables indicate that also HDMR results for almost each sub-group is more accurate than SFS method where also sequential forward selection algorithm is too dependent to classifier feedback as well. Also with compare with two ground truth groups, using machines Sysmex XE-series and also Abbott CELL-DYN range (see section 1) it can be seen from the data in confusion matrix tables that global sensitivity with Sobol on RS-HDMR expansion reveals 91% accuracy for Neutrophil, 65% rate for Lymphocyte and also 100% for Eosinophil while the expensive machines mentioned above provide 92.5%, 92.2%, and 87.7%, respectively in an ideal performance. It also provides 81%

classification rate for Monocytes and 77% for Basophils where the results obtained from machines are 75.6% and 76.3%. The following conclusions in regard to κ coefficient can be also drawn from the present confusion matrices. The Cohen's unweighted κ coefficient of the FV_{Total}, FV_{SFS}, also FV_{HDMR} are acceptable (0.81= almost perfect and $0.77, 0.79$ = substantial) in this low resolution WBC classification. Taken together, the most obvious finding to emerge from feature selection and with RS- HDMR study in particular is that all these two methods provide *substantial* performance where lessen computational time and improve model interpret-ability to enhance generalization by reducing over-fitting possibility as well.

Table 1. Confusion matrices (top to down: a,b,c) for SVM classifier, totals over testing images in invariant features & linear SVM

Linear SVM (FV_{Total}): Assigned WBC classes				
Known	Basophil	Eosinophil	Lymphocyte	Monocyte	Neutrophil
Basophil	0.72	0	0.21	0.03	0.04
Eosinophil	0	1.00	0	0	0
Lymphocyte	0.17	0	0.68	0.13	0.02
Monocyte	0.01	0	0.04	0.90	0.05
Neutrophil	0	0	0	0.03	0.97

Linear SVM (FV_{SFS}): Assigned WBC classes				
Known	Basophil	Eosinophil	Lymphocyte	Monocyte	Neutrophil
Basophil	0.72	0	0.24	0.04	0
Eosinophil	0.00	1.00	0.00	0.00	0.00
Lymphocyte	0.17	0	0.62	0.14	0.07
Monocyte	0.02	0	0.18	0.80	0.0
Neutrophil	0.01	0	0.01	0.04	0.94

Linear SVM (FV_{HDMR}): Assigned WBC classes				
Known	Basophil	Eosinophil	Lymphocyte	Monocyte	Neutrophil
Basophil	0.77	0.01	0.17	0.01	0.04
Eosinophil	0	1.00	0	0	0
Lymphocyte	0.16	0.01	0.65	0.1	0.08
Monocyte	0.04	0	0.13	0.81	0.02
Neutrophil	0.02	0.01	0.01	0.05	0.91

7 Conclusions

A machine learning approach for white blood cell classification is effective and reliable, while working under different and even unfavourable and adverse conditions. In this paper, these conditions include low resolution cytological images that are noisy digital white blood cell images. In this research, various approaches to the comprehension and analysis of invariant three main features are presented and the use of these theories is outlined. This work also concentrates on the literature concerning the usefulness of feature selection in presence of big data with high dimensional 12140 invariant features in connection with white blood cell classification. An account is provided of the widespread use of sequential feature selection (SFS) set to recent development in random sample High-dimensional model representation (RS-HDMR). It has conclusively been shown that these invariant feature collection sets are appropriate solutions as their implementations are promising strategies for representing small distorted white blood cell classifier system (see table 1a). These findings suggest that, in general, RS-HDMR emerged as a reliable input-output relationship predictor of small distorted WBCs and their own classes to allow the full feature sensitivity analysis based on Sobol sequences. It is expected that classification accuracy will be further improved by extending the data set size to reach higher performance in training and testing procedures. The findings are expected to be persuasively supported by future work considering different underdeveloped HDMR variations, i.e., Sobol HDMR using Quasi Monte Carlo, multiple subdomain random sampling HDMR, or Cut-HDMR. Briefly, the empirical findings in this study provide a better understanding of invariant feature implementation and feature selection. One of the more significant findings to emerge from this study is that the possibility of extending the use of this framework to entire field of haematology analysis or other similar medical research.

References

1. Aliş, Ö., Rabitz, H.: Efficient implementation of high dimensional model representations. Journal of Mathematical Chemistry 29(2), 127–142 (2001)
2. Bacusmber, J.W., Gose, E.E.: Leukocyte pattern recognition. IEEE Transactions on Systems, Man and Cybernetics SMC-2(4), 513–526 (1972)
3. Ben-Hur, A., Weston, J.: A user's guide to support vector machines. In: Data Mining Techniques for the Life Sciences, Methods in Molecular Biology
4. Bouatmane, S., Roula, M., Bouridane, A., Al-Maadeed, S.: Round-robin sequential forward selection algorithm for prostate cancer classification and diagnosis using multispectral imagery. Machine Vision and Applications 22(5), 865–878 (2011)
5. Buttarello, M., Plebani, M.: Automated blood cell counts -state of the art. American Journal of Clinical Pathology 130, 104–116 (2008)
6. Choi, K.S., Zeng, Y., Qin, J.: Using sequential floating forward selection algorithm to detect epileptic seizure in EEG signals. In: 11th International Conference on Signal Processing (ICSP), vol. 3, pp. 1637–1640 (2012)
7. Comaniciu, D., Meer, P.: Cell image segmentation for diagnostic pathology. In: Advanced Algorithmic Approaches to Medical Image Segmentation, pp. 541–558. Springer, New York (2002)
8. Dangott, B., Salama, M., Ramesh, N., Tasdizen, T.: Isolation and two-step classification of normal white blood cells in peripheral blood smears. Journal of Pathology Informatics 3(1), 13 (2012)
9. Dorini, L.B., Minetto, R., Leite, N.J.: Semi-automatic white blood cell segmentation based on multiscale analysis. IEEE Journal of Biomedical and Health Informatics 17(1), 250–256 (2013)
10. Fu, B., Zhou, J., Li, Y., Zhang, G., Wang, C.: Image analysis by modified legendre moments. Pattern Recognition 40(2), 691–704 (2007)
11. Grimaldi, E., Scopacasa, F.: Evaluation of the abbott CELL-DYN 4000 hematology analyzer. American Journal of Clinical Pathology 113(4), 497–505 (2000)
12. Habizadeh, M., Krzyżak, A., Fevens, T.: Analysis of white blood cell differential counts using dual-tree complex wavelet transform and support vector machine classifier. In: Bolc, L., Tadeusiewicz, R., Chmielewski, L.J., Wojciechowski, K. (eds.) ICCVG 2012. LNCS, vol. 7594, pp. 414–422. Springer, Heidelberg (2012)

13. Habibzadeh, M., Krzyżak, A., Fevens, T.: Comparative study of shape, intensity and texture features and support vector machine for white blood cell classification. Journal of Theoretical and Applied Computer Science 7, 20–35 (2013)
14. Habibzadeh, M., Krzyżak, A., Fevens, T., Sadr, A.: Counting of RBCs and WBCs in noisy normal blood smear microscopic images. In: SPIE Medical Imaging: Computer-Aided Diagnosis, Orlando, FL, USA, vol. 7963, p. 79633I (February 2011)
15. Haralick, R.M., Shanmugam, K., Dinstein, I.: Textural features for image classification. IEEE Transactions on Systems, Man and Cybernetics SMC-3(6), 610–621 (1973)
16. Hosny, K.M.: Image representation using accurate orthogonal gegenbauer moments. Pattern Recognition Letters 32(6), 795–804 (2011)
17. Hu, M.K.: Visual pattern recognition by moment invariants. IEEE Transactions on Information Theory 8(2), 179–187 (1962)
18. Jain, A., Zongker, D.: Feature selection: evaluation, application, and small sample performance. IEEE Transactions on Pattern Analysis and Machine Intelligence 19(2), 153–158 (1997)
19. Jiang, K., Liao, Q.M., Dai, S.Y.: A novel white blood cell segmentation scheme using scale-space filtering and watershed clustering. In: IEEE International Conference on Machine Learning and Cybernetics, Xi'an, China, pp. 2820–2825 (November 2003)
20. Kaya, G.T., Kaya, H., Ersoy, O.K.: Feature selection by high dimensional model representation and its application to remote sensing. In: IEEE International Geoscience and Remote Sensing Symposium (IGARSS), pp. 4938–4941 (2012)
21. Kok-Swee, S., Faizy Salleh, A., Chee-way, C., Rosli, B., Hock-Ann, G.: Translation and scale invariants of Hahn moments. International Journal of Image and Graphics 09(02), 271–285 (2009)
22. Kumar, B.R., Joseph, D.K., Sreenivas, T.V.: Teager energy based blood cell segmentation. In: 14th International Conference on Digital Signal Processing, Santorini, Greece, pp. 619–622 (July 2002)
23. Landis, J.R., Koch, G.G.: The measurement of observer agreement for categorical data. Biometrics 33(1)
24. Lezoray, O., Elmoataz, A., Cardot, H., Gougeon, G., Lecluse, M., Elie, H., Revenu, M.: Segmentation of cytological images using color and mathematical morphology. Acta Stereologica 18(1), 1–14 (1999)
25. Li, S., Lee, M.C., Pun, C.M.: Complex Zernike moments features for shape-based image retrieval. IEEE Transactions on Systems, Man and Cybernetics, Part A: Systems and Humans 39(1), 227–237 (2009)
26. Mukherjee, D.P., Ray, N., Acton, S.T.: Level set analysis for leukocyte detection and tracking. IEEE Transactions on Image Processing 13(4), 562–572 (2004)
27. Mukundan, R., Ong, S.H., Lee, P.A.: Image analysis by Tchebichef moments. IEEE Transactions on Image Processing 10(9), 1357–1364 (2001)
28. Ongun, G., Halici, U., Leblebicioglu, K., Atalay, V., Beksac, M., Beksac, S.: Feature extraction and classification of blood cells for an automated differential blood count system. In: International Joint Conference on Neural Networks, Washington, DC, USA, pp. 2461–2466 (July 2001)
29. Ping, Z., Wu, R., Sheng, Y.: Image description with Chebyshev-Fourier moments. Journal of the Optical Society of America A 19(9), 1748–1754 (2002)
30. Rahman, S.: Extended polynomial dimensional decomposition for arbitrary probability distributions. Journal of Engineering Mechanics 135(12), 1439–1451 (2009)
31. Ramoser, H., Laurain, V., Bischof, H., Ecker, R.: Leukocyte segmentation and classification in blood-smear images. In: 27th IEEE Annual Conference Engineering in Medicine and Biology, Shanghai, China, September 1-4, pp. 3371–3374 (2005)

32. Ren, H., Ping, Z., Bo, W., Wu, W., Sheng, Y.: Multidistortion-invariant image recognition with radial harmonic fourier moments. Journal of the Optical Society of America A 20(4), 631–637 (2003)
33. Rodenacker, K., Bengtsson, E.: A feature set for cytometry on digitized microscopic images. Analytical Cellular Pathology 25(1), 1–36 (2001)
34. Rowan, R., England, J.M.: Automated examination of the peripheral blood smear. In: Automation and Quality Assurance in Hematology, ch. 5, pp. 129–177. Blackwell Scientific Oxford (1986)
35. Ruzicka, K., Veitl, M., Thalhammer-Scherrer, R., Schwarzinger, I.: New hematology analyzer Sysmex XE-2100: performance evaluation of a novel white blood cell dierential technology. Archives of Pathology and Laboratory Medicine 125(3), 391–396 (2001)
36. Sheng, Y., Shen, L.: Orthogonal fourier-mellin moments for invariant pattern recognition. Journal of the Optical Society of America A 11(6), 1748–1757 (1994)
37. Shitong, W., Min, W.: A new detection algorithm (NDA) based on fuzzy cellular neural networks for white blood cell detection. IEEE Transactions on Information Technology in Biomedicine 10(1), 5–10 (2006)
38. Sinha, N., Ramakrishnan, A.G.: Automation of differential blood count. In: IEEE International Conference on Convergent Technologies for Asia-Pacific Region, Bangalore, India, pp. 547–551 (October 2003)
39. Sobol, I.M.: Global sensitivity indices for nonlinear mathematical models and their Monte Carlo estimates. Mathematics and Computers in Simulation 55(1-3), 271–280 (2001)
40. Sobrevilla, P., Montseny, E., Keller, J.: White blood cell detection in bone marrow images. In: 18th International Conference of the North American Fuzzy Information Processing Societ (NAFIPS), pp. 403–407 (1999)
41. Tamura, H., Mori, S., Yamawaki, T.: Textural features corresponding to visual perception. IEEE Transactions on Systems, Man and Cybernetics 8(6), 460–473 (1978)
42. Tang, X.: Texture information in runlength matrices. IEEE Transactions on Image Processing 7(11), 1602–1609 (1998)
43. Theera-Umpon, N., Dhompongsa, S.: Morphological granulometric features of nucleus in automatic bone marrow white blood cell classification. IEEE Transactions on Information Technology in Biomedicine 11(3), 353–359 (2007)
44. Verso, M.L.: The evolution of blood-counting techniques. Journal of Medical History 8(2), 149–158 (1964)
45. Xia, T., Zhu, H., Shu, H., Haigron, P., Luo, L.: Image description with generalized pseudozernike moments. Journal of the Optical Society of America A 24(1), 50–59 (2007)
46. Xiao-min, Y., Li-min, L., Yu, W.: Automatic classification system for leukocytes in human blood. Journal of Computer Science and Technology 17(2), 130–136 (1994)
47. Yap, P.T., Paramesran, R., Ong, S.H.: Image analysis by Krawtchouk moments. IEEE Transactions on Image Processing 12(11), 1367–1377 (2003)
48. Kang, B., Ma, Z., Ma, J.: Translation and scale invariant of Legendre moments for images retrieval. Journal of Information & Computational Science 8(11), 2221–2229 (2011)
49. Ziehn, T., Tomlin, A.S.: GUI-HDMR - a software tool for global sensitivity analysis of complex models. Environmental Modelling & Software 24(7), 775–785 (2009)

End-Shape Recognition for Arabic Handwritten Text Segmentation

Amani T. Jamal, Nicola Nobile, and Ching Y. Suen

CENPARMI (Centre for Pattern Recognition and Machine Intelligence)
Computer Science and Software Engineering Department, Concordia University
Montreal, Quebec, Canada
{am_jamal,nicola,suen}@cenparmi.concordia.ca

Abstract. Text segmentation is an essential pre-processing stage for many systems such as text recognition and word spotting. However, few methods have been published for Arabic text segmentation. In Arabic handwritten documents, separating text into words is challenging due to the enormous different Arabic handwriting styles. In this paper, we present a new segmentation methodology of an Arabic handwritten text line into words. Our proposed approach of text segmentation utilizes the knowledge of Arabic writing characteristics. This method shows promising results.

Keywords: component, Arabic Handwritten Documents, segmentation, End-Shape recognition.

1 Introduction

Extracting all the word images from a handwritten document is an essential pre-processing step for two reasons [1]. First, for text recognition methods, which can be categorized into letter-based and word-based, there is a need to work on pre-extracted word images. Secondly, for word-spotting or content-based image retrieval techniques, all the word images in the documents are required to be pre-segmented properly. Most of the techniques in handwritten document retrieval and recognition fail if the texts are wrongly segmented into words.

Few methods have been published for Arabic text segmentation. In Arabic handwritten documents, separating text into words is challenging due to the enormous different Arabic handwriting styles. In this paper, we present a new segmentation methodology of an Arabic handwritten text line into words. Our proposed approach of text segmentation utilizes the knowledge of Arabic writing characteristics.

In this Section, we provide some background of the Arabic characteristics and the previous works of text line segmentation into words. In addition, the challenges of Arabic handwritten text segmentation are given. Finally, the proposed approach is summarized with the rational of applying it and our overall methodology is explained. The secondary component removal technique is briefly explained in Section 2. The used metric-based segmentation method is explained in Section 3. The contribution of this paper is described in Section 4. The experiment is explained in Section 5. Finally the conclusion is given in Section 6.

N. El Gayar et al. (Eds.): ANNPR 2014, LNAI 8774, pp. 228–239, 2014.

1.1 Arabic Characteristics

In the Arabic script, there is a major characteristic that differentiates this language from Latin-based ones. Twenty-two letters in the Arabic language must be connected on a baseline within a word. The remaining six letters cannot be connected from the left, which we call non-left-connected (NLC) letters. In this way, NLC letters separate a word into several parts depending on how many of these letters are included in a word. In other words, NLC letters indicate a separation of Part of Arabic Word (PAW). A study shows that NLC letters represent 33% of the text [2]. The Arabic script is considered as semi-cursive [4] since each word may be composed of one or more sub-words or (PAWs). In [5], a sub-word is defined "as being a connected entity of one or several characters belonging to the word". Figure 1 shows one word with two PAWs.

Fig. 1. An Arabic word with two PAWs

1.2 Challenges

Arabic texts have two types of spacing, intra-word gaps (gaps between PAWs within a word) and inter-word gaps (gaps between words). Intra-word gaps in the Arabic language are different from the ones in Latin-based languages. In Latin, intra-word gaps refer to the spaces that arise arbitrarily between any successive letters as a result of handwriting styles. In Arabic, intra-word gaps are the ones between two PAWs, where the word must be disconnected due to NLC letters. This is part of the structure of the language.

Generally, handwritten texts lack the uniform spacing that is normally found in machine-printed texts. In Arabic machine-printed texts, the inter-word gaps are much larger than intra-word gaps. However, in Arabic handwritten documents, the spacing between the two types is mostly the same [3]. This is pointed out in Figure 2 from the CENPARMI cheque database [11]. Since the shape of most of the NLC letters are curved, with the open end to the left, they are usually written with long strokes, which shrink the distance between words. Sometimes, they caused overlapping, or touching between words.

Fig. 2. Intra and inter word gaps in Arabic text

1.3   Related Work

Word segmentation is a critical step towards word spotting and text recognition. There are many word segmentation techniques in the literature [14]. Nevertheless, it is still a challenging problem in handwritten documents. Word segmentation techniques are based mainly on the analysis of the distance between adjacent CCs. The algorithms can be categorized into gap thresholding and metric classification. In the former, the threshold is determined to distinguish between gap types. In the latter, the gaps are classified into either inter or intra word gaps.

There is little research for Arabic handwritten text segmentation. Some works apply to manual segmentation [13]. In [9], an online Arabic segmentation method was proposed. The gap types are classified based on local and global online features. The fusion of multi-classification decisions was used as a post-processing stage to verify the decisions.

J. Alkhateeb et al. proposed a method for Arabic handwritten text segmentation into words based on the distances between PAWs and words [6]. Vertical projection analysis was employed to calculate the distances. The statistical distribution was used to find the optimal threshold. Bayesian criteria of minimum classification error were used to determine the threshold. The technique was applied on a subset of the IFN/ENIT database. The correct segmentation of one-word and three-word images was 80.34% and 66.67% respectively.

In [8], an offline handwritten Arabic text segmentation technique was introduced. First, the CCs of the images were detected based on the baseline. Their bounding boxes were determined. These boxes were extended to include the dots and any small CCs. The distances between adjacent PAWs were obtained. They assumed that the distance between words is larger than the distance between PAWs. Based on that assumption, a threshold approach was used. Two conditional probabilities were determined by manually analyzing more than 200 images. A Bayesian histogram minimum classification error criteria was used to find the optimal distance. They achieved 85% of correct segmentation.

M. Kchaou et al applied scaling space to segment Arabic handwritten documents into words [7]. The techniques that were used for segmentation were scaling space and feature extraction from horizontal and vertical profiles. Two documents written by five writers were used in their experiments. Segmentation errors varied between 29.5% and 3.5%. They believe that the errors arising from different writer styles, coordinating conjunctions and distances between PAWs.

In [18], the segmentation is based on extracting several features from the adjacent clusters. The main and secondary components are merged into clusters. Nine features were extracted. The neural network is used to classify the gaps between the words. Overall performance is about 60% correct segmentation.

Due to the importance of text segmentation, four Handwriting Segmentation Contests were organized: ICDAR 2007, ICDAR 2009, ICFHR 2010, and ICDAR 2013 [15]. Therefore, a benchmarking dataset with an evaluation methodology were created to capture the efficiency of the methods. The total number of participants on these competitions was thirty research groups with different algorithms. In addition, there are plenty of methods for Latin-based languages in comparison to Arabic language that address this problem [14].

1.4   Proposed Approach

The main difference with our segmentation approach from previous methods is utilizing the knowledge of Arabic writing by shape analysis. In [6], [9], and [8], the authors pointed out the importance of using the language specific knowledge for Arabic text segmentation. In addition, in [7], the authors claim that one of the problems of Arabic text segmentation is the inconsistent spacing between words and PAWs. Our approach for segmentation is a two-stage strategy: (1) metric-based segmentation, and (2) recognition-based segmentation.

Utilizing the Knowledge of Arabic Writing. In the Arabic alphabet, twenty-two letters out of twenty-eight have different shapes when they are written at the end of a word as opposed to the beginning or middle. Two non-basic characters have different shapes at the end of a word. Therefore, analyzing these shapes can help to identify the end of a word. In fact, there are just fourteen main shapes that can be used to distinguish the end of a word, since the remaining characters have the same main part but have a different number and/or position of dots. Only NLC letter shapes, which cause the disconnection within a word, are written the same way at the beginning, the middle or the end of a word. Therefore NLC letters cannot identify the end of a word. Consequently, End Shape Letters (ESLs) can be categorized into two classes: endWord and nonEndWord. Figure 3 shows the shape of the letter Noon when it is written at the beginning of the word, the middle and the end, and this letter is part of endWord class.

Beginning   Middle   End

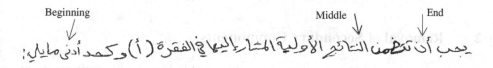

Fig. 3. Letter Noon in different positions

1.5   Our Methodology

Our methodology is composed of two stages as mentioned earlier. The first stage is called metric-based segmentation. The second stage is named ESL-based segmentation. The input of our system is a binarized text line. A method that was proposed by M. Al-Khayat et al. [16] for text line segmentation was used. First, the Connected Components (CCs) of a text line were extracted. CCs by definition, consist of connected black pixels. Normally, a PAW is composed of several CCs: a main component, diacritics, and/or directional markings. Therefore, the first main step in segmenting an Arabic handwritten word is detecting and labeling its CCs. CC analysis is the most efficient approach since the Arabic script consists of several overlapping CCs. The 8-connectivity method was used. Second, the secondary components were removed, which was explained in detail in Section 2. Then, metric-based and ESL-based segmentation were applied. The ESL-based proposed method was provided in Section 4. The overall methodology is given as a block diagram in Figure 4.

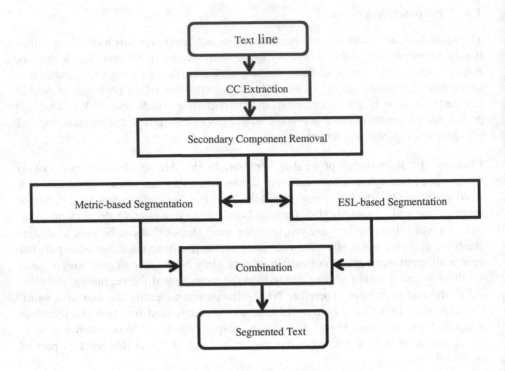

Fig. 4. Overall Methodology

2   Removal of Secondary Components

In this paper, the secondary components are removed to improve the performance of metric-based segmentation and to reduce the number of classes of Final Shape Letter recognition system. However, many algorithms also remove the secondary components to facilitate skew correction and baseline estimation. Some methods also detect the secondary components to extract more features for recognition or spotting systems. We used secondary component removal using morphological reconstruction [12].

Mathematical morphology is an essential tool in image processing that is used to process images based on its shape information. Reconstruction is a morphological transformation that involves two images. The mask image constrains the transformation. The marker image is the starting point for the transformation. Using the morphological reconstruction method that is based on a thin horizontal line facilitates main component extraction. This line is defined below the middle of the image. The reconstruction method is used to ascertain that only the main components are analyzed. We process only binary images. The word images are the masks. The marker is a generated binary image with the same size as the mask image and a horizontal line that is located below the middle of the image.

3   Metric-Based Segmentation

In this stage, the distance between adjacent components was computed using a gap metric. This method is somewhat like a writer dependent technique since the threshold was estimated based on a given text line. In fact, since spaces between words are part of a writing style, this writer dependent technique provides better result [14]. Thus, a global threshold across all documents is an inadequate solution.

3.1   Distance Computation

After extracting the main components that are ordered from left-to-right, a bounding box for each component was calculated. Then all the overlapped bounding boxes were merged. The minimum horizontal distances between pairs of adjacent bounding boxes were measured. After that, all gap metrics of the text line were sorted.

3.2   Threshold Estimation

After identifying the largest space (determined based on empirical study) between the sorted values, the threshold was determined. The threshold is the minimum value of the largest group of gap metrics. If the spaces between the gap metrics are almost the same, the threshold is calculated to be the mean of the gap metrics. Finally, the text line is segmented into words based on this threshold. The algorithm is given below:

Algorithm for Word Threshold Estimation

For each text line
Calculate the bounding boxes for each CC (Bc_i), i =1..... L
Calculate the distance between Bc_i and Bc_{i+1}
Find all gap metrics G_j
Find spaces between G_j that is denoted by S_i
If a large space is found
    The minimum value in the largest group is determined as a Threshold (T).
    $T \subset G_i$
Else
    The mean of the gap metrics is the threshold
    $T = mean(G_i)$

4   ESL-Based Segmentation

In this stage, the main idea is to recognize the ESL that helps to specify the word segment. ESL can be isolated or part of a PAW. However, the end-shape needs to be detected first before recognition can begin. Each step is described in the following sub-sections. Our method is depicted in Figure 5.

4.1 ESL Detection

At this stage, the main purpose is to detect the isolated letter or the last letter of a PAW. The last part will be extracted based on the height, width, and the baseline position.

4.2 ESL Recognition

At this stage, either the end-shape of the main component or the isolated letter is sent to an ESL recognizer. We created an ESL database and classifier to identify the end of a word. This recognizer classifies the end-shapes of main components and of an isolated letter.

The ESL database contains the shape of letters at the final position (only in its isolated form). The endWord set contains eleven classes and the nonEndWord set is composed of three classes. We used the CENPARMI Arabic isolated letter database [19]. To test the ESL recognition system before applying it to the documents, a testing model was generated using the testing set of the CENPARMI Arabic isolated letter database. We applied the method that was used by M. W. Sagheer [24]. This ESL recognizer consists of the following three phases: (1) Pre-processing, (2) Feature extraction, and (3) Recognition.

Since our concern is the main component of the letters, we removed all the secondary components that comprise less than half the area of the largest component. Then, the bounding box of the main component is calculated in order to eliminate all the white spaces around it. The image was normalized to two different sizes, 64 x 64 and 128 x 128 pixels by using aspect ratio adaptive normalization strategy [22]. Two different sizes of the image were used for different feature extraction processes. In addition, the image was skeletonized to standardize the representation of the images and facilitate feature extraction. The Zhang and Suen thinning algorithm [23] was applied. We extracted gradient features and structural features. Several experiments were conducted with different features to find the best combination of these features that produce the best results as shown in Table 1.

Gradient Features Extraction. In our gradient feature extraction phase, each image of size 128 x 128 pixels was converted into a grayscale image. Robert's filter masks were applied on the images.

Let $IM(x, y)$ be an input image; the horizontal gradient component (g_x) and vertical gradient component (g_y) were calculated as follows:

$g_x = IM(x+1, y+1) - IM(x, y)$

$g_y = IM(x+1, y) - IM(x, y+1)$

• The gradient strength and direction of each pixel $IM(x,y)$ were calculated as follows:

$$\text{Strength: } s(x, y) = \sqrt{gx^2 + gy^2} \tag{1}$$

$$\text{Direction: } \theta(x, y) = \tan^{-1}(g_y / g_x) \tag{2}$$

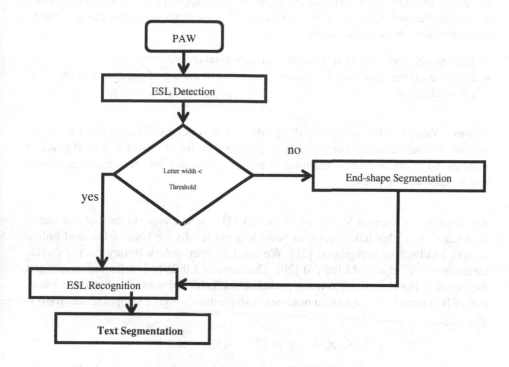

Fig. 5. ESL-based segmentation method

After calculating the gradient strength and direction for each pixel, the following steps were taken in order to calculate the feature vector:

1. The direction of a vector (g_x, g_y) in the range of [π,-π]. These gradient directions were quantized to 32 intervals of $\pi/16$ each.
2. The gradient image was divided into 81 blocks, with 9 vertical blocks and 9 horizontal blocks. For each block, the gradient strength was accumulated in 32 directions. By applying this step, the total size of the feature set in the feature vector is (9 x 9 x 32) = 2592.
3. To reduce the size of a feature vector, a 5 x 5 Gaussian filter was applied by down sampling the number of blocks from 9 x 9 to 5 x 5. The number of directions was reduced from 32 to 16 by down sampling the weight vector [1 4 6 4 1]. The size of the feature vector is 400 (5 horizontal blocks x 5 vertical blocks x 16 directions).
4. A variable transformation ($y = x^{0.4}$) was applied on all features to make the distribution of features Gaussian-like.

Structural Features Extraction. In addition to the gradient feature, other structural features were extracted. The additional features are: the number of black pixels, horizontal and vertical histograms, end and intersection points, holes, and structure of the top part of the image. However, the horizontal and vertical features were removed since they provide lower performance. Moreover, the upper profile features were used to capture the outline shape of the top part [17]. To extract the upper profile feature the following steps were followed:

- Each image was converted into a two-dimensional array.
- For each column, the distance was measured from the top of the image to the closest black pixel.

Feature Vector. After extracting the gradient and structural features from each image, all the features were merged to make a feature vector of size of 468 (400 gradient features, 64 upper profile, 4 structural features). Then, this feature vector was passed to the classification phase.

Recognition. A Support Vector Machine (SVM) is a technique in the field of statistical learning. SVMs have shown to provide good results for both offline and online cursive handwriting recognition [21]. We used an open source library for the implementation of SVM called LibSVM [20]. The input of LibSVM is a feature matrix and the output is the classification result probabilities. LibSVM uses a Radial Basis Function (RBF) kernel for mapping a nonlinear sample into a higher sample space. RBF is given by:

$$K(X_i, X_j) = \exp(-\gamma \| X_i - X_j \|2) , \gamma > 0 \tag{3}$$

For the K-class problem, K2 SVMs are trained by a pairwise approach. The probability is estimated for test sample x that belongs to class i. The probabilities are obtained from a one-against-one class probability. The two optimal parameters γ and C were chosen by using v-fold cross validation. A training model was generated for the whole collection of images with their class labels.

Table 1. Experimental Results with Different Features

Feature vector	Gradient feature	Horizontal projection	Vertical projection	Upper profile	Structural features	Feature vector	Recognition result
Fv1	x					416	89.97%
Fv2	x	x				480	89.66%
Fv3	x		x			480	89.51%
Fv4	x				x	420	90.27%
Fv5	x			x		464	90.73%
Fv6	x			x	x	468	**90.88%**

5   Experiments

We performed experiments using the IFN/EINT [10] database. It was developed by the National School of Engineers of Tunis (ENIT), and the Institute of Communication Technology (IFN) in Germany. The database contains 937 Tunisian town/village names written by 411 writers. In fact, 448 names contain two to three words. We did our experiments on the names that contain two to three words, since our gap threshold estimation is based on the distance within the images. In addition, since the CENPARMI database is composed of only the isolated shapes of the letters (which is a subset of the final shape letters) and there does not exist an available database that has all the shapes of the letters in their final positions (connected), therefore, we applied our experiments on those city names that can be segmented using the CENPARMI isolated letters. This set contains a total of 440 names. We used set-a for training and subsets from set-b, set-c, set-d, and set-e for testing. Table 2 shows the word segmentation results of metric-based, FSL-based and the result of the combined methods. In Section 1.3, the results of related work is reported.

Table 2. Word segmentation result

Set	Metric-based	ESL-based	ESL + Metric
Set-b	67.34%	82.00%	86.16%
Set-c	68.29%	84.15%	93.48%
Set-d	83.12%	98.61%	99.02%
Set-e	71.05%	86.84%	93.88%

6   Conclusion

Arabic handwriting recognition and spotting depend on accurate segmentation. In this paper, we introduced a new segmentation approach for Arabic handwritten text. Our experiments show promising results. Most of the errors are caused by the confusion between classes. An example of such an error is illustrated in Figure 6. Our future work will include creating a database of final shape letters that are connected to the PAWs, improving the segmentation method, and extract more features to improve the FSL recognition system.

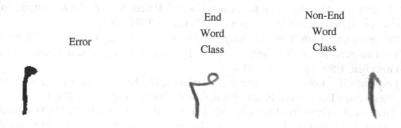

Fig. 6. Error Analysis

Acknowledgment. This work was supported by King Abdulaziz University (KAU), Jeddah, Saudi Arabia and Ministry of Higher Education in Saudi Arabia.

References

1. Huang, C., Srihari, S.: Word Segmentation of Off-line Handwritten Documents. In: Proceedings of the Document Recognition and Retrieval (DRR) XV, IST/SPIE Annual Symposium, San Jose, CA, USA, vol. 6815 (2008)
2. Olivier, G., Miled, H., Romeo, K., Lecourtier, Y.: Segmentation and Coding of Arabic Handwritten Words. In: Proceedings of 13th International Conference of Pattern Recognition (ICPR 1996), vol. 3, pp. 264–268 (1996)
3. Amin, A.: Recognition of Printed Arabic Text based on Global Features and Decision Tree Learning Techniques. Pattern Recognition 33(8), 1309–1323 (2000)
4. Miled, H., Amara, N.E.B.: Planar Markov Modeling for Arabic Writing Recognition: Advancement State. In: Proceedings of the 6th International Conference on Document Analysis and Recognition (ICDAR 2001), Seattle, USA, pp. 69–73 (2001)
5. Westall, J.M., Narasimha, M.S.: Vertex Directed Segmentation of Handwritten Numerals. Pattern Recognition 26(10), 1,473–1,186 (1993)
6. AlKhateeb, J.H., Ren, J., Ipson, S.S., Jiang, J.: Knowledge-based Baseline Detection and Optimal Thresholding for Words Segmentation in Efficient Pre-processing of handwritten Arabic text. In: Proceedings of 5th International Conference on Information Technology: New Generations (ITNG 2008), Las Vegas, Nevada, pp. 1158–1159 (2008)
7. Kchaou, M., Kanoun, S., Ogier, J.: Segmentation and Word Spotting Methods for Printed and Handwritten Arabic Texts: A Comparative Study. In: Proceedings of 13th International Conference on Frontiers in Handwriting Recognition (ICFHR 2012), Bari, Italy, pp. 274–279 (2012)
8. AlKhateeb, J.H., Jiang, J., Ren, J., Ipson, S.: Component-based Segmentation of Words from Handwritten Arabic text. In: Proceedings of World Academy of Science, Engineering and Technology (WASET), Vienna, Austria, vol. 31 (2008) ISSN: 1307- 6884
9. Elanwar, R.I., Rashwan, M., Mashali, S.: Arabic Online Word Extraction from Handwritten Text Using SVM-RBF Classifiers Decision Fusion. In: Proceedings of the 4th WSEAS International Conference on Nanotechnology, Cambridge, UK, pp. 68–73 (2012)
10. Pechwitz, M., Maddouri, S.S., M"argner, V., Ellouze, N., Amiri, H.: IFN/ENIT- Database of Handwritten Arabic Words. In: 7th Colloque International Francophone sur l'Ecrit et le Document (CIFED), Hammamet, Tunis, vol. 2, pp. 127–136 (2002)
11. Al-Ohali, Y., Cheriet, M., Suen, C.Y.: Databases for Recognition of Handwritten Arabic Cheques. Pattern Recognition 36(1), 111–121 (2003)
12. Jamal, A.T., Suen, C.Y.: Removal of Secondary Components of Arabic Handwritten Words Using Morphological Reconstruction. In: Proceedings of the 2nd International Conference on Information Technology (ICIT), Dubai, UAE (2014)
13. Manmatha, R., Rath, T.M.: Indexing Handwritten Historical Documents - Recent Progress. In: Proceedings of the Symposium on Document Image Understanding Technology, Greenbelt, USA, pp. 77–85 (2003)
14. Louloudis, G., Gatos, B., Pratikakis, I., Halatsis, C.: Text Line and Word Segmentation of Handwritten Documents. Pattern Recognition 42(12), 3169–3183 (2009)
15. Nikolaos, S., Gatos, B., Louloudis, G., Pal, U., Alaei, A.: ICDAR 2013 Handwriting Segmentation Contest. In: Proceedings of 12th International Conference on Document Analysis and Recognition (ICDAR), Washington DC, USA, pp. 1402–1406 (2013)

16. Al-Khayat, M., Lam, L., Suen, C.Y., Yin, F., Liu, C.-L.: Arabic Handwritten Text Line Extraction by Applying an Adaptive Mask to Morphological Dilation. In: Proceedings of the 10th IAPR International Workshop on Document Analysis Systems (DAS), Gold Coast, Australia, pp. 100–104 (2012)
17. Aghbari, Z., Brook, S.: HAH manuscripts: A Holistic Paradigm for Classifying and Retrieving Historical Arabic Handwritten Documents: Expert Systems with Applications. An International Journal 36(8), 10942–10951 (2009)
18. Srihari, S., Srinivasan, H., Babu, P., Bhole, C.: Handwritten Arabic Word Spotting Using the CEDARABIC Document Analysis System. In: Proceedings of Symposium on Document Image Understanding Technology (SDIUT), pp. 123–132. College Park, USA (2005)
19. Alamri, H., Sadri, J., Suen, C.Y., Nobile, N.: A Novel Comprehensive Database for Arabic Off-line Handwriting Recognition. In: Proceedings of 11th International Conference on Frontiers in Handwriting Recognition (ICFHR 2008), Montreal, Canada, pp. 664–669 (2008)
20. Chang, C., Lin, C.: LIBSVM: A Library for Support Vector Machines (2001), http://www.csie.ntu.edu.tw/cjlin/libsvm
21. Gatos, B., Pratikakis, I., Kesidis, A.L., Perantonis, S.J.: Efficient Off-Line Cursive Handwriting Word Recognition. In: The Proceedings of the 10th International Workshop on Frontiers in Handwriting Recognition (IWFHR 2006), La Baule, France, pp. 121–125 (2006)
22. Liu, C.-L., Nakashima, K., Sako, H., Fujisawa, H.: Handwritten Digit Recognition: Investigation of Normalization and Feature Extraction Techniques. Pattern Recognition 37(2), 265–279 (2004)
23. Zhang, T.Y., Suen, C.Y.: A Fast Parallel Algorithm for Thinning Digital Patterns. Communications of the ACM 27(3), 236–239 (1984)
24. Sagheer, M.W.: Novel Word Recognition and Word Spotting Systems for Offline Urdu Handwriting. Master Thesis, Concordia University, Montreal, Canada (2010)

Intelligent Ensemble Systems for Modeling NASDAQ Microstructure: A Comparative Study

Salim Lahmiri

ESCA School of Management
7 El Kindy Street, BD Moulay Youssef, Casablanca, Morocco
slahmiri@esca.ma

Abstract. In this study, four neural networks (NN) ensemble systems are presented and compared for NASDAQ returns prediction. They are the conventional feed-forward back-propagation neural network (FFNN) ensemble which widely used in the literature, time-delay neural network (TDNN) ensemble, nonlinear auto-regressive with exogenous inputs (NARX) ensemble and the radial basis neural network (RBFNN) ensemble. Each component of the NN ensemble is used to learn specific patterns related to a given NASDAQ submarket. Based on the mean of absolute errors (MAE), the experiments show that ensemble models based on advanced NN architectures such as TDNN, NARX, and RBFNN ensemble all achieve lower forecasting errors than traditional FFNN ensemble system. In addition, the RBFNN ensemble outperformed all other NN ensembles under study.

Keywords: Neural networks, ensemble, stock market microstructure, forecasting.

1 Introduction

Successful prediction models for stock market trading are of great interest for investors to make profits. Therefore, there is a need for effective stock price forecasting systems capable to limit personal bias and mistakes. However, the financial market is complex, evolutionary, and non-linear dynamical system [1,2]. As a result, financial market data are noisy and nonstationary [1,2].

Financial market prediction had received a large attention in the literature where different methods and algorithms for automated stock market prediction were proposed; including artificial neural networks, fuzzy logic, and expert systems to name a few. Indeed, such studies were surveyed in Atsalakis and Valavanis [3] and Bahrammirzaee [4]. In general, the literature [3,4] used two broad classes to predict stock market prices [5]: fundamental analysis and technical analysis. The fundamental analysis depends on knowledge of microeconomics and macroeconomics factors, whilst technical analysis is based on historical patterns to predict market prices [5]. However, fundamental analysis based knowledge is usually not readily available, and historical patterns are not always evident because of the noise [5].

N. El Gayar et al. (Eds.): ANNPR 2014, LNAI 8774, pp. 240–251, 2014.

The purpose of this study is to design a neural network (NN) committee (ensemble) to model stock market microstructure and achieve better prediction accuracy of the market returns. Indeed, neural network ensemble systems were found to be effective in forecasting stock market [5-8], and also in other time series forecasting applications such as customer purchase behavior [9], drug dissolution [10], weather [11], climate [12], and software reliability [13].

A neural network (NN) ensemble or committee is a learning paradigm where a collection of several neural networks is trained for the same task. It is expected to provide the following advantages over the traditional neural network. First, a NN is an adaptive nonlinear soft computing system that can learn from patterns and capture hidden functional relationships in a given data even if the functional relationships are not known or difficult to identify [14]. Indeed, it is capable of parallel processing of the information with no prior assumption about the model form and the process that generates the data. In addition, it is robust to noisy data; hence the network is capable to model non-stationary and dynamic data [15]. Furthermore, a NN can theoretically approximate a continuous function to an arbitrary accuracy on any compact set [16-18]. Second, a NN ensemble or committee can produce even more accurate results than any of the individual neural networks by making up the ensemble and thus intensifying discriminant capability of neural networks [19]. Third, a NN is prone to overfitting when it is too closely adjusted to the training set [19]. Therefore, its generalization error tends to increase when it is applied to previously unseen samples [19]. As a solution, a committee or an ensemble approach can make base neural networks robust to overfitting and thus reduce generalization error [19]. Fourth, a NN committee is used to deal with sampling and modeling uncertainties that may otherwise impair individual NN forecasting accuracy and robustness [20]. Fifth, previous works demonstrated the ability of NN based committee system in modeling and predicting various types of time series [5-13].

All of the aforementioned studies [5-8] only consider single neural network type for financial time series prediction. However, the performance of a neural network system significantly depends on the type of NN used to design the committee system. For instance, previous studies relied on the well known feed-forward back-propagation neural networks as basic systems. Indeed, although NN committees are essential for providing accurate forecasts the improvements in the construction of such committees is important by considering other type of neural networks.

In this paper, the performance of the NN committee is investigated depending on the type NN used to form the ensemble. In particular, four different NN architectures are considered to form the NN committee and are compared. They are the conventional feed-forward back-propagation neural network which widely used in the literature, time-delay neural network, nonlinear auto-regressive with exogenous inputs (NARX) network, and the radial basis neural network.

Despite the potential benefits of using fundamental or technical analysis knowledge for ensemble training and prediction task, we rely on information from microstructure of the stock market to train NN committees. In particular, we aim to model stock market returns based on the dynamics of each sub-market (component) that composes the whole stock market. In other words, our emphasis is on the importance of information related to price movements in each sub-market in determination of the overall stock market behavior. Indeed, we rely on microstructure information for two main reasons. First, price formation in each sub-market is key information that determines the stock market trend. Therefore, modeling the stock market microstructure from an informational point of view would be helpful to improve forecasting accuracy when predicting stock market future return. Second, unlike fundamental and technical analysis based information, microstructure information is always available and not affected with noisy information.

The rest of this paper is organized as follows. Section 2 presents the committee system, and gives a brief introduction to each type of artificial neural network adopted to form the neural network committee. Section 3 presents the empirical results from the NASDAQ market data. The paper is concluded in Section 4.

2   Methods

In this section, we present the proposed system for stock market return prediction based on neural network ensembles based on market microstructure information. Neural network ensemble was proposed by originating Hansen and Salamon [21] as a learning paradigm where several neural networks are trained for the same task. The purpose is to improve the generalization performance of NN system in comparison with using a single neural network.

The prediction system based on neural network ensemble (committee) is shown in Figure 1. The input of the system is the financial return of a given sub-market and the output of the system is the predicted return of the aggregate stock market. For instance, each NN is used as the basic prediction system making up ensemble and each is trained with a sub-market specific price returns. Finally, the output NN is used to generate a single system to produce the output of the ensemble by combining the predictions of multiple neural networks.

In this study, the neural network committee consists of K component neural networks where each component neural network is a three-layer single input single-output NN with two nodes in the hidden layer. Each component neural network is trained with different initial weights connecting three-layers. The outputs of the component neural networks are combined using the output NN as shown in Figure 1. In this study, a neural network (NN) could be the conventional feed-forward back-propagation neural network which widely used in the literature, time-delay neural network, NARX network, or the radial basis neural network. They are described in next sub-sections. The accuracy of each single NN and the NN committee is evaluated based on the mean absolute error (MAE) statistic.

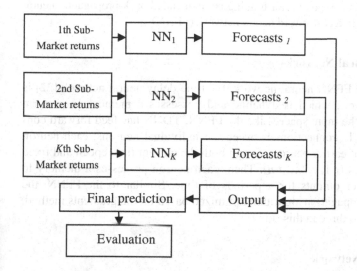

Fig. 1. A generic committee system for modeling and forecasting stock market returns

2.1   Feed-Forward Back-Propagation Neural Network

An artificial neural network (NN) [15] is a generic model for data process data that uses a brain metaphor. The feed-forward back-propagation neural network is a popular architecture that has one input layer with x predictive variables, one or more hidden layers that fulfill the input-output mapping, and an output layer with the predicted variable y. The output y is computed as:

$$y_i = f\left(\sum_{j=1}^{m} x_j w_{ij} + \theta_i\right) \tag{1}$$

where w_{ij} is a connecting weight from neural j to neural i, θ denotes the bias, and f ($\bullet$) is an activation function employed to control the amplitude of the output. In this study, the commonly used sigmoid function is used for activation. It is given by:

$$f(x) = \frac{1}{1 + e^{-x}} \tag{2}$$

The neural network is trained with the Levenberg-Marquardt (L-M) algorithm [9][22] where weights are adjusted based on the gradient method as follows:

$$\Delta w_k = -\left(J_k^T J_k + \mu I\right)^{-1} J_k e_k \tag{3}$$

where J is the Jacobian matrix (first derivatives) of weights, I is the identity matrix, μ is an adaptive learning parameter arbitrarily set to 0.001, and e is a vector of network errors.

The neural network with this type of error training is often called back-propagation neural network [15] or multi-layer feed-forward neural network (FFNN).

2.2 Time-Delay Neural Networks

Unlike the conventional FFNN neural network, the time-delay neural network [22] is a recurrent neural network which is dynamic and possesses a memory to perform temporal processing of the input space. Like the FFNN, TDNN has feed forward connections, but has multiple connections between the individual neurons: each neuron consists of the outputs of earlier neurons during both the current time epoch and fixed number of previous ones (t-1, t-2,...,t-n). Then, each neuron possesses a memory to remember previous layer outputs for n periods of time. Similar to the FFNN, the TDNN uses the back-propagation algorithm optimized based on the gradients method. The time delay n is set to three in this study.

2.3 NARX Neural Network

The Nonlinear Auto-Regressive with Exogenous inputs (NARX) neural network which was introduced by Leontaritis and Billings [23] is a class of discrete-time nonlinear systems that can represent a variety of nonlinear dynamic systems. In particular, the NARX network is a dynamic neural network that contains recurrent feedbacks from several layers of the network to the input layer. It can be mathematically represented as follows:

$$y(t+1) = f\left(y(t),\ldots,y(t-d_y),u(t-1),\ldots,u(t-d_u);W\right) \tag{4}$$

where $u(t)$ and $y(t)$ are respectively the input and the output of the system at time t, $d_u \geq 1$, $d_y \geq 1$, $d_y \geq d_u$, W is a weights matrix, and f is an unknown nonlinear function to be approximated by a FFNN neural network. As a result, the general NARX network equation can be written as:

$$y(t+1) = f_0\left(\begin{array}{c} b_0 + \sum_{h=1}^{Nh} w_{h0} f_h \\ \left(b_h + \sum_{i=0}^{d_u} w_{ih} u(t-i) + \sum_{j=0}^{d_y} w_{jh}(t-j)\right) \end{array}\right) \tag{5}$$

where $i = 1,\ldots,d_u$; $j = 1,\ldots, d_y$; $h = 1,\ldots, N_h$, f_h and f_o are the hidden and output functions, w_{ih}, w_{jh} and w_{ho} are the weights, and b_h and b_o are biases. In this study, the time delay d is set to three in this study.

2.4 Radial Basis Function Neural Network

The radial basis function neural network (RBFNN) [16] is suitable to model flexible in dynamic environment because of its ability to quickly learn data local complex patterns and adapt to changes. The RBFNN system consists of three layers; namely the input, hidden and output layer. The input layer distributes the input data among the hidden nodes (units) which are fully connected to the previous layer. In other words, the input variables are each assigned to a node in the input layer and pass directly to the hidden layer without weights. The hidden nodes contain the radial basis functions (RBF) represented by Gaussian kernels and used as transfer functions to process information contained in input layer nodes. For instance, each neuron in hidden layer computes local response to its input. Finally, the neuron in the output layer only sums up the outputs of the hidden neurons. Mathematically, the output of the jth unit, $H_j(x)$, in the hidden layer for an input x_i is computed as follows:

$$H_j(x) = H_j\left(\left\|x_i - c_j\right\|\right) = \exp\left(\frac{-\left(x_i - c_j\right)^2}{2\delta_j^2}\right) \tag{6}$$

where x_i is the first difference of natural logarithm of market price (for example the market return defined as first difference of logarithmic price), c_j represents the position of the center of the jth Gaussian function, and δ is the width parameter controlling the smoothness of the Gaussian function. Finally, the output y of the system is calculated by a linear combination of the K radial basis functions plus the bias w_0 as follows:

$$y(x_j) = \sum_{j=1}^{K} w_j H_j\left(\left\|x_i - c_j\right\|\right) + w_0 \tag{7}$$

In this paper, the width of the Gaussian kernel is set to 0.55.

In our study, the number of neurons in the input layer and hidden layer of a single NN (FFNN, TDNN, NARX, RBFNN) is set to one and two respectively. They are set to seven in both input and hidden layer in the case of the NN used to combine all forecasts. In all cases, the output layer has one neuron corresponding to the predicted return.

2.5   Performance Measure

Each single NN and NN committee accuracy is evaluated by computing the mean absolute error (MAE) which is defined as follows:

$$MAE = \frac{1}{m} \sum_{i=1}^{m} \left| y_i - p_i \right| \tag{8}$$

where y is the observed value, p is the predicted value, and m is the total number of observations in the testing data. The lower is the MAE, the better is the accuracy.

3   Data and Results

The empirical study involved the prediction of the NASDAQ price return. Submarket set includes banking (first submarket), biotechnology (second submarket), insurance (third submarket), other finance (fourth submarket), Telecom (fifth submarket), transport (sixth submarket), and computers (seventh submarket). The data were daily price values from 3 January 2007 to 15 November 2013. The purpose is to predict the aggregate market (NASDAQ) return series. They are computed as first difference of log-price. Figures 2 to 8 provide banking, biotechnology, insurance, other finance, Telecom, transport, and computers submarket return series respectively. The aggregate market NASDAQ return series are depicted in Figure 9. The learning phase consisted of 80% of the observations, while the testing phase consisted of the remaining 20%. All the obtained results were compared and evaluated by the mean absolute error (MAE) statistic. The empirical results are depicted in Table 1. It indicates that in all cases the NN ensemble (committee) generated the lowest errors than single NN. This result is in accordance with previous works found in the literature [5-13].

Comparing accuracies between FFNN ensemble, TDNN ensemble, NARX ensemble, and RBFNN ensemble, there are differences in favor of the latter when looking at MAE statistic. For instance, they respectively achieved 0.0036, 0.033, 0.0028, 0.0016 MAE. This result indicates that TDNN ensemble, NARX ensemble, and RBFNN ensemble all outperformed the conventional FFNN ensemble used as the main reference NN ensemble. Thus, ensemble system composed of more advanced NN architecture yield to lower prediction error. Among the NN ensembles considered in this study, the RBFNN achieved the lowest forecasting error. This could be explained by the fact that

RBF neural networks have advantages of easy design, good generalization, strong tolerance to input noise, and online learning ability in comparison with traditional neural networks including sophisticated fuzzy inference systems [24]. In addition, contrary to the traditional neural networks trained with back-propagation algorithm each hidden unit of the RBFNN acts locally by computing a score for the match between the input vector and its centers. As a result, the basis units are highly specialized to detect patterns in the underlying data.

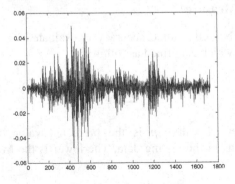

Fig. 2. Banking sub-market return series

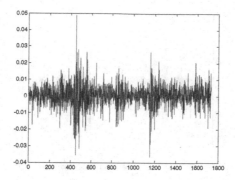

Fig. 3. Biotechnology sub-market return series

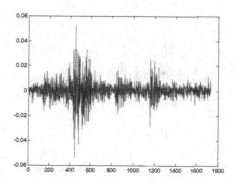

Fig. 4. Insurance sub-market return series

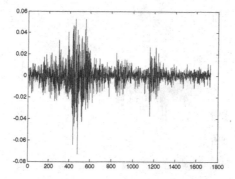

Fig. 5. Other finance sub-market return series

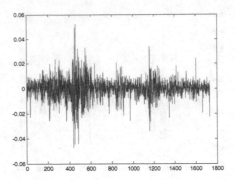

Fig. 6. Telecom sub-market return series

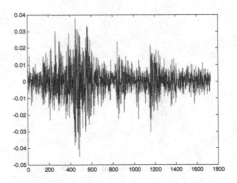

Fig. 7. Transport sub-market return series

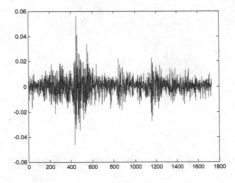

Fig. 8. Computers sub-market return series

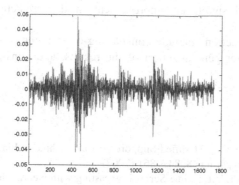

Fig. 9. NASDAQ return series

Table 1. MAE results

	FFNN	TDNN	NARX	RBFNN
Component 1	0.0152	0.0416	0.0080	0.0025
Component 2	0.0189	0.0176	0.0325	0.0025
Component 3	0.0379	0.0431	0.0095	0.0019
Component 4	0.0248	0.0349	0.0084	0.0023
Component 5	0.0220	0.0327	0.0410	0.0021
Component 6	0.0279	0.0288	0.0063	0.0023
Component 7	0.0192	0.0264	0.0252	0.0018
Ensemble	**0.0036**	**0.0033**	**0.0028**	**0.0016**

4   Conclusion

This paper evaluates four neural network ensembles each with different neural network architecture with application in forecasting NASDAQ returns. They are the conventional feed-forward back-propagation neural network which widely used in the literature, time-delay neural network, NARX network, and the radial basis neural network. Each ensemble component is used to model the relationship between NASDAQ returns and a given NASDAQ submarket returns and to provide a forecast. The outputs of the component neural networks are combined together by the combination module to produce the final output of the ensemble. The combination module is represented by a neural network. This result indicated that TDNN ensemble, NARX ensemble, and RBFNN ensemble all outperformed the conventional FFNN ensemble which was used as the main reference NN ensemble. In addition, the RBFNN achieved the lowest forecasting error.

In summary, our findings are in accordance with those of the literature: ensemble neural networks perform better than single neural networks. In addition, this work demonstrated that ensemble systems composed of more advanced NN architectures yield to lower prediction error than neural network ensemble composed of conventional feed-forward neural networks widely used in the literature. Indeed, such sophisticated architectures provided significant improvement in the prediction accuracy of the NASDAQ return series.

Future research directions include considering other types of time series forecasting problems for better generalization of the results, and considering classification problems.

References

1. Francis, E.H., Chao, L.J.: Modified Support Vector Machine in Financial Time Series Forecasting. Neurocomputing 48, 847–861 (2002)
2. Zhang, G.P., Berardi, V.L.: Time Series Forecasting with Neural Network Ensembles: An Application for Exchange Rate Prediction. Journal of the Operational Research Society 52, 652–664 (2001)
3. Atsalakis, G.S., Valavanis, K.P.: Surveying Stock Market Forecasting Techniques – Part II: Soft Computing Methods. Expert Systems with Applications 36, 5932–5941 (2009)
4. Bahrammirzaee, A.: A Comparative Survey of Artificial Intelligence Applications in Finance: Artificial Neural Networks, Expert System and Hybrid Intelligent Systems. Neural Computing & Applications 19, 1165–1195 (2010)
5. Xiao, Y., Xiao, J., Lu, F., Wang, S.: Ensemble ANNs-PSO-GA Approach for Day-Ahead Stock Exchange Prices Forecasting. International Journal of Computational Intelligence Systems 6, 96–114 (2013)
6. Kim, M.J., Min, S.H., Han, I.: An Evolutionary Approach to The Combination of Multiple Classifiers to Predict a Stock Price Index. Expert Systems with Applications 31, 241–247 (2006)
7. Tsai, C.-F., Lin, Y.-C., Yen, D.C., Chen, Y.-M.: Predicting Stock Returns by Classifier Ensembles. Applied Soft Computing 11, 2452–2459 (2011)
8. Wang, L., Wu, J.: Neural Network Ensemble Model using PPR and LS-SVR for Stock Market Forecasting. In: Huang, D.-S., Gan, Y., Bevilacqua, V., Figueroa, J.C. (eds.) ICIC 2011. LNCS, vol. 6838, pp. 1–8. Springer, Heidelberg (2011)
9. Kim, E., Kim, W., Lee, Y.: Combination of Multiple Classifiers for The Customer's Purchase Behavior Prediction. Decision Support Systems 34, 167–175 (2003)
10. Goh, W.Y., Lim, C.P., Peh, K.K.: Predicting Drug Dissolution Profiles with An Ensemble of Boosted Neural Networks: A Time Series Approach. IEEE Transactions on Neural Networks 14, 459–463 (2003)
11. Langella, G., Basile, A., Bonfante, A., Terribile, F.: High-Resolution Space–Time Rainfall Analysis using Integrated ANN Inference Systems. Journal of Hydrology 387, 328–342 (2010)
12. Fildes, R., Kourentzes, N.: Validation and Forecasting Accuracy in Models of Climate Change. International Journal of Forecasting 27, 968–995 (2011)
13. Zheng, J.: Predicting Software Reliability with Neural Network Ensembles. Expert Systems with Applications 36, 2116–2122 (2009)
14. Zhang, G., Patuwo, B.E., Hu, M.Y.: Forecasting with Artificial Neural Networks: The State of The Art. International Journal of Forecasting 14, 35–62 (1998)
15. Rumelhart, D.E., Hinton, G.E., Williams, R.J.: Learning Representations by Back-Propagating Errors. Nature 323, 533–536 (1986)
16. Cybenko, G.: Approximation by Superpositions of Sigmoidal Function. Math. Contr. Signals Syst. 2, 303–314 (1989)

17. Funahashi, K.-I.: On The Approximate Realization of Continuous Mappings by Neural Networks. Neural Networks 2, 183–192 (1989)
18. Hornik, K.: Approximation Capabilities of Multilayer Feedforward Networks. Neural Networks 4, 251–257 (1991)
19. Kim, M.-J., Kang, D.-K.: Ensemble with Neural Networks for Bankruptcy Prediction. Expert Systems with Applications 37, 3373–3379 (2010)
20. Kourentzes, N., Barrow, D.K., Crone, S.F.: Neural Network Ensemble Operators for Time Series Forecasting. Expert Systems with Applications 41, 4235–4244 (2014)
21. Hansen, L.K., Salamon, P.: Neural Network Ensembles. IEEE Trans. Pattern Analysis and Machine Intelligence 12, 993–1001 (1990)
22. Kim, S.S.: Time-Delay Recurrent Neural Network for Temporal Correlations and Prediction. Neurocomputing 20, 253–263 (1998)
23. Leontaritis, I.J., Billings, S.A.: Input–Output Parametric Models for Non-Linear Systems. Part I: Deterministic Non-Linear Systems. International Journal of Control 41, 303–328 (1985)
24. Hao, Y., TianTian, X., Paszczynski, S., Wilamowski, B.M.: Advantages of Radial Basis Function Networks for Dynamic System Design. IEEE Transactions on Industrial Electronics 58, 5438–5450 (2011)

Face Recognition Based on Discriminative Dictionary with Multilevel Feature Fusion

Hongjun Li[1,2], Nicola Nobile[2], and Ching Y. Suen[2]

[1] School of Electronic Information Engineering, Nantong University, Nantong, 226019, China
[2] Centre for Pattern Recognition and Machine Intelligence, Concordia University, Montreal, Quebec H3G 1M8, Canada
hongjun@encs.concordia.ca

Abstract. In order to alleviate the influence of illumination, pose, expression and occlusion variations in face recognition, in this paper, an effective face recognition method based on discriminative sparse representation is proposed. To solve the problem of these variations, we extract discriminative features which represent for each of the training images, and propose a novel dictionary by learning discriminative features. Firstly, we decompose a test image by using nonsubsampled contourlet transform (NSCT), and then fuse the information according to the features from each subband and their contributions. Finally, we obtain the discriminative features of training images and construct a discriminative dictionary. Fuse these multiple features can improve the efficiency and effectiveness of face recognition, especially when training samples are limited and the dimension of feature vector is low. Experimental results on two widely used face databases are presented to demonstrate the efficiency of the proposed approach.

Keywords: Face recognition, Sparse representation, Discriminative, NSCT.

1 Introduction

Face recognition is always an attractive topic in computer vision and pattern recognition [1, 2]. For face recognition, image features are firstly extracted and then matched to those features in a gallery set. The task is to find the class to which a test sample belongs by given training samples from multiple classes. Recently, there has been an increasing interest in classification problem where the data across multiple classes comes from a collection of low dimensional linear subspace. An important method that deals with data on multiple subspaces relies on the notion of sparsity. During the past few years, sparse representation theories have been applied to face recognition and were paid much attention. It has been one of the most successful applications of image analysis and understanding. Although many technologies have been proposed to perform tasks of classifying facial images well, face recognition problem is still challenging. Wright et al. [3] used sparse representation for face recognition and the performance is impressive. This method constructs dictionary by training all kinds of samples, however, when the training samples are limited and the dimension of feature

N. El Gayar et al. (Eds.): ANNPR 2014, LNAI 8774, pp. 252–263, 2014.
© Springer International Publishing Switzerland 2014

vector is low, it becomes less efficient. Yang et al. [4] further extended SRC based framework to a sparsity constrained robust regression problem for handling outliers such as occlusions in facial images. In addition, low-rank representation method [5, 6] which has been established recently is based on the hypothesis that the data is approximately spanned by some low-rank subspaces. Liu and Yan [5] further proposed a latent low-rank representation approach. In latent LRR, the hidden data can be regarded as the input data matrix after being transposed. This idea has been used to design a classifier for image classification [6]. Recently, some works considering multi-resolution information have been proposed for face recognition [7, 8]. Most of these methods realize multi-resolution face recognition by extracting Gabor features in different scales and orientations, which are fused to form multi-resolution features.

Inspired by previous works, we aim to find a discriminative dictionary for face recognition. Most existing discriminative dictionary learning algorithms try to learn a common dictionary shared by all classes, as well as a classifier of coefficients for classification. However, the shared dictionary loses the correspondence between the dictionary atoms and the class labels, and hence performing classification based on the reconstruction error associated with each class is not allowed. Different from these works, Yang et al. [4] proposed a discriminative dictionary learning framework which employs fisher discrimination criterion to learn a structured dictionary. This method uses the reconstruction error associated with each class as the discriminative information for classification, but it does not enforce discriminative information analysis in dictionary construction.

Our method considers NSCT features of facial images in different scales and orientations. NSCT is shift invariant and can reduce the effect of posture variations in the process of face recognition. In addition, in order to obtain robust features, we use contribution criterion calculated by structure similarity when fusing features from different scales and orientations. The structural similarity criterion is imposed on the latent feature images to make them discriminative. We try to propose a new discriminative dictionary by learning fused discriminative features to improve the classification efficiency.

The remainder of the paper is organized as follows. Preliminary works are presented in Section 2. Section 3 presents the proposed method. Section 4 is devoted to experimental results and analysis, and Section 5 concludes the paper.

2   Multi-scale Geometry Analysis

Contourlet transform offers a high directionality but due to the up-sampling and down-sampling, it is shift-variant. However, image analysis applications such as edge detection, image enhancement etc. requires a transform which is shift-invariant. This requirement is fulfilled by nonsubsampled contourlet transform (NSCT) [9, 10]. The NSCT consists of nonsubsampled pyramid (NSP) that ensures multiscale decomposition and nonsubsampled directional filter bank (NSDFB) that offers directionality. NSP is constructed by using a low pass filter and a high pass filter, and it can decompose an image into a low frequency subband and a high frequency subband. NSDFB is then applied to high frequency subband whereas the low frequency subband is used for next finer pyramidal decomposition.

NSCT is shift-invariant so that each pixel of the transformed subbands corresponds to that of the original image in the same spatial location. Therefore we gather the geometrical information pixel by pixel from NSCT coefficients. There are three classes of pixels: strong edges, weak edges and noise. Strong edges correspond to those pixels with large magnitude coefficients in all subbands. Weak edges correspond to those pixels with large magnitude coefficients in some directional subbands but small magnitude coefficients in other directional subbands within the same scale. Noise corresponds to those pixels with small magnitude coefficients in all subbands.

We decompose a facial image under 3 scales by NSCT. In **Fig. 1**, there is an obvious difference between **Fig. 1** (c) and **Fig. 1** (d). In subimages from different scales, the amounts of coefficients are different. According to the characteristic of NSCT, we know that strong coefficient keeps a big value and do not varies when the scale changes, the value of weak coefficient varies as the scale changes and holds different value in different subbands. However, noise signal keeps a small value in all subbands. As a result, the recognition by strong coefficients is more robust than that by other coefficients. Combine coefficients from different scales and directions, acquire discriminative features may improve the efficiency of face recognition.

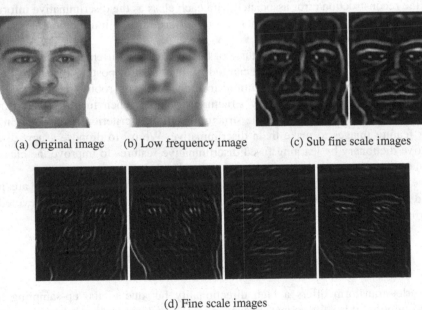

(a) Original image   (b) Low frequency image   (c) Sub fine scale images

(d) Fine scale images

Fig. 1. Image decomposed by NSCT

3   Face Recognition Algorithm

3.1   Discriminative Feature Fusion

Unsupervised feature extraction is a fundamental and critical step in face recognition. Raw facial images usually contain a high dimensional data structure. It is desirable to utilize feature extraction for the sake of avoiding the curse of dimensionality.

We propose a novel approach based on discriminative sparse representation. Firstly, we decompose an image sparsely by NSCT and acquire subimages in different scales and directions. As shown in **Fig. 2**, we can see that subimages in different scales contain different information, some information from key points of the face mainly concentrates on fine scale. As a result, extracting key information from different subimages to construct discriminative features is very important.

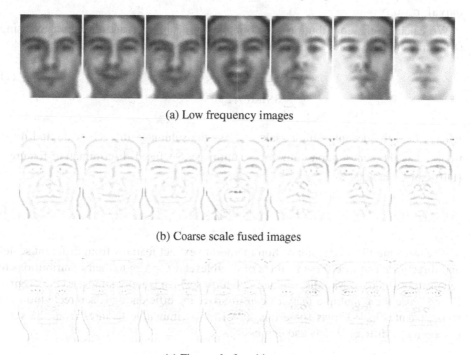

(a) Low frequency images

(b) Coarse scale fused images

(c) Fine scale fused images

Fig. 2. Facial feature images of different training samples

The features in different scales can be easily extracted by multi-scale geometry analysis. If we use a simple and effective feature fusion method to combine these features, the efficiency and the performance of face recognition should be improved. Images which are used to reconstruct should be similar to the original image, so we use Structural SIMilarity (SSIM) as a criterion [11] to calculate the contribution of each subimage over each scale and fuse discriminative features according to their contributions. SSIM is easy to calculate and applicable to various image processing applications. SSIM is defined as:

$$SSIM(u,v) = \frac{\sigma_{uv}}{\sigma_u \sigma_v} \cdot \frac{2\overline{u}\,\overline{v}}{(\overline{u})^2 + (\overline{v})^2} \cdot \frac{2\sigma_u \sigma_v}{\sigma_u^2 + \sigma_v^2} \qquad (1)$$

The first component is the correlation coefficient between u and v. Even if u and v are linearly related, there still might be relative distortions between them, which are evaluated in the second and third components. The second component, with a value range in [0, 1], measures how close the mean luminance is between u and v. It equals 1 if and only if $\overline{u} = \overline{v}$. σ_u and σ_v can be viewed as the estimation of contrast of u and v, so the third component measures how similar the contrasts of the image are. The values range from 0 to 1, where the best value 1 is achieved if and only if $\sigma_u = \sigma_v$.

We calculate the contribution of subimages in each scale and direction by using structural similarity theory. The contribution of each subimage can be defined as:

$$c_d = \frac{1}{\sum_{d=1}^{2^J} SSIM(y, z_d)} SSIM(y, z_d) \qquad (2)$$

where J is the decomposition scale, z_d is the subimage in each scale and direction, c_d is the contribution value of a subimage. So fused discriminative feature (DF) can be defined as:

$$DF = \sum_{d=1}^{2^J} c_d * z_d \qquad (3)$$

We use a multilevel decomposition method to extract features from different scales and directions, and fuse these features of each level according to their contributions to reconstruct an image with discriminative features which is more similar to the original image. Fuse these multiple features can improve the efficiency of face recognition to some extent. **Fig. 3** shows some images with discriminative features from the same person under different lights and expressions.

Fig. 3. Images with discriminative features

3.2   Multilevel Discriminative Dictionary Construction

In this section, we construct a multilevel discriminative dictionary by learning discriminative features. Define matrix Y as the entire training set which consists of n training samples from all c different classes: $Y = [Y_1, Y_2, Y_3, ..., Y_c]$ where $Y_i \in R^{d*n_i}$ is all the training samples from i-th class, d is the dimension of samples, and n_i is the sample from i-th class.

We construct dictionary by training set, using NSCT analysis method. The training samples are decomposed by NSCT, and then using formula (2) and (3) to obtain fused discriminative features. We construct sub-dictionaries class by class, and finally form a dictionary $D = [D_1, D_2, ..., D_k] \in R^{m \times N}$, where D_i is the vector of discriminative features of face images in each class.

```
Algorithm 1: Dictionary Construction
Input: Training Sample Y
Output: Dictionary D
1. Decompose training sample Y by using NSCT.
2. Obtain fused discriminative features by formula (2-3).
3. Construct sub-dictionary of each class and gain
dictionary D.
```

3.3   Face Recognition Algorithm

In practical, occlusion exists in both training and testing samples, the dictionary constructed via nonsubsampled contourlet transform can reduce this influence. **Fig. 4** shows the procedure of our face recognition algorithm. Constructing dictionary $D = [D_1, D_2, ..., D_N]$ via Algorithm 1 on face database, where D_i is the constructed sub-dictionary of i-th class. Algorithm 2 below summarizes the complete recognition procedure. Assume that y is a test sample, and we can get the sparse coefficients vector by solving:

$$x = \arg \min_x \{\|y - Dx\|_2^2 + \lambda \|x\|_1\} \tag{4}$$

Denoted by $x = [x_1; x_2; ...; x_c]$, where x_i is the vector of coefficients in sub-dictionary D_i. We can calculate the residual associated with i-th class by:

$$e_i = \| y - D_i x_i \|_2^2 \tag{5}$$

The identity of testing sample y is determined according to:

$$identify(y) = \arg \min_i \{e_i\} \tag{6}$$

Algorithm 2: Face Recognition
Input: Dictionary D, a test sample y
Output: $identify(y) = \arg\min_{i}\{e_i\}$

1. Decompose training sample y by using NSCT.
2. Obtain contributions of subimages in each scale by formula (2).
3. Fuse discriminative features by formula (3).
4. Minimize x_i to problem (4).
5. Compute the residual $e_i = \parallel y - D_i x_i \parallel_2^2$.

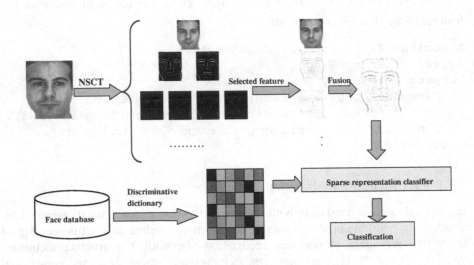

Fig. 4. Face recognition algorithm

4   Experiments and Analysis

In this section, we conduct experiments on AR Database [14] and Extended Yale B Database [15] which are widely used to test the efficacy of our method. The experiments are implemented by MATLAB R2013b on a computer with Intel(R) Xeon(R) CPU X3450@2.67GHz 2.66 GHz, and windows 7 operating system. We repeat each experiment 10 times and the accuracy is averaged. We examine the performance of our method when dealing with different illuminations, expressions, occlusions and different amounts of training samples. Compared with several state-of-the-art methods, such as SRC [3], LRC [12], and FDDL [13], our proposed approach achieves a better performance in terms of high recognition accuracy and robustness.

4.1   AR Face Database

The AR database consists of over 4000 frontal images from 126 individuals. For each individual, 26 pictures were taken in two separate sessions. A subset that contains males and females was chosen from AR database. For each class, seven images are randomly selected as training images, and seven images as testing images. **Fig. 5** shows some facial images from the AR database. We select four feature space dimensions: 30, 54, 120, and 300, according to the down sample ratios 1/24, 1/18, 1/12, and 1/8, respectively. **Table 1** shows the recognition rates of four algorithms when feature dimension changes. Our method gains the highest recognition accuracy of four algorithms in same feature dimension and it outperforms SRC about 3% on average. The experimental results illustrate the accuracy and the robustness of our proposed method when illumination or expression changes. The reason for this improvement is due to the fusion of discriminative features extracted by NSCT which can enhance the representation power.

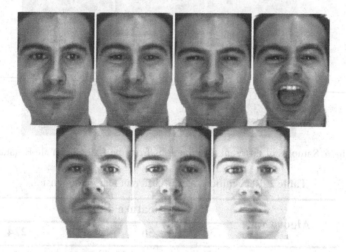

Fig. 5. Samples of different illuminations and expressions from the AR database

Table 1. Recognition rates of different feature dimensions

Algorithm	Feature dimension			
	30	54	120	300
LRC[12]	60.6%	65.8%	71.2%	75.2%
SRC[3]	66.8%	83.0%	88.6%	93.2%
FDDL[13]	64.6%	79.2%	86.5%	93.0%
Our method	69.5%	85.4%	91.5%	94.7%

4.2   Extended Yale B Face Database

The Extended Yale B database consists of 2414 frontal-face images of 38 individuals. The images were captured under various laboratory-controlled illumination conditions. For each class, there are about 64 images, half of the images are randomly selected as training images, and the rest as testing images. **Fig. 6** shows some facial images from the Extended Yale B database. We compute the recognition rates in the feature space dimensions 30, 56, 120, and 224, according to the down sampling ratios of 1/32, 1/24, 1/16, and 1/12, respectively. **Table 2** shows the recognition rates of different feature dimensions, our method outperforms SRC about 8% on average.

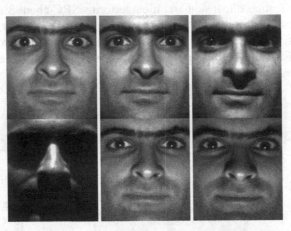

Fig. 6. Samples of different illuminations from the Extended Yale B database

Table 2. Recognition rates of different feature dimensions

Algorithm	Feature dimension			
	30	56	120	224
LRC[12]	49.6%	61.9%	71.2%	75.2%
SRC[3]	53.8%	67.0%	81.6%	88.2%
FDDL[13]	25.0%	40.0%	81.5%	92.7%
Our method	67.5%	77.4%	88.5%	91.7%

4.3   Robustness to Occlusion Samples

In this part of the experiments, we choose a subset of the AR database consisting of *both* neutral and corrupted images. We consider the following scenarios:

Sunglasses: We consider corrupted training images due to the occlusion of sunglasses. We use seven neutral images plus one image with sunglasses for training, and the remainder as test images. Note that the presence of sunglasses occludes about 20% of the facial image.

Scarf: We consider the training images corrupted by scarves, occluding roughly 40% of the facial images. We apply a similar training/test set choice, and have a total of eight training images (seven neutral plus one randomly selected image with scarf) and the remainder as test images.

Sunglasses+Scarf: We consider the case that both sunglasses and scarves are presented in the images during training. We choose all seven neutral images and two corrupted images (one with sunglasses and the other with scarf) for training. A total of two test images (one with sunglasses and the other with scarf) are available for this case.

Comparing our method with SRC and FDDL under three different scenarios, **Table 3** summarizes the comparison of three approaches. Our approach achieves the best result and outperforms other approaches more than 7% for the sunglass scenario, 2% for scarf scenario, and 10% for the mixed scenario when the sizes of dictionaries are the same.

Table 3. Recognition rates of different occlusion samples

Dimension300	Sunglass	Scarf	Sunglass+scarf
SRC[3]	82%	69%	76%
FDDL[13]	83%	60%	75%
Our method	90%	71%	86%

4.4 Influence of Training Samples

We conduct experiments according to different amounts of training samples, compare the recognition rate of our algorithm with that of SRC and FDDL. **Table 4** illustrates the results. The recognition rate of our method is higher than that of SRC and almost equal to FDDL by the same amount of training samples, but our method performs better when there are few samples.

Table 4. Recognition rates of different amounts of training samples

Algorithm	1	2	3	4	5	6	7
SRC[3]	45.0%	60.2%	68.5%	77.1%	79.9%	83.5%	87.0%
FDDL[13]	----	67.4%	73.4%	79.9%	88.4%	90.7%	92.0%
Our method	53.0%	68.9%	80.2%	83.8%	87.1%	90.0%	91.2%

In all the experiments, we can find that our method outperforms others. The dictionary constructed via NSCT can optimize the discriminative features and is more robust. Compared with SRC and FDDL, our method is higher in recognition rate, especially, when the training samples are limited. It can reduce the training time and also suitable for lack samples situation.

5   Conclusion

In this paper, we propose an efficient discriminative sparse representation face recognition method. The proposed method obtains fused discriminative features according to the characteristics of coefficients in different subbands and their contributions, and constructs a dictionary by learning discriminative features. The dictionary constructed by our method has two characteristics: Firstly, the subdictionary can optimize discriminative features in each class. Secondly, the dictionary has low-rank optimized features. The discriminative power of the dictionary comes from minimizing the characteristic of shift invariable, multi-direction and multi-scale of NSCT. It can reduce the influence of illumination, pose, expression and occlusion in both training and testing samples and performs well when there are few training samples. We apply our algorithm to face recognition and the experimental results clearly demonstrate its superiority to numerous other state-of-the-art methods.

Acknowledgment. We thank Wei Hu for the help and discussion. This work was supported by the National Natural Science Foundation of China (NO.61171077), University Science Research Project of Jiangsu Province (NO.12KJB510025), and the Natural Sciences and Engineering Research Council of Canada.

References

1. Yang, M., Zhang, L., Shiu, S.C.-K., Zhang, D.: Monogenic Binary Coding: An Efficient Local Feature Extraction Approach to Face Recognition. IEEE Trans. on Information Forensics and Securit 7(6), 1738–1751 (2012)
2. Ma, L., Wang, C., Xiao, B., Zhou, W.: Sparse representation for face recognition based on discriminative low-rank dictionary learning. In: Proc. CVPR 2012, pp. 2586–2593 (2012)
3. Wright, J., Yang, A.Y., Ganesh, A., Sastry, S.S., Ma, Y.: Robust face recognition via sparse representation. IEEE Transactions on Pattern Analysis and Machine Intelligence 31(2), 210–227 (2009)
4. Yang, M., Zhang, D., Yang, J.: Robust sparse coding for face recognition. In: Proc. CVPR, pp. 625–632 (2011)
5. Liu, G., Yan, S.: Latent low-rank representation for subspace segmentation and feature extraction. In: Proc. ICCV, pp. 1615–1622 (2011)
6. Bull, G., Gao, J.B.: Transposed low rank representation for image classification. In: Proc. DICTA, pp. 1–7 (2012)
7. Xu, Y., Li, Z., Pan, J.S., Yang, J.Y.: Face recognition based on fusion of multi-resolution Gabor features. Neural Computing and Application 23(5), 1251–1256 (2013)
8. Pong, K.H., Lam, K.M.: Multi-resolution feature fusion for face recognition. Pattern Recognitio 47(2), 556–567 (2014)
9. Cunha, A.L., Zhou, J., Do, M.N.: The nonsubsampled Contourlet transform: theory, design and applications. IEEE Transactions on Image Processin 15(10), 3089–3101 (2006)
10. Li, H.J., Zhao, Z.M., Yu, X.L.: Grey theory applied in non-subsampled Contourlet transform. IET Image Processin 6(3), 264–272 (2012)
11. Wang, Z., Bovik, A.C.: A universal image quality index. IEEE Signal Processing Letters 9(3), 81–84 (2002)

12. Naseem, I., Togneri, R., Bennamoun, M.: Linear regression for face recognition. IEEE Trans. on Pattern Analysis and Machine Intelligenc 32(11), 2106–2112 (2010)
13. Yang, M., Zhang, D., Feng, X.: Fisher discrimination dictionary learning for sparse representation. In: Proc. ICCV, pp. 543–550 (2011)
14. Martinez, A., Benavente, R.: The AR face database. CVC Tech. Report No. 24 (1998)
15. Lee, K.C., Ho, J., Kriegman, D.: Acquiring linear subspaces for face recognition under variable lighting. IEEE Trans. on Pattern Analysis and Machine Intelligence 27(5), 684–698 (2005)

Investigating of Preprocessing Techniques and Novel Features in Recognition of Handwritten Arabic Characters

Ahmed T. Sahlol[1,*], Ching Y. Suen[2],
Mohammed R. Elbasyoni[1], and Abdelhay A. Sallam[3]

[1] Computer Teacher preparation Dept., Damietta University, New Damietta, Damietta, Egypt
asahlol@encs.concordia.ca, mmrefaat@hotmail.com
[2] (CENPRMI) Centre for Pattern Recognition and Machine Intelligence, Concordia University, Canada
parmidir@encs.concordia.ca
[3] Department of Electrical Engineering, Port-Said University, Port-Said, Egypt
aasallam@ucalgary.ca

Abstract. There are many difficulties facing a handwritten Arabic recognition system such as unlimited variation in character shapes. This paper describes a new method for handwritten Arabic character recognition. We propose a novel efficient approach for the recognition of off-line Arabic handwritten characters. The approach is based on novel preprocessing operations, structural statistical and topological features from the main body of the character and also from the secondary components. Evaluation of the importance and accuracy of the selected features was made. Our method based on the selected features and the system was built, trained and tested by CENPRMI dataset. We used SVM (RBF) and KNN for classification to find the recognition accuracy. The proposed algorithm obtained promising results in terms of accuracy; with recognition rates of 89.2% for SVM. Compared with other related works and also our recently published work we find that our result is the highest among them.

Keywords: Arabic OCR, Noise removal, Secondaries.

1 Introduction

The Arabic alphabet is used by a wide variety of languages besides Arabic (especially in Africa and Asia) such as Persian, Kurdish, Malay and Urdu. Thus, the ability to automate the interpretation of written Arabic would have widespread benefits. The calligraphic nature of the Arabic script is distinguished from other languages in several ways.

Optical character recognition (OCR) problems can be distinguished into two domains. Off-line recognition; which deals with the image of the character after it inputs to the system for instant scanning. On-line recognition which has different input way,

---

* Corresponding author.

N. El Gayar et al. (Eds.): ANNPR 2014, LNAI 8774, pp. 264–276, 2014.

where the writer writes directly to the system using, for example, light pen as a tool of input. These two domains (offline & online) can be further divided into two areas according to the way that the character itself has been written (by hand or by machine) to handwritten or printed character .In this paper we deal with the Off-line handwritten OCR. Offline recognition of handwritten cursive text is more difficult than online recognition because more information is available in online recognition, like the movement of the pen may be used as a feature of the character, on the contrary the Offline recognition systems must deal with two-dimensional images of the text after it has already been written. Although there are a few commercial Arabic OCR systems for printed text (like Sakhr, IRIS, ABBYY, etc.), there is no commercial product for handwritten Arabic OCR available in the market.

There are many other applications for analysis of human handwriting such as writer identification and verification, form processing, interpreting handwritten postal addresses on envelopes and reading currency amounts on bank checks etc. The main problems encountered when dealing with handwritten Arabic characters are:

— The letters are joined together along a writing line. This big difference between Arabic handwriting and English handwriting, is that the English characters are easier to separate but Arabic are not.
— More than half the Arabic letters are composed of main body and secondary components. The secondary components are letter components that are disconnected from the main body. That secondary component s should be taken into account by any computerized recognition system. Also the type and position of the secondary components are very important features of Arabic letters.
— Each character is drawn in three or four forms when it is written connected to other characters in the word depending on his position of the word. The same letter at the beginning and end of a word can have a completely different appearance.

Various approaches have been proposed to deal with this problem. Many approaches have been adopted in various ways to improve accuracy and efficiency.

In our literature review, we focus on offline Arabic handwritten characters. As for printed Arabic text recognition, some of the recently used techniques can be found in Benjelil et al. [1], Ben Cheikh et al. [2], Kanoun et al. [3], Khan et al. [4], Ben Moussa et al. [5], Prasad et al. [6], Saeeda and Albakoor [7], and Slimane et al. [8].

Also for Recent attempts for online recognition of Arabic characters can be seen in Kherallah et al. [9], [10], Mezghani and Mitiche [11], Saabni and El-Sana [12], and Sternby et al. [13].

Benouareth et al. [14] described an offline Arabic handwritten word recognition system based on segmentation-free approach and hidden Markov models.

Abandah et al. [15] extracted 96 features from the letter's secondary components, main body, skeleton, and boundary. These features are evaluated and best subsets of varying sizes are selected using five feature selection techniques. The evolutionary algorithm has the highest time complexity but it selects feature subsets that give the highest recognition accuracies.

Abdelazeem et al [16] used vertical and horizontal projections which gave more valuable information to capture the distribution of ink along one of the two dimensions in the character. Another kind of useful feature is topological features.

Aburas [17] presented new construction of OCR system for handwriting Arabic characters using the technique similar to that is used in wavelet compression.

The proposed algorithm obtained promising results in terms of accuracy (reaches 97.9% for some letters at average 80%) as well as in terms of time consuming.

Bluche and Ney [18] and [19].made a combination of a convolutional neural network with a HMM gave better results compared with recurrent neural networks, instead of using only HMM in [20].

Prum et al [21] introduced a novel discriminative method that relies, in contrast, on explicit grapheme segmentation and SVM-based character recognition. In addition to single character recognition with rejection, bi-characters are recognized in order to refine the recognition hypotheses. In particular, bi-character recognition is able to cope with the problem of shared character parts. Whole word recognition is achieved with an efficient dynamic programming method similar to the Viterbi algorithm.

Chowdhury et al [22] formulated a distance function based on Levenshtein metric to compute the similarity between an unknown character sample and each training sample. He studied also the effect of pruning the training sample set based on the above distance between individual training samples of the same character class. The proposed approach has been simulated on different publicly available sample databases of online handwritten characters. The recognition accuracies are acceptable.

Chherawala et al [23] built a recognition model is based on the long short-term memory (LSTM) and connectionist temporal classification (CTC) neural networks. This model has been shown to outperform the well-known HMM model for various handwriting tasks, In its multidimensional form, called MDLSTM, this network is able to automatically learn features from the input image. The IFN/ENIT database has been used as benchmark for Arabic word recognition, where the results are promising. A more recent survey on Arabic handwritten text recognition can be found in was presented in [24].

The goal of this work is to develop a reliable offline OCR system for handwritten Arabic characters. In order to overcome the writing variations described before:

First we make different kind of noise removal then we used different kind of features (Whole body features, Main body features and Secondary component features):

Support vector machine and K-nearest neighbor are then used to classify the characters based on the features that were extracted from the input character. Figure 1 summarizes the methodology adopted in this paper.

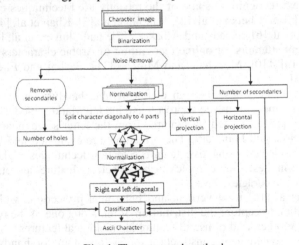

Fig. 1. The proposed method

2   Materials and Methods

2.1   Binarization

We use Otsu's method [25] to convert the grey character image to a binary image which is a normalized intensity value that lies in the range [0, 1]. We don't use this default value of binarization (0.5) because by experiment we find out that useful information have been lost from the character, so we compute the level of intensity for each character and then replace all pixels in the input image with luminance greater than level with the value 1 (white) and replaces all other pixels with the value 0 (black).

2.2   Noise Removal

Although that noise removal techniques have the effect of slightly distorting the actual image, but often this is a small price to pay for the removal of distracting noise and also we were so circumspect when choosing suitable techniques and their parameters.

We remove from character all connected components (objects) that have fewer than 5 pixels. By experience we find that less than 5 connected pixels can be determined as noise and this operation has no bad effect on character main shape or any secondary components.

Median Filtering

Median filtering [26] is an image processing filter used to reduce the effects of random noise. We adopt a 3×3 median filter was because it gave us the best result.

Dilation [27]

It is an operation that grows or thickens objects in a binary image the specific manner and extent of this thickening is controlled by a shape referred to as a structuring element. In this paper we use a square of 2x2 of ones as a structuring element as it gives us the best dilation job.

Morphological Noise Removal

- Filling: fill isolated interior pixels, for each pixel p if the number of non-zero neighbors are 7 pixels.
- Cleaning: Remove isolated pixels. For each pixel p if all neighbors are zeros.
- Adjacent neighbors and Diagonal neighbors: For each pixel we check each pixel diagonal ($\{\{(x-1,y-1\ (x-1,y+1), (x+1,y+1), (x+1,y-1)\}\}$) and   adjacent ($\{\{(x-1,y), (x,y+1), (x+1,y), (x,y-1)\}\}$) neighbors.

If three from its four neighbors are zeroes, so it become zero.

2.3   Normalization

Size normalization is an important pre-processing technique in character recognition because the character image is mapped onto a predefined size so as to give a representation of fixed dimensionality for classification.

We use the Linear Backward mapping method [28].

2.4   Feature Extraction

We divide our features into 3 groups in terms of the kind of information we want to extract:

Features from Whole Character (main body and secondaries)

Vertical and Horizontal Projections.
Vertical profile is the sum of white pixels perpendicular to the y axis. Similarly, the horizontal projection profile is sum of black pixels but it is perpendicular to the x axis.

Right and Left Diagonal of Each Part of the Four Triangular Character Parts.
We divide each character into four triangular and crop each part by determining the boundaries for the last non-zero pixel as shown in Figure 1.

Fig. 2. a. Upper triangle b. Right triangle c. Left triangle d. Lower triangle

Then we get the right and the left diagonal for each triangle of the character by:

The columns of the first output matrix contain the nonzero diagonals of the character. See Figure2 "the dark blue arrow".

The longest nonzero diagonal in the character is determined.

For the nonzero diagonals below the main diagonal of the character, extra zeros are added at the tops of columns.

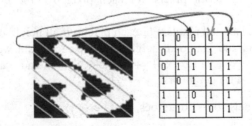

Fig. 3. a. Right diagonal of "Haa" b. Matrix for non-zero diagonals

Number of Secondaries.
This feature recognizes the connected components and number of them like Hamza and dots.

We use the connected component labeling techniques [29].

We identify the main body easily as it is usually the largest component so any other connected components are considered as secondaries.

Features from Only Main Body of the Character
Which represent only the character body without any secondaries.

Number of Holes.
We see that this feature can give accurate results if we eliminate all secondaries with the character correctly. So we keep only the main body for this feature and remove any secondaries connected with the character:

We use again the algorithm [29] but in another way:
1. After getting labels of all connected components.
2. Sort the secodaries by size so that the largest is the first.
3. Keep only the largest connected-components; at this point we eliminate all secondaries.
4. Trace the boundaries of holes inside the character by using Moore-Neighbor tracing algorithm modified by Jacob's stopping criteria [30].

By experiments we find some defects resulted from quick hand written as shown in the figure below

Fig. 4. Number of holes for "Haa" a. Num =1 b. Num =2 c. Num = 3 d. Num =4

Although the above figures represent the same character (ﻪ) after preprocessing yet we unfortunately extract different number of holes from each of them.

Feature from the biggest secondary component of the character (dots and hamzas).

Position of Secondaries.
As we said before that secondary position is the only way to distinguish between a character and another. Those groups of characters (ﺡ , ﺥ ,ﻨ , ﺐ , ﻰ, ﻲ) can be distinguished by machine or by human eye only by the position of the secondary component

We utilize again the connected component labeling techniques [29].

We get easily the largest component so any other connected components are considered as secondaries then eliminate the largest component which is considered the main body of the character then eliminate all the other smallest components after sorting them by size so we start by the smallest, except for the last one B (which is considered the big connected component after the main body) then determine the row and the column for this component, then divide height of B by width of B to get the height/width feature, then count the total number of white and black pixels of B to get Density feature.

Normalization of Feature Data: The attribute data which might have different ranges (min to max) is scaled to fit in a specific range 0, 1. We use Min–max method [31] for normalization.

Handwritten Arabic Characters Dataset: Our database of handwritten Arabic samples is CENPRMI dataset [32]. It includes Arabic off-line isolated handwritten characters. The database contains 11620 characters. These characters were written according to 12 different templates by 13 writers, with each template adopted by 5–8 writers.

2.5   Classification

Support Vector Machine

The Support Vector Machine (SVM) was proposed by Vapnik in [33]. SVM classifies data by finding the best hyperplane that separates all data points of one class from those of the other classes. The best hyperplane for an SVM means the one with the largest margin among the classes. The RBF kernel is a measure of similarity between two examples (training and testing data).We use SVM package called LIBSVM [34].

The SVM uses RBF kernel parameters C and γ where C (cost) is a regularization parameter which controls the penalty for imperfect fit to training labels, and gamma (γ) controls the shape of the separating hyperplane. Increasing gamma usually increases number of the support vectors. Using grid search several experiments were carried out. After several trials of tuning parameters we find that c= 12 and gamma parameter γ= 0.04 give the best results; Accuracy = 89.2 %. Figure 4 below shows the relation between gamma value and recognition rate.

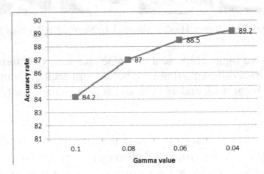

Fig. 5. The relationship between gamma γ and the classification rate

It is obvious from the previous figure that Gamma γ affects significantly in the classification rate.

K-Nearest Neighbor (k-NN) [35]

K-NN Calculates distances of all training vectors and picks k closest vectors Calculate average/majority. Classification using an instance-based classifier can be a simple matter of locating the nearest neighbor in instance space and labeling the unknown instance with the same class label as that of the located (known) neighbor.

The k-nearest-neighbor classifier is commonly based on the Euclidean distance between a test sample and the specified training samples. More robust models can be achieved by locating k, where k > 1, neighbors and letting the majority vote decide the outcome of class labeling. A higher value of k results in a smoother, less locally sensitive, function. We tried to tune neighbor parameter on our Arabic handwritten database and from the figure below we can see the effect of number of neighbors on accuracy.

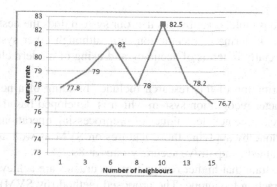

Fig. 6. The relationship between number of neighbors and recognition rate

As shown above. K=10 gives us the best results, but more than 10 neighbors the results get worst.

3   Results and Discussions

The recognition results of KNN were compared to those of the SVM classifier. Table I shows the recognition rate of KNN and SVM.

Table 1. Comparison between the Used Classifiers

Classifier	KNN	SVM (RBF)
Accuracy Rate (%)	82.5	89.2

Table II shows summarized the most recent work in the isolated handwritten Arabic characters. As we can see they are sorted according to the published date. The data used, the feature set and the improvement of recognition rate can also be seen.

Table 2. Comparison between Previous Results and Ours

Previous study	Approach	Results
A.A. Aburas et al [36]	Haar Wavelet transform	70%
M. Z. Khedher, et al [37]	Not mentioned	73.4%
G. Abandah et al [15]	Combination of multi-objective genetic algorithm and SVM	not mentioned exactly
A. T. Al-Taani et al [38]	Decision tree	75.3%
G. A. Abandah et al [39]	Linear Discriminant Analysis	87%
Our previous work [40]	Feed forward neural network.	88%
Proposed method	SVM	**89.2%**

From the previous table, it is obvious that our system does the best when compared with other systems in terms of recognition rate, although other systems make great contributions especially in terms of accuracy and using of modern classification techniques.

The main contribution of this research includes building of a new offline Arabic handwritten character recognition system which is developed based on the novel extracted feature after some new techniques of preprocessing operations. The evaluation of our system is done by applying those features on SVM as well as KNN. The proposed method obtained competitive accuracy rates at 89.2%.

The results illustrate that higher recognition accuracies are achieved using the proposed feature extraction technique. The proposed method (by SVM) gives a recognition rate of about 100 % for (أ, ي , ت,ن,م,ل,خ,ج, ح , هـ).

The worst recognized characters was (ف) by SVM and also by KNN. It was misclassified as (ن and ق) this is because the similarities between those two characters in some writing styles and also they all have upper secondaries and holes. The second misclassified character is (ع) was misclassified as (ح) by SVM as well as by KNN. We think that this is because the similarities in their shapes especially at the lower part. The third misclassified character is (س) was misclassified as (ص) by SVM as well as by KNN. We think that this is because the similarities in their left part.

We think that preprocessing operations as well as selecting most proper feature can minimize classification error. For example we use different kinds of noise removal (statistical and morphological) for erasing useless parts of the character which can occur during hand writing process, ink stain or even by digitizing the image. We make also dilation for fixing damaged pixels of the character occurred as a result of preprocessing operations (binarization- noise removing) or during the digitizing process. Any of those preprocessing operations could have bad effects on character shape if they don't used properly and this can reflect on the quality of the extracted features for example if we overuse of noise removal techniques we can easily remove a dot if it was written slightly and consequently we lose very important information of this character dots.

We extract features from the whole character, as well as, its main body and secondary components themselves which provide more valuable features that exploit the recognition potential of the secondary components of handwritten Arabic letters. These results also confirm the importance of the secondary components of the handwritten Arabic characters. For example if we make a comparison between س and ش, ص and ض we will find no differences between each pair of them except for the secondary component.

We use not only different kinds of features (structural features, statistical features, topological features) which represent different aspects of the character's characteristic, but also (after many trials) we choose the most significant features for distinguishing between characters. After careful examination of the samples that were incorrectly recognized, we concluded that most of these samples are hard to recognize by native or even by a human expert reader. However, we think that the door is open to search for extracting new features that capture subtle differences in loop shapes and secondary types.

4   Conclusion

This paper presents a novel approach for extracting features to achieve high recognition accuracy of handwritten Arabic characters. We tune the used parameters during the preprocessing phase including binarization, normalization and some noise removal methods accurately to preserve all useful information that can be extracted from the character.

Selecting proper features for recognizing handwritten Arabic characters can give better recognition accuracies, therefore we included statistical, morphological and topological features. Also we pay more attention to the secondaries like secondary position, ratios and density because we think that may overcome some of handwritten characters variations. Although, there are some challenges with some characters, the overall recognition rate is encouraging especially when compared to other handwritten Arabic character systems.

After examining the recognition accuracy of each character using SVM and KNN we found that the best accuracy is given by SVM which is 89.2%. The other misrecognized characters such as (ف, ع) we think that this is because those characters similarities between those characters and others in some writing styles and also they have secondaries and holes. Our future work includes increasing the efficiency of the proposed approach especially for the characters that were not recognized well by finding out more powerful features, also including variations in writing the main body of the character and also the secondaries.

References

1. Benjelil, M., Kanoun, S., Mullot, R., Alimi, A.M.: Arabic and latin script identification in printed and handwritten types based on steerable pyramid features. In: Proceedings of the 10th International Conference on Document Analysis and Recognition (ICDAR), pp. 591–595 (2009)

2. Cheikh, I.B., Bela, A., Kacem, A.: A novel approach for the recognition of a wide Arabic handwritten word lexicon. In: Proceedings of the 19th International Conference on Pattern Recognition (ICPR), pp. 1–4. IEEE (2008)
3. Kanoun, S., Slimane, F., Guesmi, H., Ingold, R., Almi, A.M., Hennebert, J.: Affixal approach versus analytical approach for off–line arabic decomposable vocabulary recognition. In: Proceedings of the 10th International Conference on Document Analysis and Recognition (ICDAR), pp. 661–665 (2009)
4. Khan, T.K., Azam, S.M., Mohsin, S.: An improvement over template matching using k-means algorithm for printed cursive script recognition. In: Proceedings of the 4th IASTED International Conference on Signal Processing, Pattern Recognition, and Applications, pp. 209–214 (2007)
5. Benmoussa, S., Frissard, Q., Zahour, A., Benabdelhafid, A., Alimi, A.M.: New features using fractal multi-dimensions for generalized Arabic font recognition. Pattern Recogn. Lett. 31(5), 361–371 (2010)
6. Prasad, R., Saleem, S., Kamali, M., Meermeier, R., Natarajan, P.: Improvements in hidden markov model based Arabic ocr. In: Proceedings of the 19th International Conference on Pattern Recognition, ICPR (2008)
7. Saeeda, K., Albakoor, M.: Region growing based segmentation algorithm for typewritten and handwritten text recognition. Appl. Soft Comput. 9(2), 608–617 (2009)
8. Slimane, F., Ingold, R., Kanoun, S., Alimi, A.M., Hennebert, J.: A new Arabic printed text image database and evaluation protocols. In: Proceedings of the 10th International Conference on Document Analysis and Recognition (ICDAR), pp. 946–950 (2009)
9. Kherallah, M., Haddad, L., Alimi, A.M., Mitiche, A.: On-Line handwritten digit recognition based on trajectory and velocity modeling. Pattern Recogn. Lett. 29(5), 580–594 (2008)
10. Kherallah, M., Bouri, F., Alimi, A.M.: On-line Arabic handwriting recognition system based on visual encoding and genetic algorithm. Engin. Appl. Artif. Intell. 22(1), 153–170 (2009)
11. Mezghani, N., Mitiche, A.: A gibbsian kohonen, "Network for online arabic character recognition". In: Bebis, G., et al. (eds.) ISVC 2008, Part II. LNCS, vol. 5359, pp. 493–500. Springer, Heidelberg (2008)
12. Saabni, R., El-sana, J.: Hierarchical on-line Arabic handwriting recognition. In: Proceedings of the 10th International Conference on Document Analysis and Recognition (ICDAR), pp. 867–871 (2009)
13. Sternby, J., Morwing, J., Andersson, J., FRriberg, C.: On-Line Arabic handwriting recognition with templates. Pattern Recogn., New Frontiers Handwrit. Recogn. 42(12), 3278–3286 (2009)
14. Benouareth, A., Ennaji, A., Sellami, M.: Arabic Handwritten Word Recognition Using HMMs with Explicit State Duration. Journal on Advances in Signal Processing 2008, 1–13 (2008)
15. Abandah, G.A., Anssari, N.: Novel moment features extraction for recognizing handwritten Arabic letters. Journal of Computer Science 5(3), 226–232 (2009)
16. Abdelazeem, S., EL-Sherif, E.: Arabic handwritten digit recognition. International Journal on Document Analysis and Recognition (IJDAR) 11(3), 127–141 (2008)
17. Aburas, A.A., Rehiel, S.M.: Off-line Omni-style handwriting Arabic character recognition system based on wavelet compression. In: Arab Research Institute in Sciences & Engineering ARISER, pp. 123–135 (2007)

18. Bluche, T., Ney, H., Kermorvant, C.: Feature extraction with convolutional neural networks for handwritten word recognition. In: 12th International Conference on Document Analysis and Recognition, pp. 285–289 (2013)
19. Kozielski, M., Doetsch, P., Ney, H.: Improvements in RWTH's system for off-line handwriting recognition. In: 12th International Conference on Document Analysis and Recognition, pp. 935–939 (2013)
20. Rothacker, L., Vajda, S., Fink, G.A.: Bag-of-features representations for offline handwriting recognition applied to Arabic script. In: Proceedings of the 3rd International Conference on Frontiers in Handwriting Recognition (ICFHR), Bari, Italy, pp. 149–154 (2012)
21. Prum, S., Visani, M., Fischer, A., Ogier, J.M.: A Discriminative Approach to On-Line Handwriting Recognition Using Bi-Character models. In: 12th International Conference on Document Analysis and Recognition (ICDAR), pp. 364–368 (2013)
22. Chowdhury, S.D., Bhattacharya, U., Parui, S.K.: Online Handwriting Recognition Using Levenshtein Distance Metric. In: 12th International Conference on Document Analysis and Recognition (ICDAR), pp. 79–83 (2013)
23. Chherawala, Y., Roy, P.P., Cheriet, M.: Feature design for offline Arabic handwriting recognition: handcrafted vs automated?. In: 12th International Conference on Document Analysis and Recognition (ICDAR), pp. 290–294 (2013)
24. Parvez, M.T., Mahmoud, S.A.: Offline Arabic handwritten text recognition: A Survey. Journal of the Association for Computing Machinery 45(2), 23:1–23:35 (2013)
25. Otsu, N.: A Threshold Selection Method from Gray-Level Histograms. IEEE Transactions on Systems, Man, and Cybernetics 9(1), 62–66 (1979)
26. Jae, S.L.: Two dimensional signal and image processing. PRENTICE HALL PTR, Upper Saddle River, New Jersey, 07458 (1990)
27. Rosenfeld, A., Kak, A.: Digital Picture Processing, 1st edn. Academic Press, New York (1976) ISBN: 10: 0125973608
28. Cheriet, M., Kharma, N., Liu, C.L., Suen, C.Y.: Character Recognition Systems, pp. 36–38. John Wiley (2007)
29. Rosenfeld, A., Kak, A.: Digital Picture Processing, 1st edn. Academic Press, New York (1976) ISBN: 10: 0125973608
30. Gonzalez, R.C., Woods, R.E., Eddins, S.L.: Digital Image Processing Using MATLAB. Gatesmark Publishing (2009)
31. Hann, J., Kamber, M.: Data Mining: concepts and techniques, 3rd edn. Morgan Kaufman (2011)
32. Alamri, H., Sadri, J., Suen, C.Y., Nobile, N.: A novel comprehensive database for Arabic off line handwriting recognition. In: The 11th International Conference on Frontiers in Handwriting Recognition (ICFHR), pp. 664–669 (2008)
33. Vapnik, V.: Statistical Learning Theory. Wiley, Chichester (1998)
34. Hsu, C.W., Lin, C.J.: A comparison of methods for multi-class support vector machines. IEEE Trans. Neural Network 13, 415–425 (2002)
35. Mitchell, T.: Machine Learning. McGraw-Hill, New York (1997)
36. Aburas, A.A., Rehiel, S.M.: Off-line Omni-style handwriting Arabic character recognition system based on wavelet compression. In: Arab Research Institute in Sciences & Engineering ARISER, pp. 123–135 (2007)
37. Khedher, M.Z., Abandah, G.A., Al-Khawaldeh, A.M.: Optimizing feature selection for recognizing handwritten Arabic characters. World Academy of Science, Engineering and Technology 4, 81–84 (2005)

38. Al-Taani, A.T., Al-Haj, S.: Recognition of on-line Arabic Handwritten Characters Using Structural Features. Journal of Pattern Recognition Research, 23–37 (2010)
39. Abandah, G.A., Younis, K.S., Khedher, M.Z.: Handwritten Arabic character recognition using multiple classifiers Based on letter form. In: Conf. on Signal Processing, Pattern Recognition& Applications, Austria, pp. 128–133 (2008)
40. Sahlol, A.T., Suen, C.Y., Sallam, A.A., Elbasyoni, M.R.: A proposed OCR Algorithm for cursive Handwritten Arabic Character Recognition. Journal of Pattern Recognition and Intelligent Systems (PRIS), 90–104 (2014)

A Time Series Classification Approach for Motion Analysis Using Ensembles in Ubiquitous Healthcare Systems

Rana Salaheldin[1], Mohamed ElHelw[1], and Neamat El Gayar[2]

[1] Center for Informatics Science, Nile University, Giza, Egypt
[2] Faculty of Computers and Information, Cairo University, Giza, Egypt

Abstract. Human motion analysis is a vital research area for healthcare systems. The increasing need for automated activity analysis inspired the design of low cost wireless sensors that can capture information under free living conditions. Body and Visual Sensor Networks can easily record human behavior within a home environment. In this paper we propose a multiple classifier system that uses time series data for human motion analysis. The proposed approach adaptively integrates feature extraction and distance based techniques for classifying impaired and normal walking gaits. Information from body sensors and multiple vision nodes are used to extract local and global features. Our proposed method is tested against various classifiers trained using different feature spaces. The results for the different training schemes are presented. We demonstrate that the proposed model outperforms the other presented classification methods.

Keywords: human motion analysis, time series classification, multiple classifier systems.

1 Introduction

Human health monitoring continues to be an increasingly active research area. Ubiquitous healthcare systems provide information necessary to recognize emerging physical problems. This is useful for monitoring and controlling the elderly and chronically ill patients inside their homes [1]. In general, human activity can be captured within a home environment. This can automatically provide an online analysis of the user's health status [2]. One of the most promising health care areas is human motion analysis. Understanding user walking patterns and identifying changes in everyday behavior can reveal the onset of adverse health problems. Moreover, capturing walking abnormalities is important for assessing people who may have a greater risk of falling [3]. A set of sensors is used to capture information of human activity patterns. Recognizing various activities requires different sensors at different locations and time. Among sensors that are helpful in context recognition tasks are Body Sensor Networks and Visual Sensor Networks. Body Sensor Networks (BSNs) are wireless wearable sensors that capture continuous data over extended periods of time [4].

N. El Gayar et al. (Eds.): ANNPR 2014, LNAI 8774, pp. 277–288, 2014.

BSNs can be easily worn with minimal inconvenience. Wearable sensors prove to be helpful in health monitoring of patients in ambulatory settings [5,6] and in measuring gait parameters [7,8]. Visual Sensor Networks (VSNs) are ambient sensors that contain a number of low cost vision sensor nodes [9]. Information computed from distributed multiple vision nodes can monitor the movement of human body.

In this paper we address the problem of human motion analysis from a time series perspective. Sensor data is a typical form of time series observations captured along a period of time. We propose a new methodology for classification of human motion activity by retaining the temporal aspect found in sensor captured measurements. The proposed architecture performs automated differentiation between impaired and normal walking gaits. Real time motion monitoring and recognition is implemented for gait analysis. The objective is to identify walking patterns for unseen individuals using a training set of different subjects. We previously tested the model on character and sign language recognition applications and produced satisfactory results [10].

The proposed model uses multiple classifiers to integrate feature and distance based methods extracted from body sensor and multiple vision nodes. The aim of this study is to investigate the different classifier integration methods for the problem of human motion analysis. Different types of local and global features are explored. Our model is mainly though for real time classification, however we also investigate the performance in an offline setting and discuss impact of preprocessing and feature effectiveness in this case.

The data set used represents information captured by an ear worn body sensor node and four wireless cameras. A home care environment is simulated to record motion information for different targets.

The paper is organized as follows: The following section highlights important background related to human motion analysis, time series classification and multiple classifier systems. Section 3 presents the proposed ensemble. Section 4 introduces the data set, experimental setup and results. The discussion is presented in section 5. The final section concludes the paper and discusses future work.

2   Background

2.1   Human Motion Analysis

The process of human motion analysis can be classified into three parts: human detection, tracking and behavior recognition [11]. In this work, we are concerned with human behavior understanding. Most techniques for activity recognition using sensor data follow a number of steps [12]: First the captured signal is divided into windows. The windowing technique segments the signal sequentially into smaller parts with or without overlap [13-16]. The second step is to extract features from each window. These features should be able to discriminate between different classes of action. Widely used features include mean, variance, entropy, energy, skewness and kurtosis [16-20]. Also the frequency content of a signal is analyzed using extracted frequency domain features. Some of these features such as the fast Fourier transform entropy can be used to differentiate between actions with highly varying acceleration patterns [13].

Finally, the generated features are used as input to a classification process. The last step is applying a classification algorithm to distinguish between different human activities. Comparisons of different classifiers for activity recognition are found in [13] and [21].

2.2   Time Series Classification

Time series data is a sequence of observation values ordered with respect to time in ascending order [22]. Time series analysis studies the structural dependencies between the observations. Among the challenging tasks in time series analysis is time series classification. Similar to conventional supervised classification, each series is associated with a class label. During training phase, examples of series with known classes are presented. The goal is to learn patterns and assign unlabeled time series into predefined classes.

Three main approaches are used along the literature for time series classification; distance based methods, feature extraction followed by a classification method, and finally model based classification [23]. In the distance based method approach a distance function is used to define the similarity between time series data [24]. Many methods have been proposed to define similarity between time series data. Some of these techniques are listed in [24]. The most widely popular techniques are Euclidean distance (ED) and Dynamic Time warping (DTW) [25]. Another approach for time series classification is transforming the observations into a feature vector thus allowing the usage of a conventional pattern recognition scheme [26,27]. Global and local features are extracted from each time series sample. These features represent the global characteristics and the temporal aspect of a time series respectively [26]. Other methods include classifying time series using modeled based algorithms such as Recurrent Neural Networks (RNN) [28,29] and Hidden Markov Models (HMM) [30].

The field of Multiple Classifier Systems (MCS) has attracted great interest in pattern recognition research. The main objective is based on the continuous need for improving the classification accuracy. The idea of MCS is combining learners to generate more precise results than individual classifiers [31]. The decision aggregation is dependent on using competitive experts as single classifier, and combining their different predictions to provide complementary information about the problem. MCS works best when base classifiers produce accurate and diverse results [32].

3   Proposed Ensemble Model

A multiple classifier systems approach is proposed for human motion recognition using time series analysis. The proposed architecture is a two layer ensemble; it combines classifiers trained with different features and distance measures. The decision fusion is performed using a trainable combiner that learns the class from the outputs of classifiers in the first layer.

Figure 1 shows the model architecture. Initially three base classifiers are trained independently with a different set of features. The first classifier is trained using local features while the second classifier is trained using global features. The third classifier is a K Nearest Neighbor classifier with Dynamic Time Warping similarity distance function. Next, a fusion layer is trained to perform mapping of the classifiers' outputs into the set of desired class labels.

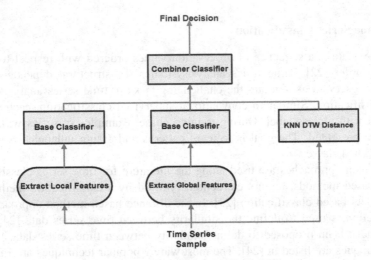

Fig. 1. Ensemble Architecture Block Diagram

The success of the ensemble depends on the use of different models and feature spaces for training the base classifiers. This provides complementary and diverse information about each subject's walking pattern. The model captures differing information from sensors, and uses them to train individual classifiers to produce independent errors. The outputs of the three base classifiers are used as training data for the fusion classifier to make an improved estimate of the activity pattern. The combiner classifier is adaptive enough to learn the weights of different classifiers and the best combination of base classifiers' decisions. As follows we present the details of the proposed ensemble model.

3.1   Feature Extraction

The precision of the classification process is highly related to the selection of attributes. Both local and global features are used in order to capture the fundamental trends in the motion activity. Each type will represent a different aspect of the structure of patterns, thus generate accurate approximations.

Local features are features extracted from interval regions of a time series [33]. Sliding time windows are used to divide the sensor signals into segments. A sliding window covers a small portion and moves along the series, extracting a set of features from each window. The number of windows varies from a time series to another. For time series T, a sliding window W^J is moved along the series dividing it into parts $T(i)_{i=1}^I$ for each window $W^j, j = 1, ... J$. Starting from the first observation, the sliding window extracts the features until the whole time series is covered. This method will reveal temporal information in BSN and VSN sequences. Spikes, edges or sudden abrupt changes in time domain of a walking pattern are defined. Local features extracted are the average of raw values, minimum, maximum, amplitude and standard deviation.

Global features are based on the global characteristics and information of the whole time series instead of the temporal property [26]. Global features give a measure of the overall properties of a subject's complete walking activity. By following the general trend of the entire movement, valuable gait information is extracted. This information is not affected by the contrasting sub regions in the pattern. Each sequence is analyzed independently and also the association between different series is examined. The total distance covered by each time series is calculated. Also we compute the Euclidean distance between each pair of series, the minimum and maximum of each attribute and the mean of each sequence.

3.2   Similarity Measure

Dynamic time warping (DTW) [25] is an algorithm that measures the similarity among sequences of data. The algorithm computes the best alignment between two walking patterns that can be of different length. DTW can find the likeness between sequences that are warped non-linearly in time dimension. Unlike Euclidean distance, if two similar gait measurements are not exactly timely aligned, DTW algorithm can map them to the same class. In other words, the algorithm is able to examine two series very much like the way humans may compare and recognize the similarity between them.

The DTW algorithm can be formulated as follows:
Given two sequences X and Y of length m and n respectively:

$$X = x_1, x_2, \dots, x_m \quad Y = y_1, y_2, \dots, y_n$$

The goal is to find a path which minimizes the total distance between the two sequences.

The sequences are used to form a matrix M of size [m, n]. Each cell in M denotes an alignment between elements $x_i \in X$ and $y_j \in Y$ where:

$$0 \leq i < m \text{ and } 0 \leq j < n$$

This alignment is denoted by the squared distance between the points x_i and y_j.

$$d(x_i, y_j) = (x_i - y_j)^2$$

The matrix is searched to find an optimal distance path W between the two sequences.

$$W = w_1, w_2, \dots, w_K$$

Each w_K corresponds to a d(i, j) point in the matrix M. The optimal path is the one that minimizes the distance:

$$DTW(X, Y) = \min\left[\sum_{k=1}^{K}(w_k)\right]$$

Where (w_k) is the kth matrix element in the warping path W.

Dynamic programming is used to find the minimal warping path W. Dynamic Programming divides the problem into sub problems, and uses the solutions repeatedly to solve the original problem [34].

The path is discovered using the following recurrence:

$$\gamma\,(i,j) = d(i,j) + \min\{\ \gamma(i\text{-}1,j\text{-}1)\ ,\ \gamma(i\text{-}1,j)\ ,\ \gamma(i,j\text{-}1)\}$$

For each element in the matrix d(i,j), the three adjacent elements are examined, and the minimum cumulative distance $\gamma\,(i,j)$ is selected in a recursive way.

The proposed ensemble uses KNN classifier with DTW algorithm applied as a similarity measure. When minimal distance is found between a given sequence and a class label, the pattern is assigned to this particular label. A warping path is calculated between each series in the training data and in the test sequence. An unknown input pattern is compared to all labeled series. The classification is based on ranking the labeled instances by their similarity measure to the unlabeled pattern.

3.3   Ensemble Fusion

The fusion layer for the architecture is a classifier which adaptively combines the outputs of the three classifiers to produce the final class label. The fusion classifier is trained on a separate data set that has not been used to train the base classifiers. This means that a training set is used to build the base classifiers, and a validation set is used to train the combiner classifier using the outputs from the base classifiers as features. This two layered ensemble architecture allows the combiner classifier to learn the mapping between the labels produced from the base classifiers and the desired class labels. Since the base classifiers are trained using different feature spaces, each classifier makes different mistakes and produces independent errors. Thus, the trainable combiner learns the different outputs of each classifier (including their individual errors). After training level one and two of the architecture, a separate test set is used to evaluate the classification process.

4   Data and Experiments

4.1   Data Set

The data set represents motion information of different targets. It is obtained from [35] and experiments were carried out in a lab-based home monitoring environment. The data set contains accelerometer information from a wearable body sensor and information from multiple vision nodes. The cameras simulate visual information from vision sensor nodes by capturing and sending images at 10 frames per second. The proposed framework in [35] employs ubiquitous sensing to acquire non redundant, complimentary features for improved motion analysis. The data set consists of two classes; impaired walking (limping) and normal walking patterns. Ten subjects are used, and for each subject four different examples for each limping and walking patterns are recorded.

4.2   Setup and Results

Data from body and visual sensors are captured in frames per second. Thus information from each sensor is considered a time series sample. The temporal aspect in the measured data is used to build the classification model. A set of experiments are carried out for evaluating the proposed model. Experiments are performed for real time and offline motion classification.

For the DTW approach, data abstraction is performed for reducing the size of the input time series. This helps in speeding up the algorithm. The time series is reduced and the warping path is found by DTW on lower resolution time series. Each pair of adjacent observations in the series is averaged, this way the size is reduced by the factor of two every time resolution is decreased.

In the first experiment we test the efficiency of our proposed ensemble compared to conventional single classification techniques using both local and global features. The effect of these features on classification accuracy is analyzed. In this experiment, our objective is real time classification; to recognize the motion pattern instantly. We also test the impact of different sensor nodes on motion analysis and present the results for training with BSN and VSN data. Next we explain the set up for the first experiment followed by the results.

The following single classifiers are used: support vector machines, decision trees, K - Nearest Neighbor and naïve bayes classifiers. As for the proposed ensemble, it's diversity depends on the different feature representations. Two base classifiers are trained using local and global features, and a KNN classifier that uses DTW as the similarity measure. To train the combiner classifier the output labels from the base classifiers are used as features. These features along with the actual labels from the training set are the input to the second fusion layer. Support vector machine classifier with polynomial kernel is used for combination due to its generalization capability. WEKA [36] is used for our implementation. All experiments are conducted using leave one out method. The classification accuracy of the proposed framework is tested using unseen subjects.

A window of three seconds is used for training. This decreases the delay, and also increases the number of training samples used in classification, since each subject's recordings are divided into many training samples, three seconds each. We test whether the time series model can represent the short window well enough to recognize the subject's motion pattern. Global features are extracted from the whole motion sample. Windows representing one second each are defined concurrently, these windows are used to extract local features.

Below are the findings of the first experiment. Tables 1 and 2 present the results for classification using BSN and VSN respectively. The tables demonstrate the results for classification using local and global features. Also, the accuracy of the proposed ensemble architecture (section 2) is presented. The mean and standard deviation for the different classification accuracies are shown.

Table 1. Percentage accuracy for single classifiers and proposed ensemble - BSN data set – real time experiments

Single Classifiers	Local Features	Global Features
Naïve Bayes	72.03% ±8.42	84.78% ±2.11
Decision Tree	78.89% ±6.34	81.20% ±2.53

Table1. (*Continued*)

KNN	80.01% ±3.52	92.94% ±1.90
SVM	80.66% ±3.54	93.94% ±0.9
Proposed Ensemble	*96.48±0.47*	

Table 2. Percentage accuracy for single classifiers and proposed ensemble - VSN data set – real time experiments

Single Classifiers	Local Features	Global Features
Naïve Bayes	74.64±12.76	81.45±1.89
Decision Tree	79.34±10.40	93.75±0.9
KNN	84.2±9.48	93.23±0.89
SVM	**86.43±8.58**	**94.11±0.88**
Proposed Ensemble	*98.48±0.52*	

In general, results reveal that using global features outperforms local features. Additionally, Support Vector Machine is the winning 'single' model using local and global features for both data sets. As for the sensor recordings, training using BSN and VSN data yield similar results. We should note that the result of training single classifiers using both BSN and VSN data yields close results to when only BSN data is used. Finally, the proposed ensemble architecture outperforms the performance of single classifiers. The classification is improved significantly by combining feature and distance based techniques and introducing the trainable fusion layer.

The second experiment is presented next. To verify the usefulness of our approach, we test offline classification using the whole sensor recordings. In some applications it is useful to use the whole subject's motion pattern as input data, this provides more information about the motion pattern, but does not allow online classification. Similar to real time classification, smaller windows are moved along each sensor recording to extract the local features. The number of windows varies for each series under consideration; and depends on the number of instances produced from each subject's recordings.

Table 3. Percentage accuracy for single classifiers trained using different features – BSN and VSN combined – offline experiments

Single Classifiers	Local Features	Global Features
Naïve Bayes	91.78±6.2	**85.34±4.81**
Decision Tree	**93.52±5.95**	81.89±7.29
KNN	89.65± 7.23	76.72±9.25
SVM	90.89±6.11	84.91±5.29

Table 3 shows the accuracies of single classifiers trained using local and global features from both sensors. Here, global features are clearly performing worse in offline experiments than in real time results. The table also indicates that Decision Tree produces best results for local features while Naïve Bayes outputs the best results for global features. This is different from previous experiments where Support Vector Machines outperformed other single classifiers.

Finally, studies have indicated the need for preprocessing the data before classification [35]. In the next experiment we evaluate the usefulness of preprocessing. We note that the preprocessing is performed for the offline experiment, which makes this step unsuitable for real time continuous classification.

The average and variance of the signals are used as features instead of the raw data. For each sensor signal, a time window of size 4 seconds and a step of one eighth of a second are used to calculate the average and variance of each time series. The average and variance of the features from the four cameras and from the BSN node are augmented together to form the feature vector used for classification.

Consistent with the previous experiments, the results of training using individual sensors is not different from combining them together. We choose to present the result of concatenating BSN and VSN data. Below are the results for single classifiers using data from both sensors.

Table 4. Percentage accuracy for single classifiers trained using different features – BSN and VSN combined – offline experiments- with preprocessing

Single Classifiers	Local Features	Global Features
Naïve Bayes	100.00±0.00	91.25±13.93
Decision Tree	100.00±0.00	95.00±10.05
KNN	97.50±7.54	77.00±20.93
SVM	100.00±0.00	92.25±14.08

Table 4 presents the results for training single classifiers using the offline recognition scheme. Most classifiers trained using local and concatenated features achieve accuracies close to 100%. Also, in this experiment the local features produce better results than global features for all classifiers. This result will be further analyzed in the discussion section. The table displays the results for using both BSN and VSN data combined. The K Nearest Neighbor yields worse results than other three single models using local features.

5   Discussion

The results show that the proposed ensemble architecture outperforms the single classification methods in case of real time experiments. The classification is improved because of the multiple representation of information extracted from each sensor.

In particular, the classification is boosted by combining feature and distance based techniques and introducing the trainable fusion layer. The combiner classifier can effectively learn the errors of the base classifiers. The choice of diverse base classifiers produces independent errors and the process of aggregating the decisions results in better accuracy.

It is worth noting that our experiments reveal that there is a relation between the time series length and the impact of both local and global feature vectors. As the size of the data set decreases the global features become more effective than the local features. This happens because as the width of the local feature windows decreases, the features become less meaningful and do not truly discriminate among the extracted subsets, thus the trained model is unable to describe the classes at hand. To illustrate this observation, we note that the offline experiments are performed with the whole data series, while the real time experiments use a shorter window. In offline experiments, local features seem to outperform global features since the temporal aspect is fully maintained in the time series data. Alternatively, in the real time experiments, when only a significantly smaller portion of the series is used for training, global features have better impact on classification accuracy over other features.

Combining the two sensor data together does not boost the accuracy of classifiers. There is no significant difference over the results using BSN or VSN individually.

It is clear that classification accuracy increases significantly when preprocessing the data before classification instead of using the raw series. The processed values provide more useful information about the motion pattern over the individual samples.

6   Conclusion and Future Work

The focus of this work is using pattern recognition techniques to analyze human motion patterns and classify an unknown motion sequence. In this paper, an efficient multiple classifier design is proposed. The ensemble is capable of recognizing the difference between normal and impaired walking gaits. We show the results for training classifiers using local and global features extracted from sensor data. Future work includes speeding up the distance calculations for reducing the computational cost of real time experiments [37]. Also we are going to analyze the effect of different components of the ensemble on the results and compare it to other techniques such as Recurrent Neural Networks. Also the dataset will be extended to cover a wider range of subjects. Finally, the proposed method is intended to be tested in a wider range of applications in the ubiquitous computing field, such as activity recognition.

References

1. Aziz, O., Atallah, L., Lo, B., ElHelw, M., Wang, L., Yang, G.Z., Darzi, A.: A pervasive body sensor network for measuring postoperative recovery at home. Surgical Innovation 14(2), 83–90 (2007)
2. Lukowicz, P., Anliker, U., Ward, J., Tröster, G., Hirt, E., Neufelt, C.: Amon: A wearable medical computer for high risk patients. In: 2012 16th International Symposium on Wearable Computers, p. 0133. IEEE Computer Society (October 2002)
3. Gurley, R.J., Lum, N., Sande, M., Lo, B., Katz, M.H.: Persons found in their homes helpless or dead. New England Journal of Medicine 334(26), 1710–1716 (1996)
4. Yang, G.Z., Yacoub, M.: Body sensor networks (2006)
5. Jovanov, E., Milenkovic, A., Otto, C., De Groen, P.C.: A wireless body area network of intelligent motion sensors for computer assisted physical rehabilitation. Journal of Neuro Engineering and Rehabilitation 2(1) 6 (2005)
6. Istepanian, R.S., Jovanov, E., Zhang, Y.T.: Guest editorial introduction to the special section on m-health: Beyond seamless mobility and global wireless health-care connectivity. IEEE Transactions on Information Technology in Biomedicine 8(4), 405–414 (2004)

7. Bamberg, S.J.M., Benbasat, A.Y., Scarborough, D.M., Krebs, D.E., Paradiso, J.A.: Gait analysis using a shoe-integrated wireless sensor system. IEEE Transactions on Information Technology in Biomedicine 12(4), 413–423 (2008)
8. Ramachandran, R., Ramanna, L., Ghasemzadeh, H., Pradhan, G., Jafari, R., Prabhakaran, B.: Body sensor networks to evaluate standing balance: interpreting muscular activities based on inertial sensors. In: Proceedings of the 2nd International Workshop on Systems and Networking Support for Health Care and Assisted Living Environments, p. 4. ACM (June 2008)
9. Akyildiz, I.F., Melodia, T., Chowdhury, K.R.: A survey on wireless multimedia sensor networks. Computer Networks 51(4), 921–960 (2007)
10. Salaheldin, R., El Gayar, N.: Multiple Classifiers for Time Series Classification Using Adaptive Fusion of Feature and Distance Based Methods. In: UKCI 2011, p. 114 (2011)
11. Aggarwal, J.K., Cai, Q.: Human motion analysis: A review. In: IEEE Proceedings of the Nonrigid and Articulated Motion Workshop, pp. 90–102. IEEE (June 1997)
12. Preece, S.J., Goulermas, J.Y., Kenney, L.P., Howard, D., Meijer, K., Crompton, R.: Activity identification using body-mounted sensors—a review of classification techniques. Physiological Measurement 30(4), R1 (2009)
13. Bao, L., Intille, S.S.: Activity recognition from user-annotated acceleration data. In: Ferscha, A., Mattern, F. (eds.) PERVASIVE 2004. LNCS, vol. 3001, pp. 1–17. Springer, Heidelberg (2004)
14. Ge, X., Smyth, P.: Segmental semi-markov models for endpoint detection in plasma etching. IEEE Transactions on Semiconductor Engineering (2001)
15. Huynh, T., Schiele, B.: Analyzing features for activity recognition. In: Proceedings of the 2005 Joint Conference on Smart Objects and Ambient Intelligence: Innovative Context-Aware Services: Usages and Technologies, pp. 159–163. ACM (October 2005)
16. Kern, N., Schiele, B., Schmidt, A.: Multi-sensor activity context detection for wearable computing. In: Aarts, E., Collier, R.W., van Loenen, E., de Ruyter, B. (eds.) EUSAI 2003. LNCS, vol. 2875, pp. 220–232. Springer, Heidelberg (2003)
17. Thiemjarus, S.: A device-orientation independent method for activity recognition. In: 2010 International Conference on Body Sensor Networks (BSN), pp. 19–23. IEEE (June 2010)
18. Atallah, L., Lo, B., King, R., Yang, G.Z.: Sensor placement for activity detection using wearable accelerometers. In: 2010 International Conference on Body Sensor Networks (BSN), pp. 24–29. IEEE (June 2010)
19. Heinz, E.A., Kunze, K.S., Sulistyo, S., Junker, H., Lukowicz, P., Tröster, G.: Experimental evaluation of variations in primary features used for accelerometric context recognition. In: Aarts, E., Collier, R.W., van Loenen, E., de Ruyter, B. (eds.) EUSAI 2003. LNCS, vol. 2875, pp. 252–263. Springer, Heidelberg (2003)
20. Zheng, Y., Wong, W.K., Guan, X., Trost, S.: Physical Activity Recognition from Accelerometer Data Using a Multi-Scale Ensemble Method. In: IAAI (July 2013)
21. Ravi, N., Dandekar, N., Mysore, P., Littman, M.L.: Activity recognition from accelerometer data. In: AAAI, vol. 5, pp. 1541–1546 (July 2005)
22. Palit, A.K., Popovic, D.: Computational intelligence in time series forecasting: theory and engineering applications. Springer (2006)
23. Xing, Z., Pei, J., Keogh, E.: A brief survey on sequence classification. ACM SIGKDD Explorations Newsletter 12(1), 40–48 (2010)
24. Wang, X., Mueen, A., Ding, H., Trajcevski, G., Scheuermann, P., Keogh, E.: Experimental comparison of representation methods and distance measures for time series data. Data Mining and Knowledge Discovery 26(2), 275–309 (2013)

25. Berndt, D.J., Clifford, J.: Using Dynamic Time Warping to Find Patterns in Time Series. In: KDD Workshop, vol. 10(16), pp. 359–370 (1994)
26. Dietrich, C., Schwenker, F., Riede, K., Palm, G.: Classification of bioacoustic time series utilizing pulse detection, time and frequency features and data fusion, pp. 2001–2004. Univ., Fak. für Informatik (2001)
27. Ghosh, J., Beck, S.D., Chu, C.C.: Evidence combination techniques for robust classification of short-duration oceanic signals. In: Aerospace Sensing, pp. 266–276. International Society for Optics and Photonics (August 1992)
28. Elman, J.L.: Finding structure in time. Cognitive Science 14(2), 179–211 (1990)
29. Williams, R.J., Zipser, D.: A learning algorithm for continually running fully recurrent neural networks. Neural Computation 1(2), 270–280 (1989)
30. Rabiner, L.: A tutorial on hidden Markov models and selected applications in speech recognition. Proceedings of the IEEE 77(2), 257–286 (1989)
31. Kuncheva, L.I.: Combining pattern classifiers: methods and algorithms. John Wiley & Sons (2004)
32. Kittler, J., Hatef, M., Duin, R.P., Matas, J.: On combining classifiers. IEEE Transactions on Pattern Analysis and Machine Intelligence 20(3), 226–239 (1998)
33. Dietrich, C., Palm, G., Riede, K., Schwenker, F.: Classification of bioacoustic time series based on the combination of global and local decisions. Pattern Recognition 37(12), 2293–2305 (2004)
34. Sakoe, H., Chiba, S.: Dynamic programming algorithm optimization for spoken word recognition. IEEE Transactions on Acoustics, Speech and Signal Processing 26(1), 43–49 (1978)
35. ElSayed, M., Alsebai, A., Salaheldin, A., El Gayar, N., ElHelw, M.: Body and Visual Sensor Fusion for Motion Analysis in Ubiquitous Healthcare Systems. In: 2010 International Conference on Body Sensor Networks (BSN), pp. 250–254. IEEE (June 2010)
36. Hall, M., Frank, E., Holmes, G., Pfahringer, B., Reutemann, P., Witten, I.H.: The WEKA data mining software: An update. ACM SIGKDD explorations newsletter 11(1), 10–18 (2009)
37. Xi, X., Keogh, E., Shelton, C., Wei, L., Ratanamahatana, C.A.: Fast time series classification using numerosity reduction. In: Proceedings of the 23rd International Conference on Machine Learning, pp. 1033–1040. ACM (June 2006)

Author Index